Batch Adsorption Process of Metals and Anions for Remediation of Contaminated Water

Batch Adsorption Process of Metals and Anions for Remediation of Contaminated Water

Edited by
Deepak Gusain and Faizal Bux

CRC Press is an imprint of the
Taylor & Francis Group, an **informa** business

First edition published 2021
by CRC Press
6000 Broken Sound Parkway NW, Suite 300, Boca Raton, FL 33487-2742

and by CRC Press
2 Park Square, Milton Park, Abingdon, Oxon, OX14 4RN

CRC Press is an imprint of Taylor & Francis Group, LLC

Library of Congress Cataloging-in-Publication Data
Names: Gusain, Deepak, editor. | Bux, F. (Faizel), editor.
Title: Batch adsorption process of metals and anions for remediation of contaminated water / edited by Deepak Gusain and Faizal Bux.
Description: First edition. | Boca Raton : CRC Press, 2021. | Includes bibliographical references and index.
Identifiers: LCCN 2020047300 | ISBN 9780367436483 (hardback) | ISBN 9781003006367 (ebook)
Subjects: LCSH: Water—Purification—Adsorption.
Classification: LCC TD449.5 .B38 2021 | DDC 628.1/64—dc23
LC record available at https://lccn.loc.gov/2020047300

ISBN: 978-0-367-43648-3 (hbk)
ISBN: 978-0-367-70496-4 (pbk)
ISBN: 978-1-00-300636-7 (ebk)

Typeset in Times
by codeMantra

Contents

Acknowledgement

First, I would like to thank my parents, Bharat Singh Gusain and Kamla Gusain; brother, Himanshu Gusain; and sister, Kavita Gusain; for their support, constant loving encouragement and being kindly patient with my academic career.

I thank Himanshu Gusain and Swatantra Tripathi for their help and valuable input during the creation of Figure 8.3 (Economic cost graph for sustainable use of adsorbent).

This book is based on batch adsorption experiments conducted over the years. I am indebted to peers, organizations and academic institutions whose research publications, reports and books provided much-needed information and act as building block for text, figure and table of this book. My special thanks to persons who helped me learn and practice the batch adsorption experiments and analysis of its data (especially, Prof. Yogesh Chandra Sharma and Dr. Varsha Srivastava).

Thanks to everyone on my CRC Press publishing team especially Jessica Poile (Editorial assistant) and Hilary Lafoe (Senior Editor) for their kind guidance and patience during the review process. I thank CRC Press for kindly considering our document as worthwhile to publish. I thank, Prof. Stanley E. Manahan and anonymous the reviewers for their valuable comments and suggestions in spite of their busy schedules, which enhanced this edition of the book.

I appreciate the support of co-contributors during the work, with special thanks to Dr. Shikha Dubey, without whom the work would not have been completed on time.

Deepak Gusain
New Delhi

Acknowledgement

List of Abbreviations

%	Percentage
°C	Celsius
μg/d	Microgram per day
μg/l	Microgram per liter
Å	Angstrom
AC	Activated carbon cloth
ACC	Activated carbon cloth
AMD	Amidoxime
APTES	3-Aminopropyl triethoxysilane
ATR-FT-IR	Attenuated total reflection Fourier transform infrared spectroscopy
ATSDR	Agency for toxic substances and disease registry
BET	Brunauer–Emmett–Teller
Bq/g	Becquerel per gram
Bq/L	Becquerel per liter
c.a.	Circa, or approximately
CA	Carboxyl
Ce	Concentration of adsorbate at equilibrium
cm^{-1}	Wavenumber
cm^3	Cubic centimeter
cm^3/g or $cm^3\ g^{-1}$	Cubic centimeter per gram
CNTs	Carbon nanotubes
COF-S-SH	Thiol-functionalized covalent organic framework
Conc.	Concentration
CPCB	Central Pollution Control Board (India)
CTAB	Cetyltrimethylammonium bromide
DAPF	2,3-diaminophenol and formaldehyde
d_{basal}	Basal spacing
DEFRA	Department for Environment, Food and Rural Affairs (UK)
d-PDF	Differential atomic pair distribution function analysis
DTA	Differential thermal analysis
DTPA	Diethylenetriaminepentaacetic acid
e.g.	For example
ECH	Epichlorohydrin
ED/EDA	Ethylenediamine
ED-MIL-101	Ethylene diamine- Matérial Institut Lavoisier-101 (ethylene diamine modified chromium-based metal organic framework)
EDS/EDX	Energy-dispersive X-ray spectroscopy
EDTA	Ethylenediaminetetraacetic acid
EELS	Electron energy loss spectroscopy

e_g	Triply degenerate set of three orbitals (d_{xy}, d_{yz}, and d_{xz})
eV	Electron-volt
EXAFS	Extended X-ray absorption fine structure
Fe/RGO	Iron-reduced graphene oxide composite
FITEQL	Computer program for the determination of chemical equilibrium constants
FT-IR	Fourier transform infrared spectroscopy
g	Gram
g/l	Gram per liter
g/t	Gram per ton
GO	Graphene oxide
GOMNP	Graphene oxide magnetic nanoparticles
h	Hour
HF	Hydrogen fluoride
HPLC	High performance liquid chromatography
HSAB principle	Hard/soft/acid/base principle
IARC	International Agency for Research on Cancer
ICP	Inductively coupled plasma spectroscopy
K	Kelvin
kJ/mol	Kilo Joule per mole
kJ/mol/K	Kilo Joule per mole per kelvin
K_p or K_c	Thermodynamic equilibrium constant derived from partition method
L	Liter
LDH	Layered double hydroxides
LEED	Low energy electron diffraction
M	Molar
m^2/g or $m^2\ g^{-1}$	Square meters per gram
MCM-41	Mesoporous material (Mobil composition of matter No. 41)
mg	Milligram
Mg Al LDH	Magnesium- and aluminum-based layered double hydroxide
mg/g	Milligram per gram
mg/l	Milligram per liter
mgN/L	Milligram of nitrogen per liter of solution
MIL-101	Matérial Institut Lavoisier-101 (chromium-based metal organic framework)
min	Minute
ml/g	Milliliter per gram (distribution coefficient unit)
mmol/g	Millimoles per gram
MOF	Metal-organic framework
mol	Mole
Mol/l or mol l^{-1}	Mole per liter

MPS	Metal chalcophosphate
mV	Millivolt
MWCNTs	Multiwalled carbon nanotubes
NaHS	Sodium hydrosulfide
nm	Nanometer
nm^3	Cubic nanometer
NMR	Nuclear magnetic resonance
NU-1000	Zirconium-based metal organic framework (Northwestern University)
OTMAC	Octodecyltrimethylammonium chloride
PAA-MGO	Poly (allylamine)-modified magnetic graphene oxide
PAN	1-(2-pyridylazo)-2-napththol
PEG_{2000}	Polyethylene glycol 2000
PE-MA-NN	Porous chelating fiber
PGMA-MAn	Poly(glycidyl methacrylate-maleic anhydride) copolymer
pH_{IEP}	Isoelectric point
pH_{pzc}	Point of zero charge
pH_{pzc}/pH_{zpc}	pH at zero point of charge
pKa	Acid dissociation constant
PMM	Phosphate-modified montmorillonite
PP-g-PGMA	Polypropelene-graft-poly(glycidyl methacrylate)
ppm	Parts per million
PVA/EDTA	Polyvinyl alcohol/ethylenediaminetetraacetic acid
PZC	Point of zero charge
Q3/Q4	29Si magic angle spinning NMR spectra peaks
q_e	Amount of adsorbate adsorbed on per unit gram of adsorbent at equilibrium
q^t	Amount of adsorbate adsorbed at time "t" on per unit gram of adsorbent
R^2	Coefficient of determination
RAFT-IIP	Hydrophilic ion-imprinted polymer based on graphene oxide
Raman SPECTRA	Light scattering spectra named after C.V. Raman
rGO	Reduced graphene oxide
rpm	Rotations per minute
SAC	Sulfurized activated carbon
SBA-15	Santa Barbara Amorphous-15 (Mesoporous silica)
SEM	Scanning electron microscopy
SEM-HADF (High angular dark field)	High-angular dark-field imaging in scanning electron microscopy
STEM-EDS	Scanning transmission electron microscopy—energy dispersive X-ray spectroscopy
STEM-XDS	Scanning transmission electron microscopy—X-ray dispersive spectrometry

SWCNTs	Single-walled carbon nanotubes
t	Time
$t^{1/2}$	Square root of time (Intraparticle diffusion)
t_2g	Doubly degenerate set of two orbitals (dx^2-y^2) and dz^2
TCLP	Toxicity characteristic leaching procedure
TEM	Transmission electron microscope
TMU-16-NH_2	Amino functionalized zinc-based metal organic framework (TMU = Tarbiat Modares University)
UiO-66	Zirconium based metal organic framework (Universitetet i Oslo)
USEPA	United States Environmental Protection Agency
UV-Visible spectroscopy	Ultraviolet–visible spectroscopy
WHO	World Health Organization
XANES	X-ray absorption near edge structure
XCMP	Xanthate-modified cross-linked magnetic chitosan/poly(vinyl alcohol) particles
XPS	X-ray photoelectron spectroscopy
XRD	X-ray diffraction
ZVI or Fe^0	Zero valent iron
γ	Gamma
ΔG^o	Change in standard free energy
ΔH^o	Change in standard enthalpy
ΔS^o	Change in standard entropy
χ^2	Chi-square

Editors

Dr Deepak Gusain is an assistant professor in the Department of Environmental Studies in PGDAV (Evening) College, University of Delhi. He earned his master's degree in environmental studies from the University of Delhi and completed his Ph.D. at the Department of Chemistry, Indian Institute of Technology (Banaras Hindu University), Varansiin Varanasi, followed by a postdoctoral fellowship at the Institute for Water and Wastewater Technology in Durban University of Technology. His Ph.D. research focused on the remediation of chromium and cadmium from aqueous solution using nanocrystalline material. In addition, he investigated the effect of linear and nonlinear curve fitting on the estimation of isotherm and kinetic and thermodynamic parameters during the adsorption process.

Professor Faizal Bux is a professor and director of the Institute for Water and Wastewater Technology, Durban University of Technology. His affiliations also include Fellow of the International Water Association, Royal Society of Chemistry, and Water Institute of Southern Africa and he is a member of the Academy of Science of South Africa. He has won several national and international awards and published extensively. Professor Faizal Bux's group works on wastewater treatment and reuse, wastewater beneficiation, and health-related microbiology, bioremediation, algal biotechnology, and constructed wetlands.

Contributors

Faizal Bux
Institute for Water and Wastewater Technology
Steve Biko Campus
Durban University of Technology
Durban, Republic of South Africa

Shikha Dubey
Department of Chemistry
IIT(BHU)
Varanasi, India

Deepak Gusain
Institute for Water and Wastewater Technology
Steve Biko Campus
Durban University of Technology
Durban, Republic of South Africa

Yogesh Chandra Sharma
Department of Chemistry
IIT(BHU)
Varanasi, India

1 Introduction

Deepak Gusain
Durban University of Technology

Shikha Dubey and Yogesh Chandra Sharma
IIT(BHU), Varanasi

Faizal Bux
Durban University of Technology

CONTENTS

1.1 INTRODUCTION

The rapid pace of development, or industrialization, resulted in an increased demand for metals. The widespread use of metals leads to their presence in the pristine environment. Over time, the natural remediation by the environment was not quite sufficient to maintain the level of contaminants below the safety levels recommended by health and environment organizations. In addition, lack or high cost of alternatives leads to oxymoron conditions in the use of metal industries. The concentration of contaminants can also present due to geological reasons, as in the case of arsenic in Bangladesh and West Bengal (India) where the concentration of the arsenic was sufficient to act as a pollutant (Nordstrom 2002). One of the major areas where the contaminants are present for a significant period of time is the aquatic ecosystem.

Living organisms are highly dependent on the aquatic ecosystem for their life sustenance. Consumption of contaminants present in the aquatic ecosystem causes health effects for the living organisms. Contaminants like chromium, manganese, iron, cobalt, nickel, copper, zinc, arsenic, lead, mercury and cadmium are known to affect human health (Klaassen 2013). In particular, chromium, cadmium, nickel and arsenic are reported to be carcinogenic in nature (WHO 2017). In addition, chromium causes ulceration and perforation of nasal septum, allergies, proteinuria, hematuria and anuria (Wilbur et al. 2012). Arsenic causes hyperkeratosis, liver injury (IARC 2012), black foot disease (Yu et al. 2002) and interference in heme synthesis, with an increase in urinary porphyrin excretion (Ng et al. 2005). Cadmium causes renal injury, osteoporosis and cardiovascular disease (Faroon et al. 2012). Lead causes peripheral neuropathy, chronic nephropathy and hypertension (Goyer 1990), proximal tubular dysfunction and renal failure (Goyer 1989).

In addition to heavy metals, other inorganic pollutants like anions also have adverse health effects; for example, a high intake of fluoride leads to dental or skeletal fluorosis (Death et al. 2015; Choubisa 2012) and a high intake of nitrate leads to methemoglobinemia (Gilchrist et al. 2010).

The need to deal with the adverse effects of metals and inorganic contaminants and the inability of the natural process to maintain the concentration of contaminants below the safety levels, above which they act as pollutants, lead to the development of methods for the treatment of these pollutants. Water can be remediated with a number of processes like precipitation, ion exchange, reverse osmosis, nanofiltration, coagulation–flocculation, electrocoagulation and adsorption. These technologies have their own benefits and limitations. Limitations include sludge generation during precipitation and the coagulation–flocculation process (Fu and Wang 2011), inability to handle high metal ion concentrations in ion exchange processes (Barakat 2011), membrane fouling in reverse osmosis (Greenlee et al. 2009; Kurniawan et al. 2006) and high operational and maintenance cost in electrochemical methods (Fu and Wang 2011). Adsorption also has limitations like the absence of a universal adsorbent and variable adsorption capacity for different materials (Dubey et al. 2017). In spite of this, adsorption is known for its cost-effectiveness, energy efficiency and complete removal of pollutants even at trace levels from dilute solutions, which makes it an attractive process for the removal of contaminants.

1.2 ADSORPTION: A BRIEF HISTORY

Adsorption is a surface phenomenon where change in concentration of the adsorbate occurs as compared to surrounding phases or accumulation or increase in concentration occurs at the interface between phases such as liquid-gas, liquid-liquid, and solid–gas; and the case we are discussing in this chapter is solid-liquid interface (Dąbrowski 2001; Summers et al. 2011; Swenson and Stadie 2019).

Adsorption process was observed in 1773 and 1777 by Scheele and Fontana, respectively (Dąbrowski 2001; West 1945; Swenson and Stadie 2019) during uptake of gases by charcoal. This was followed by Lowitz in 1785 when charcoal used for decolorization of tartaric acid solution. The term *adsorption* was first coined by du Bois-Reymond and introduced into the literature by Kayser (Dąbrowski 2001; West 1945; Patel 2019). Later, adsorption found its application in industries like sugar refinery and in chromatography.

Adsorption also forms the basis of chromatography developed by Mikhail Semyonovich Tsvet (Dąbrowski 2001). In addition, surface science yielded Nobel laureates like Irving Langmuir in 1932 and Gerhard Ertl in 2007.

1.3 ABSORPTION

The term *absorption* was proposed by McBain while monitoring the slower uptake of hydrogen by carbon (Dąbrowski 2001; Swenson and Stadie 2019; Tan 2014). He also proposed the term *sorption* in cases where adsorption and absorption are difficult to distinguish, which led to the use of other terms like *sorbate*, *sorbent* and *sorptive* rather than *adsorbate*, *adsorbent* and *adsorptive*, respectively (Dąbrowski 2001).

Absorption is a bulk, or volumetric, phenomenon rather than a surface phenomenon, and it is limited by the degree of solubility of the adsorbate. Adsorption of gas in liquid depends on its solubility, whereas adsorption of gas depends on the adsorption isotherm. In absorption, the absorbate or material or gases penetrate into the structure of the solid or liquid rather than remaining on the surface and are mostly controlled by diffusion (Tan 2014; Worch 2012e).

1.4 ADSORPTION PROCESS

Adsorption process can be carried out in any of the following ways: batch adsorption, continuous fixed-bed adsorption, continuous-flow tank adsorption, continuous moving bed, continuous fluidized bed and pulsed bed. The two most commonly used adsorption procedures are the batch and continuous mode for the removal of contaminants from an aqueous medium. These two adsorption procedures can be distinguished from each other in many ways (Table 1.1).

Batch adsorption: In this process, a fixed amount of the adsorbent (solid phase) and the adsorbate (liquid phase) is agitated in a closed vessel. The detailed description of the process is given below.

Batch adsorption process in most of the studies is conducted in an Erlenmeyer flask or reagent bottles. Agitation is carried out in a temperature-controlled water bath shaker or orbital shaker. First, a requisite amount of the adsorbent is weighed and put into the reagent bottle. The temperature of the adsorbate solution and the water bath should be the same to minimize any energy exchange due to temperature difference. Thin-walled bottles are more prone to energy transfer than thick-walled reagent bottles. The adsorbate solution setup at the requisite temperature is then poured into the reagent bottle containing the adsorbent. The solution is then stirred for the requisite time or until equilibrium time is achieved. The adsorbent is then separated from the aqueous solution by filtration or centrifugation process. The solution is then analyzed for adsorbate concentration. The adsorbent post adsorption can then be regenerated by desorbing agents for reuse. The detailed batch adsorption process is illustrated in Figure 1.1.

Data collected during batch adsorption were used to test the new adsorbents and their mechanism of adsorption. Batch adsorption experiments are not useful to treat voluminous amount of wastewater (Tien 2018) or for industrial scale (Tien 2018; Ray et al. 2020a).

Continuous-flow tank: In this system, the adsorbent is directly added to the wastewater treatment process. Adsorbents in powdered form are directly added to the wastewater treatment process (Tien 1994, 2018).

TABLE 1.1
Merits and Demerits of Batch and Continuous Mode of Adsorption Process

	Batch Adsorption Process	Continuous Adsorption Process	References
1	Batch adsorption process is classified as a discontinuous process.	Column adsorption experiments are identified as continuous experiments.	Worch (2012c)
2	Low amounts of adsorbent dispersed in large volumes of liquid. So, bulk density has the characteristics of a mass concentration of the solid particles rather than that of a conventional density.	Adsorbent particles are arranged in an adsorbent bed. Consequently, the proportion of the void volume is much lower than in the case of a batch reactor. For fixed-bed absorbers, the term *bed density* is often used instead of *bulk density*.	Worch (2012a)
3	In batch adsorbers, the adsorbent is in contact with the adsorbate solution until equilibrium is reached. The batch adsorber design for single-solute adsorption is, therefore, very simple and requires only combining the material balance with the isotherm equation.	A comparable material balance equation can be used under the assumption that the contact time in the adsorbent is longer than the time needed for establishing the equilibrium.	Worch (2012c)
4	In batch reactors, to minimize the influence of film diffusion, higher stirring speed is a prerequisite. High stirring speed leads to an increase in the risk of destruction of adsorbent particles. This can lead to errors in the determination of mass transfer coefficients, due to their particle size dependence.	In column experiments, the solution flows through a fixed bed of adsorbent. High flow velocities avoid the influence of film diffusion. In column experiments, adsorbent particles are shielded by the destruction due to high velocity.	Worch (2012d)
5	The mass transfer driving force decreases along with the adsorption rate due to declining concentration in the reactor.	In a fixed-bed reactor, the adsorbent faces a high driving force over the whole process	Worch (2012b)
6	In a batch reactor, very low residual concentrations can be achieved only if very high adsorbent doses are applied.	In a fixed-bed reactor, the adsorbate will be totally removed until the breakthrough occurs.	Worch (2012b)

Fixed-bed adsorption: This system consists of passing a solution of adsorbate at a constant rate through a column packed with adsorbents (Tien 2018; Ray et al. 2020a; Patel 2019).

Moving-bed adsorption: In this method, both adsorbent and adsorbate are in continuous motion and the bed of adsorbent part remains constant (Tien 1994, 2018; Ray et al. 2020a; Patel 2019).

Pulsed-bed adsorption: In this adsorption, the column consists of a number of adsorbent beds and the bed from the bottom can be replaced with a new fresh bed (Ray et al. 2020a).

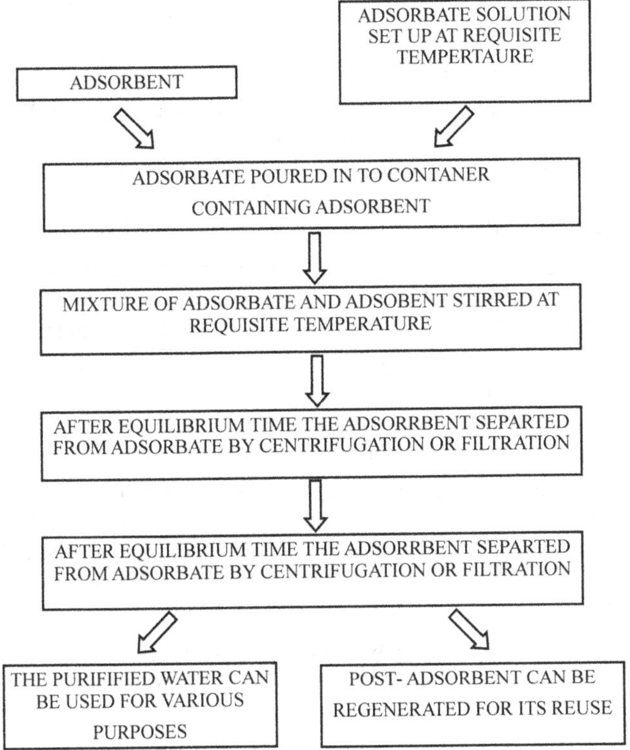

FIGURE 1.1 Illustration of batch adsorption process.

Column experiments usually show a higher efficiency than batch adsorption experiments (Hu et al. 2016). The column method or continuous adsorption methods are used for performance enhancement. This book focuses on the mechanism of adsorption process in which new adsorbents can be easily used in batch adsorption process. So, we focused our study specifically on batch adsorption process.

1.5 WATER TREATMENT BY BATCH ADSORPTION PROCESS

Batch adsorption experiments are scale-down technologies that help in short-listing the factors for adsorption at full scale (Ye et al. 2019). Batch adsorption experiments are used before the pilot study (Leiviskä et al. 2017) and a full-scale study for the determination of optimized parameters as well as isotherm and kinetic data.

In the case of vanadium removal from industrial wastewater ferric oxyhydroxide (a commercial iron sorbent), AQM PalpowerK10PM (a commercial sorbent), blast furnace sludge, steel converter sludge, Fe Cr granulated slag and steel foundry slag were used for batch experiments (Leiviskä et al. 2017). Among these, ferric oxyhydroxide exhibited best results for removal of vanadium. Batch experiment results were followed by column experiments, and ferric oxyhydroxide was used based on the results of batch experiments. Here, batch experiments saved a lot of time and energy.

In addition, stand-alone batch experiments were conducted in real wastewater for removal of pollutants and checking the efficiency in new pollutants (Hagemann et al. 2020) or adsorbents. Wood-based activated biochar is used to remove organic micropollutants (4/5-methylbenzotriazole and benzotriazole, amisulpride, candesartan, carbamazepine, citalopram, clarithromycin, diclofenac, hydrochlorothiazide, irbesartan, metoprolol, venlafaxine, as well as sulfamethoxazole and its human metabolite N4-acetylsulfamethoxazole) from biologically treated wastewater for checking the efficiency. The original wastewater was spiked with organic micropollutants, and the results showed that locally produced activated biochar has a similar and higher efficiency in the removal of organic micropollutants as compared to activated biochar produced from fossil precursors.

Batch adsorption was also used to study the adsorption characteristics of pilot-scale-constructed wetland substrate (sand, rock, quartz or clay) for removal of benzotriazole (Wagner et al. 2020). The batch experiments helped in estimating the stake of adsorption process along with photodegradation and biodegradation in the abatement of benzotriazole. The mixture of sand and granular activated carbon leads to an improved adsorptive removal of benzotriazole and can help in future natural remediation methods.

Thus, it is useful to conduct batch adsorption experiments before undertaking full-scale adsorption experiments (Maccotta et al. 2016; Vacca et al. 2016).

1.6 SCOPE AND OUTLINE OF THE CHAPTERS

The process of adsorption is affected by a number of factors like pH of the solution (Figures 3.1, 4.1 and 5.1), surface modification of the adsorbent and temperature of the solution. There is a lack of books on factors affecting adsorption of various inorganic species, which encouraged us to write this book. In this book, the effect of various parameters along with the reasons for their influence is discussed element by element. The elements discussed in this book are scandium, titanium, vanadium, chromium, manganese, iron, cobalt, nickel, copper, zinc, gallium, germanium, arsenic, selenium, strontium, cadmium, cesium, mercury, lead and uranium and anions like fluorides, nitrates, perchlorates and sulfates. The effects of initial concentration, adsorbent dose, ionic strength and contact time as well as thermodynamics, kinetics and isotherm studies have been discussed in separate sections.

Chapter 2 covers the properties of major adsorbents like activated carbon, nano-adsorbents, metal organic frameworks, carbon nanotubes and graphene oxides. The role of functional groups present on the surface of adsorbents is also discussed in this chapter.

The contaminants discussed in this book are arranged in five categories:

1. Major toxic metals: Elements that do not have any biological role and that are known to have a critical effect on humans are discussed in Chapter3. The elements in this chapter include vanadium, chromium, nickel, arsenic, strontium, cadmium, mercury, lead and uranium.

Essential elements but toxic on excessive exposure: Elements that are known to have a biological role in the human body, but whose excessive exposure leads to health problems are discussed in Chapter 4. The elements included in this category are manganese, cobalt, copper, zinc and selenium.

2. Miscellaneous: Adsorption behavior of cesium and iron ions is discussed in Chapter 5. Nonradioactive cesium does not have any severe effect on humans. Adverse health effects of excess iron are also not known.
3. Nontoxic elements: The elements that do not have any demonstrated toxicity toward human beings are also discussed in Chapter 5. The elements included under this category are scandium, titanium, gallium and germanium.
4. Anions: Anionic water contaminants like fluorides, nitrates, perchlorates and sulfates are discussed in Chapter 6.

Chapter 7 covers the impact of parameters like initial concentration, adsorbent dose and ionic strength during batch adsorption process of metals and anions.

Kinetics, isotherm and thermodynamic study of metals and anions are included in Chapter 8. In addition, Chapter 8 also includes the literature on disposal and reusability of spent adsorbents.

2 Adsorbents
Classification, Characteristics, Chemical Nature, and Interaction with Contaminants

Shikha Dubey
IIT(BHU), Varanasi

Deepak Gusain
Durban University of Technology

Yogesh Chandra Sharma
IIT(BHU)

Faizal Bux
Durban University of Technology

CONTENTS

This chapter presents detailed information on a variety of materials employed as adsorbents for the remediation of water containing hazardous organic and inorganic contaminants. The adsorbents based on their wide applicability such as carbon-based materials, nano-adsorbents, inorganic–organic hybrids (MOF), green adsorbents, and zeolites for water treatment have been selected and are discussed here. The effectiveness of the remediation process is solely dependent upon the physical and chemical properties of the adsorbents; hence, the physical and chemical properties of the adsorbents are also discussed. The adsorption of contaminants occurs through a variety of mechanisms such as physical adsorption, ion exchange, electrostatic interaction, surface complexation, and precipitation that are explained precisely. Further, the effect of functionalization of adsorbents and their remediation performance was evaluated for various metals and anions. Emphasis has been given to the functionalization with oxygen, nitrogen, and sulfur-containing functional groups.

2.1 CLASSIFICATION OF ADSORBENTS

In the water treatment processes, the potential of a wide range of solid materials as adsorbents has been investigated so far. These adsorbents are either natural materials or engineered materials (Barakat 2011; Renu and Singh 2016; Worch 2012a). A further classification of these adsorbents has been done by various groups as (a) naturally occurring materials (e.g. diatomaceous earth, fuller's earth, clay), (b) modified/activated natural materials for intended applications (e.g. activated carbon, activated alumina), (c) synthetic materials (e.g. aluminosilicates, polymers, resins, zeolites), (d) agricultural solid wastes and industrial by-products (e.g. rice husk, wheat bran, orange peels, sugarcane bagasse, corn cob, fly ash, red mud), and (e) biosorbents (e.g. chitosan, medicinal herbs, fungi, weeds, bacterial biomasses; Crini et al. 2018). Dabrowski introduced another classification as follows: (a) carbon adsorbents (e.g. activated carbons, molecular carbon sieves, fullerenes, carbonaceous materials),

(b) mineral adsorbents (e.g. activated alumina, silica gel, metal oxides and hydroxides, inorganic nanomaterials, zeolites), and (c) other adsorbents (synthetic polymers, composite adsorbents, mixed adsorbents) (Dąbrowski 2001). Another group classified the adsorbents as conventional and nonconventional where commercially available activated carbons, ion-exchange resins, activated alumina, silica gel, zeolites, and molecular sieves were enlisted as conventional adsorbents. The nonconventional adsorbents constitute activated carbons prepared from natural materials, biosorbents, agricultural solid wastes, industrial by-products, and some miscellaneous adsorbents such as alginates and hydrogels (Crini 2005, 2006; Crini and Badot 2007; De Gisi et al. 2016).

Considering the various classifications present in the literature, we provide a detailed discussion of adsorbents such as carbon-based adsorbents, nano-adsorbents, inorganic–organic hybrids, green adsorbents, and zeolites (Figure 2.1) in this section.

2.1.1 Carbon-Based Materials

Being one of the most readily available elements, carbon finds extensive applications in various fields, including targeted drug delivery, environmental remediation, and energy storage, due to certain fascinating properties such as hardness, low density, remarkable mechanical durability, outstanding chemical and thermal stability, and corrosion resistance (Gusain et al. 2020). Similarly, the synthesis and advanced applicability of carbon-based materials are gaining momentum worldwide due to their distinctive properties (e.g. high surface area, flexibility, superior directionality) that enable them to be more efficient than bulk materials (Stankovich et al. 2007; Chen et al. 2016). The recent decades have observed an escalation in the utilization of different types of carbon-based materials/nanomaterials for water treatment due to their considerable abundance, ease of preparation and handling, easy modification, high surface area and porosity, stable structure, environmentally benign nature, and

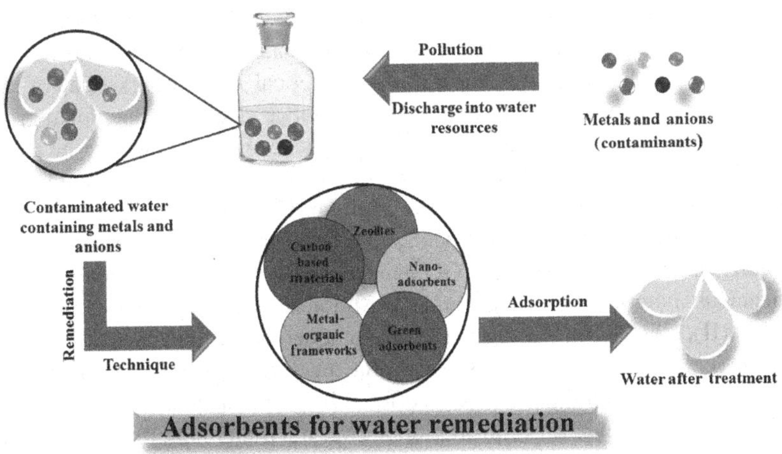

FIGURE 2.1 Adsorption of contaminants through various adsorbents from aqueous solutions.

high sorption capacities (Perreault et al. 2015; Upadhyayula et al. 2009, 2014; Sarkar et al. 2018). A variety of carbon-based materials have been investigated for water treatment by various groups such as activated carbon, graphene, and graphene-based adsorbents such as graphene oxides, reduced graphene oxides, and carbon nanotubes.

2.1.1.1 Activated Carbon

Over the years, activated carbon has been the most extensively used material for elimination of inorganic and organic pollutants from water streams. It can be prepared from readily available carbonaceous materials such as agricultural wastes, wood, and coconut coir (Adeleke et al. 2019) and possesses high surface area and high porosity that make it an excellent adsorbent. In addition, it has a hydrophobic surface, so it gets less influenced by water and holds a significantly weak acidic ion-exchange character that empowers it to take out metallic pollutants from wastewater (Alothman et al. 2016). Activated carbon has been efficiently and extensively used for the removal of a number of metallic species via different adsorption mechanisms (Kuroki et al. 2019). Several factors govern the adsorption of metal ions through the activated carbon where the solution pH is the most influential factor (Bouhamed et al. 2016) favoring the adsorption of metal ions Cu^{2+}, Ni^{2+}, and Zn^{2+} at higher pH values. Surface chemistry (i.e. surface area and pore size/volume) is another dominant factor and has a direct effect on the adsorption where higher surface area and pore size/volume favor the adsorption (Borhan et al. 2016). The chemical properties of activated carbons (e.g. surface charge and functional groups) also influence the adsorption of metal ions where an increase in the number of acidic functional groups on the surface increases the removal performance (Yoshihara et al. 2009).

2.1.1.2 Graphene

Graphene is an essential nonclassical carbon adsorbent of a flat and single-atom-thick honeycomb structure (Kyzas and Kostoglou 2014). The framework consists of nanosheets of sp^2 hybridized carbon atoms in a two-dimensional densely packed lattice structure. It has a variety of applications in various fields, including energy storage and separation technology, due to its certain extraordinary characteristics such as chemical stability and inertness and mechanical strength. The relatively large specific surface area-to-volume ratio (theoretical ~2675 m^2/g), relative abundance, economic viability, eco-friendliness, abundance of surface functional groups, chemical stability, and better conductivity of graphene sheets enable it to be a promising material for the comprehensive removal of organic and inorganic contaminants through the pre-concentration of aqueous medium (Gusain et al. 2020; Verma et al. 2011). Graphene and its derivatives such as graphene oxide, reduced graphene oxide, chemically functionalized GO, and GO nanocomposites have evidenced themselves as efficacious materials in the dynamic adsorption of heavy metals such as Cr(VI), Cd(II), Pb(II), Co(II), Zn(II), and Fe(III) from aqueous medium (Yusuf et al. 2015; Carmalin Sophia et al. 2016; Cortés-Arriagada and Toro-Labbé 2015) via physisorption. Regardless of a number of benefits offered by graphene when used as an adsorbent, it possesses a tendency to form graphite via agglomeration that can reduce its specific surface area, hinder the rapid mass transport of adsorbate, and influence the overall performance of the material (Ersan et al. 2017; Ali et al. 2019).

2.1.1.3 Graphene Oxides

Graphene oxide is the oxidized form of graphene functionalized by the functional groups containing reactive oxygen, viz. hydroxyl, carboxylic, and epoxy groups. The oxygen-containing functionalities make graphene oxide hydrophilic and significantly contribute to its hydrophilicity, leading to its solubility in both polar and nonpolar solvents, while the graphene domain makes graphene oxide hydrophobic, and the high negative charge density on its surface favors the adsorption of positively charged contaminants from water (Kumar et al. 2019). Mostly, the functional groups containing negatively charged groups present on the edges and on the basal planes of graphene oxide are involved in the removal of positively charged pollutants via electrostatic interactions between them (Bao et al. 2013). Besides being one of the most important graphene derivatives and precursors for almost all graphene-based materials, it possesses numerous outstanding features such as large specific area, excellent mechanical strength, chemical stability, high flexibility, less weight, and abundant oxygen-containing functional groups that not only make it a suitable candidate for wastewater treatment but also provide sensitive sites for certain surface alteration reactions for the functionalization of graphene oxide and graphene-based materials for their wide applications (Hao et al. 2012; Schedin et al. 2007; Liu et al. 2019). Although the adsorption selectivity of graphene oxide is poor, it shows strong affinities for various metal ions having higher electronegativity values, leading to their stronger attraction on the negatively charged graphene oxide surfaces. The adsorption of various heavy metallic species from aqueous medium is efficiently accomplished, for example, Cr(VI) (Zheng et al. 2019), Cd(II) (Pakulski et al. 2018), Cu(II) (Fu and Huang 2018), Co(II) (Wang et al. 2018), and Pb(II) (Weng et al. 2019). The adsorption occurred by the formation of metal complexes with the oxygen-containing functional groups (viz. hydroxyl and carbonyl groups) and target metal ions and is mainly chemical in nature (Sitko et al. 2013).

2.1.1.4 Reduced Graphene Oxides

Reduced graphene oxide is obtained by the reduction of GO, i.e. removal of the ionic oxygen groups present in it by hydrothermal, chemical reaction, photocatalytic, or thermal annealing methods. The chemical reduction is the most common method for reducing GO to rGO by means of a variety of reducing agents such as sodium borohydride, ascorbic acid, hydrazine hydrate, pyrrole, amino acids, and urea (Gusain et al. 2020). It has a quasi-two-dimensional structure formed by carbon atoms via sp^2 hybridization (Zhang et al. 2018). The efficiency and selectivity of adsorption onto the surface of reduced graphene oxide can be improved by its functionalization, for which metals, metal chalcogenides, biopolymers, and dopants (N, S, etc.) are commonly used (Gusain et al. 2020). Similarly, the properties of reduced graphene oxide can be tailored by incorporating certain oxygen-containing functional groups, defects/edges, and dopants in its structure. The oxygen-containing functional groups such as –OH, –COO–, –C-O–C–, and –COOH present on the surface of rGOs interact with various water contaminants and help in their removal from the water resources (Zhang et al. 2018; Wang, Li, et al. 2017; Song et al. 2017; Zheng, Wu, et al. 2017).

2.1.1.5 Carbon Nanotubes

Carbon nanotubes (CNTs) are special kinds of adsorbents having one-dimensional hollow tube-like structures that result from rolling-up of thin carbon walls (Kurwadkar et al. 2019). They are categorized into single-walled and multiwalled carbon nanotubes on the basis of number and arrangement of graphene sheets where single-layered graphene sheets rolled into a cylindrical tube form single-walled carbon nanotubes (SWCNTs) and multi-layered graphene sheets rolled into an array of concentric cylinders form multiwalled carbon nanotubes (MWCNTs) (Mubarak et al. 2014). The distinctiveness in the structure of CNTs offers them excellent physicochemical properties such as high surface area (150–1500 m^2/g), highly porous and hollow structure, layered architecture, diameter ranging from 1 nm to several nanometers, a wide spectrum of surface functional groups, and a strong interaction with the pollutant molecules, which make them an ideal candidate for pollutant adsorption. The mesoporous structure of CNTs empowers their wide application in water treatment via adsorption (Upadhyayula et al. 2009; Ihsanullah, Al-Amer, et al. 2016). The mechanism with which metal ions adsorbed onto CNTs is very complicated, and the driving force for their adsorption on CNTs involves electrostatic interaction, sorption–precipitation, charge transfer, and p–p interactions between the metal ions and the attached functional groups on the surface of CNTs (Liu, Wang, et al. 2013; Song et al. 2018; Lu and Liu 2006).

CNTs offer four possible active adsorption sites for the adsorption of different organic and inorganic contaminants (Figure 2.2): (a) "internal sites" – the hollow interior tubular structure that is accessible only if the caps are removed and the open ends are unblocked; (b) "interstitial channels" – the space between the individual nanotubes in the nanobundles; (c) "groove sites" – the grooves between the peripheral nanobundles and the exterior surface of the outermost nanotubes where two adjacent parallel tubes meet; and (b) "exterior surface" – the curved exposed active

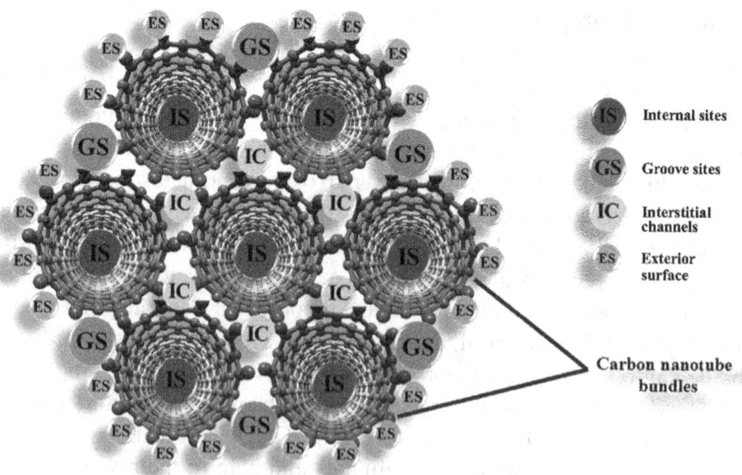

FIGURE 2.2 Illustration of four active adsorption sites on carbon nanotubes (CNTs).

surface of individual nanotubes on the outside of the nanotube bundles (Ren, Chen, et al. 2011; Das et al. 2014).

The functionalization of carbon of CNTs via oxidation, grafting, and physical adsorption of molecules of various functional groups greatly improves its selectivity and adsorption capacity. Similarly, by alteration in the structural and physical properties of CNTs, the affinity and mechanism of adsorption of contaminants can be controlled (Maleki et al. 2017). During the removal of contaminants, they are either trapped into the pores of CNTs or adsorbed onto their surface via diverse interactions (Maleki et al. 2017; Saber-Samandari et al. 2017). The heavy metal ions get adsorbed onto CNT surfaces via both chemical and physical sorption processes (Yu et al. 2014; Bankole et al. 2019). In some cases, more than a single force is involved in the adsorption; for instance, advanced adsorption of Cd(II) ions on Al_2O_3-decorated MWCNT hybrids involved a variety of forces, namely physical adsorption, cationic metal attraction, surface precipitation, electrostatic interactions, inner-sphere complexation, and complex formation between the hybrid MWCNTs and Cd(II) ions (Liang, Liu, et al. 2015). The oxidation process of CNTs introduces oxygen functionalities and hydrophilicity onto CNT surfaces making them partially negatively charged, which significantly improves their capacity toward adsorption of cationic pollutants.

2.1.2 Nano-Adsorbents

In the last two decades, nanotechnology has revolutionized all branches of science and technology such as medical, electronics, environment, catalysis, energy generation, and energy storage. In this sequence, water treatment procedures are not deprived of nanotechnology where it affords a special platform for the decontamination of water by offering the facility of manufacturing the materials down to the nanometer scale. In contrast to conventional materials for decontamination of water, nanomaterials due to their extraordinarily high surface area, capability of chemical modification and easier regeneration exhibited higher adsorption efficiencies and faster kinetic rates (Ray and Shipley 2015; Sadegh et al. 2017). To date, the potential of a number of low-cost, sustainable, and highly efficient nanomaterials possessing distinctive properties has been explored for reclamation of water resources. There are certain requirements that any ideal and efficient adsorbent must possess in order to be applied in water treatment processes such as higher surface areas for more interaction with the contaminants, selectivity, higher sorption capacity, faster reaction rates, capability to remove contaminants at low concentrations, low-dose requirements making the overall process economical, regeneration, and reuse (Hua et al. 2012). Nanomaterials proved themselves as the ideal and promising adsorbing materials for the decontamination of water containing a variety of contaminants such as organics, biological contaminants, and metallic species (Santhosh et al. 2016). The materials down to the nanometer scale often exhibit some fascinating properties, such as enhanced surface area-to-volume ratio to provide more sites for more surface interactions, and ease of functionalization to increase their functionality toward specific targets. For the application in the adsorption of heavy metals, these adsorbents can be easily functionalized to increase their affinity for particular metals (Santhosh et al. 2016; Ali 2012).

Among a variety of available nano-adsorbents, carbon-based nanomaterials, metal/metal oxide nanoparticles, and polymer-supported adsorbents are qualified as the promising ones for decontamination of water containing heavy metals.

2.1.2.1 Carbon-Based Nano-adsorbents

The carbon-containing nanomaterials such as graphene, graphene oxide, carbon nanotubes, and fullerenes have been successfully exploited in the decontamination of water laden with heavy metals. The detailed study of some of these carbon-based nanomaterials has been already discussed earlier in this section. Their intriguing properties such as high surface area, ease of functionalization resulting in multifunctional nano-adsorbents, ease of desorption of adsorbed metal ions, regeneration, and reuse, in addition to biocompatibility, make them potential adsorbents for removal of various hazardous pollutants from water resources (Ali et al. 2016; Gusain et al. 2020; Yang, Wan, et al. 2019; Baby et al. 2019).

2.1.2.2 Metal Oxide Nano-adsorbents

Transition metal oxide nanoparticles possess many fascinating properties due to which they have been extensively studied and exploited in areas related to energy and environment (Hua et al. 2012). In decontamination of water, these metal oxides play a crucial role as they provide a large surface area for greater interaction with various types of contaminants and high activities due to surface quantization effect. In addition, they exhibit advanced removal capacity, fast kinetics, and selectivity toward heavy metals, which would result in meeting the stringent regulations imposed by effectively removing the toxic metals. Thus, all these properties enable them to be widely explored as a potential candidate for adsorption of heavy metals. The commonly used metal oxide nanomaterials consist of nano-sized oxides of aluminum, cerium, copper, iron, manganese, zinc, titanium, magnesium and zirconium, cobalt ferrites, and zinc sulfides (Ray and Shipley 2015; Ali 2012; Hua et al. 2012).

Although metal-based nanoparticles find extensive application in water treatment processes, there exist some bottlenecks such as their poor stability due to agglomeration leading to decreased capacity and selectivity, and difficulty in their separation from aqueous medium. To overcome these drawbacks, these adsorbents are then impregnated onto some porous supports of large size, such as natural materials, activated carbon, and polymeric hosts, to achieve composite adsorbents. In addition, their facile separation from aqueous medium has been made possible by making their surface magnetic, i.e. magnetic metal oxide nano-adsorbents. Thus, magnetic metal oxide-based nanocomposites seem to be an efficient alternative in order to respond to these issues (Yang, Hou, et al. 2019; Hua et al. 2012).

2.1.2.3 Nanocomposite Adsorbents

Nanocomposites are the materials comprising various phases among which at least one phase must be in the nanometer scale. Based on their matrix type, nanocomposites are classified into three different categories: ceramic-matrix nanocomposites consisting of metals or ceramic as dispersed phase, metal-matrix nanocomposites having irregular reinforced materials as dispersed phase, and polymer-matrix nanocomposites comprising nanoparticles added into the polymer matrix (Yadav et al.

2019). The polymer-based nanocomposites possess perfect skeleton strength, tunable surface functional groups, viable regeneration, environmentally benign nature, biodegradability, very high surface area, and high porosity that make them the epitome for the application as potential adsorbents. Further, their properties such as active surface area, porosity, pore volume, and pore size are enhanced by functionalization of one or more components of the nanocomposites (Tesh and Scott 2014; Opoku et al. 2017).

Up to now, various nanocomposites have been extensively exploited for the removal of heavy metals from water through adsorption. As present in the literature, the following nanocomposites have been investigated for various metal ions; sulfonated magnetic nanocomposite based on reactive PGMA-MAn copolymer@ Fe_3O_4 nanoparticle for removal of copper ions (Hasanzadeh et al. 2016); poly(-aniline-co-m-phenylenediamine)@Fe_3O_4 nanocomposite for removal of Pb(II), Cd(II), and Co(II) (Zare and Lakouraj 2014); poly(N-vinylpyrrolidone-co-maleic anhydride)@eggshell/Fe_3O_4 for removal of Cd(II) and Pb(II) (Zare, Motahari, et al. 2018); poly(m-phenylenediamine)-grafted dextrin for removal of Pb(II) (Zare, Lakouraj, et al. 2018); poly(aniline-co-3-aminobenzoic acid)-based magnetic core-shell nanocomposite for removal of Pb(II) (Zare et al. 2016); and thiacalix[4] arene-functionalized chitosan for removal of Pb(II), Cd(II), Co(II), Ni(II), Cu(II), and Cr(III) are reported in the literature (Lakouraj, Mojerlou, et al. 2014; Lakouraj, Hasanzadeh, et al. 2014).

2.1.3 Inorganic–Organic Hybrids (Metal–Organic Frameworks)

A new class of adsorbing materials consisting of a three-dimensional lattice having a central metal ion or clusters and organic linkers known as metal–organic frameworks (MOFs) is gaining immense attention all over the world for water treatment processes within the past few decades. The most frequently used metal ions in MOFs are Zr(II), Cu(II), Fe(II), and Zn(II), and 2-methylimidazole, terephthalic acid, and trimesic acid are most commonly used as organic linkers (Drout et al. 2019; Yuan, Zou, et al. 2017; Cao et al. 2017; Xie et al. 2017). These synthetic porous materials are self-assembled through strong coordination bonds and use the combined effects of their organic and inorganic moieties to attain extraordinary properties such as ultrahigh and homogeneous porosity, tunable pore sizes, high surface area-to-mass ratio, and high thermal and chemical stability (Kobielska et al. 2018). The pores of crystalline MOFs are highly ordered, and their sizes as well as their shapes can be altered via a post-synthetic modification strategy to modify the ways of their interaction with target molecules by selectively choosing the metal ions and the linkers for intended applications. The high surface area and porosity of MOFs can assist in the accessibility of adsorption sites and diffusion of contaminants through the three-dimensional frameworks while considering their application as adsorbent materials. These porous solid materials can be altered to possess extremely high surface area that is a prerequisite condition for any material to be used as adsorbent in water treatment procedures, but due to their instability in aqueous medium, a very limited number of MOFs are reported for the decontamination of contaminated water (Bedia et al. 2019; Efome et al. 2018). However, efforts have been made to overcome the problem of low water stability associated with the application of MOFs in water treatment. The stability of MOFs in aqueous medium depends upon the strength of the metal–organic linker coordination bonds, and the problem of their collapse in the water arises when there exists competitive coordination between water and organic linkers with the metal ions. The stability of the framework is also associated with certain other factors such as the geometry of the coordination between metal–organic linker, hydrophobicity of the surface, crystallinity, and presence of defective sites. The use of certain additives or the formation of their composites such as graphene, carbon nanotubes, quantum dots, metal nanoparticles, and metal oxides for fabricating water-stable MOFs can have a positive effect on the framework stability and exhibit significantly improved properties (Feng et al. 2018). Further, for the removal of metallic species utilizing MOFs and their stability, the pH and temperature of the medium must be considered. MOFs proved themselves as excellent adsorbents by successfully decontaminating water containing heavy metals such as Hg, Pb, Pd, Cr, Cd, Cu, Th, U, Zn, Ni, Co, and Se (Li, Wang, et al. 2018; Peng et al. 2018; Okoro et al. 2018; Gu et al. 2018).

2.1.4 Green Adsorbents

Green adsorbents are the materials that are synthesized according to the principles of green chemistry, which encourages the use of nontoxic, biocompatible, and biodegradable precursors, and utilize energy-efficient as well as moderate reaction conditions for synthesizing the adsorbents (Ali et al. 2016; Kyzas and Kostoglou 2014).

There are various kinds of biological components (such as bacteria, algae, fungi, and plants) that have been extensively exploited as efficient resources for the synthesis of materials. Amid these naturally occurring components, plant biodiversity has been broadly considered by reason of their cost-effectiveness, environmental compatibility, widespread plethora, and sustainability. The phytochemicals and biomolecules found in various plant extracts, especially in leaves, such as polysaccharides, vitamins, carboxylic acids, aldehydes, ketones, flavones, amides, polyphenols, ascorbic acids, and terpenoids, act as capping agents, reducing agents, and/or complexing agents (Iravani 2011). The cooperative effect of these naturally procured phytochemicals may perhaps result in the development of adsorbents with precise size or shape, distinct morphologies, and improved stability (Renu and Singh 2016).

So far, a plethora of nano-sized metal oxides such as Al_2O_3, ZrO_2, CuO, Fe_xO_y, MnO_2, MgO, and ZnO are efficiently synthesized by the green approach and efficiently applied in the removal of numerous metal ions from water resources (Yuliarto et al. 2019; Singh et al. 2018; Reinsch 2016).

2.1.5 Zeolites

Zeolites are naturally occurring crystalline, environmentally and economically acceptable hydrated aluminosilicate materials consisting of an infinite three-dimensional framework of alumina and silica tetrahedron connected together by common oxygen atoms (Margeta et al. 2013). Further, this alumina–silica tetrahedron connection in different ways results in various types of zeolitic frameworks of different compositions. These are also called molecular sieves as they contain an organized distribution of micropores in molecular dimensions (Fernńdez-Reyes et al. 2020; Wang and Peng 2010).

In the primary structural unit of zeolites, the isomorphous substitution of silicon by aluminum causes an excess of negative charges on its surface that get accompanied by exchange with cations such as Na^+, Ca^{2+}, and K^+ ions in the framework (Bosso and Enzweiler 2002). These cations get further exchanged by other cations, making zeolites as cation exchangers and a most suitable alternative for elimination of lethal metallic ions from water resources (Belova 2019).

The chemical modification of zeolitic surface by inorganic or organic salts/surfactants results in the formation of positively charged oxyhydroxides or surfactant micelles, as a consequence of which the zeolite is able to bind anions too, in addition to cations, such as arsenates or chromates. The decomposition of zeolite frameworks into various rings of diverse sizes attributes to the opening of the pores that further lays the foundation of their categorization into small-, medium-, large-, and extra-large-pore zeolites accordingly (Li, Li, et al. 2017; Li and Yu 2014).

The utilization of natural zeolites in the treatment of water/wastewater containing several contaminants is one of the most established areas of their application. Zeolites exhibit various fascinating properties such as high porosity, high ion-exchange capability, high sorption capacity due to sieving properties, cost-effectiveness, and outstanding selectivity for diverse cationic species that commonly go together with a release of some exchangeable cations (e.g. K^+, Na^+, Ca^{2+}, and Mg^{2+}) that are relatively innocuous (Martínez and Corma 2011). In addition, they are easy to handle and allow cost-effective full-scale applications. These properties led zeolites to be explored for the treatment of water laden with heavy metals.

To improve the metal removal capacity and efficiency, zeolites have recently been modified by quite a few techniques such as acid treatment, ion exchange, and surfactant functionalization with metals or metal oxides (Li, Li, et al. 2017). The altered zeolites exhibit improved capacities; for instance, iron-coated clinoptilolite showed higher removal capacity, especially for Pt, in comparison with uncoated clinoptilolite; the apparent reason for this enhancement was attributed to the reduction of Pb(II) ions to Pb(0) on the surface of clinoptilolite caused by iron (Nguyen et al. 2015). Similarly, the advantage of reductivity of iron was utilized in the formation of nanoscale zerovalent iron/zeolite composites. The combined effect of large surface area possessed by Fe^0 nanoparticles and the stabilization influence of zeolites for Fe^0 nanoparticles improved the adsorption of Pb(II) ions (Arancibia-Miranda et al. 2016). In another case, the core-shell ZnO/Y particles depicted higher removal of Pb(II) ions than that of pristine zeolite (Y) and it can be explained by ZnO nanoflake formation on the surface of zeolite, which resulted into production of net negative charge on its surface that attracts the positively charged Pb(II) ions (Shaw et al. 2016).

The overall effectiveness of water decontamination processes by natural as well as modified zeolites entirely rests on several aspects such as the type, quantity, and size distribution of zeolite particles, and initial concentration, pH, and ionic strength of the solution containing contaminants. Temperature, pressure, contact time of the system zeolite/solution, and the presence of other organic compounds and anions/cations also affect the treatment procedures (Margeta et al. 2013).

2.2 CHARACTERISTICS OF ADSORBENTS

The effectiveness of the treatment of toxic and heavy metal-bearing water and wastewater through the process of adsorption is solely dependent upon the physical and chemical characteristics/properties of the adsorbents. The particle size, surface area (external/internal), porosity, pore size distribution, pore volume, pore structure, bulk density including the hydrophobic and hydrophilic behavior, chemical and thermal stability, surface morphology, presence of functional groups on the surface, and surface chemistry are some of the physical and chemical characteristics of the adsorbents that are reported to affect the adsorption type. The nature of the properties of the adsorbents forms the basis of their categorization into physical and chemical characteristics (Worch 2012a; Wang, Shi, Wang, et al. 2020).

2.2.1 Physical Characteristics

The morphological features of the adsorbent constitute the physical characteristics such as surface area, particle size, pore size, and density. The porosity of any solid is the part of the space available in the total volume of the material. Generally, the more porous the material, the more will be its adsorption capacity (Worch 2012a).

The external surface area is supposed to influence the mass transfer rate during the adsorption process. For porous adsorbents, the external mass transfer occurs via the formation of hydrodynamic sheath around the adsorbent, whereas internal mass transfer happens by way of intra-particle diffusion processes. The internal surface area is considered as the most important parameter of any adsorbent as the adsorption capacity of the adsorbents totally depends upon it. Sometimes, it exceeds several fold the external surface area of the porous adsorbents. This is the reason that led research groups worldwide to give emphasis on the synthesis of materials with extremely large internal surface areas.

The materials possessing large internal surface areas are endowed with a plethora of pores of variable shapes and sizes that form the basis of their classification into macropores, mesopores, and micropores based on their sizes. Macropores and mesopores play a significant role in the mass transfer of adsorbate into the internal part of the adsorbent, and the volume possessed by the micropores aids in the determination of size of the internal surface and, in turn, the adsorption capacity of the adsorbent. Higher micropore volume exhibits the larger quantity of the material that will be adsorbed. However, in the case of minute pores and large adsorbate molecules, the size exclusion limits the extent of adsorption.

The particle size is of extreme importance in the case where nanomaterials are used as adsorbents. The unique features that endow them as ideal adsorbents are all size-dependent, particularly large surface area and surface area-to-volume ratio leading to their improved adsorption capacity. The high surface area provides a greater number of active sites for interaction with a diverse class of contaminants. The particles having their sizes in the nanoscale range are highly reactive and possess catalytic potential, enabling them as a better alternative than conventional materials (Worch 2012a; Hua et al. 2012; Ali 2012).

2.2.2 Chemical Characteristics

The chemical composition, existence of diverse functional groups on the surface-active sites of the adsorbent and their interaction with a wide range of contaminants, and surface chemistry constitute the chemical characteristics of the adsorbent.

The interaction between the adsorbate and the adsorbent is largely affected by the surface chemistry of the adsorbent, especially for the case where adsorption of ions occurs onto oxidic surfaces (Worch 2012a). In this perspective, the pH_{pzc} of the adsorbent assists in the comprehension of the influence of pH on the overall adsorption of charged species on their surfaces. The point of zero charge (pH_{pzc}) evaluates the point at which any adsorbent is electrically neutral in nature and is referred to as a region in the pH scale where the surface bears a net zero charge due to the equal sum of positive and negative charges on the surface of the adsorbent. When the pH of the medium is lower than the pH_{pzc} of the adsorbent – the surface becomes positively charged and the medium becomes acidic due to donation of protons more than the hydroxide groups by the acidic water (Roushani et al. 2017). Thus, attraction of negatively charged species becomes feasible toward positive adsorbent surface. On the contrary, at a pH value higher than the pH_{pzc}, the surface bears negative charges elicited by the attraction of cations and vice versa. The point of zero charge (pH_{pzc}) is the characteristic feature of adsorbents. Generally, the electrostatic attraction/repulsion forces strongly affect the adsorption of charged adsorbate molecules on the charged adsorbent surfaces (Worch 2012a; Hua et al. 2012; Ali 2012).

In recent years, considerable effort has been made in the modification of adsorbent surfaces or functionalization of adsorbents in order to overcome their inherent limitations and enhance their low adsorption capacities and selectivity of both organic and inorganic moieties from water. The adsorbent surfaces are being modified with a variety of different compounds to alter their wide range of surface characteristics such as stability, functionality, reactivity, biocompatibility, dispensability, and inertness in intolerable environments (Manyangadze et al. 2020). Surface modification or functionalization has been carried out with a multiplicity of agents; for example, some carbon materials are activated by H_2SO_4, H_3PO_4, $ZnCl_2$, H_2O_2, KOH, etc., and certain ligands are also used, for example, surfactants, polymers, dendrimers, small molecules, and biomolecules. Surface modification also aids in controlling the aggregation of nanoparticles, which is their perennial property, assists in their easy separation when the surface is magnetically functionalized, results in high density of reactive sites, and assures high and rapid adsorption rates. Oftentimes, it is witnessed that the degree of modification/functionalization of adsorbent surface leads to a decrease in specific surface area but eventually results in an increased adsorption capacity (Yang, Wan, et al. 2019; Worch 2012a).

2.2.3 Adsorption

Adsorption is a surface phenomenon defined as the accumulation of any substance from the bulk to the surface of any material in contact. This accumulation results from the surface energy generated by the imbalanced forces of attraction experienced by the molecules of any substance over the material surface (Dąbrowski 2001).

The surface-active materials, i.e. the materials capable of adhering other substance onto their surfaces, are called adsorbents, and the materials that are accumulated on the adsorbent surfaces are referred to as adsorbates. The strength with which the adsorbate molecules are adhered or attached on the surface of adsorbents dictates the nature of adsorption as physical or chemical (Worch 2012a).

2.2.3.1 Physical Adsorption

Physical adsorption or physisorption is an adsorption process in which the forces operating amid adsorbate and adsorbent are weak intermolecular forces (van der Waals forces) of attraction. This is characterized by the formation of multiple layers of adsorbed molecules over the adsorbent surface via van der Waals forces, electrostatic interaction, hydrogen bonds, weak covalent bonding, or dipole–dipole interaction (Lima et al. 2015).

2.2.3.1.1 Features of Physisorption

The salient features of physisorption are as follows (Grassi et al. 2012):

- **Energetics and kinetics:** Due to weak van der Waals force of attraction, physisorption is described by lower enthalpy values (i.e. 20–80 kJ/mol) and is generally an exothermic process. The activation energy is also low and the sorption process is reversible in nature (Piccin et al. 2017).
- **Temperature:** Because of the exothermic nature of physisorption, adsorption declines with an increase in temperature.
- **Specificity:** The presence of universal van der Waals forces of attraction makes the adsorbent surface indifferent for any adsorbate; i.e. physical adsorption is inexplicit in nature.
- **Surface area of adsorbent:** Generally, the surface area of the adsorbent increases the magnitude of physical adsorption. Accordingly, finely divided porous substances, metals, etc., are supposed to be good candidates as adsorbents.

2.2.3.2 Chemical Adsorption

Chemical adsorption or chemisorption is an adsorption process in which the operational force amid the adsorbate–adsorbent is a valence force similar to that involved in the formation of any chemical compound. Due to the involvement of valence bonds, the adsorbed molecules of the adsorbate get linked to the adsorbent surface, usually reside in certain adsorption sites available on the surface, and eventually lead to the formation of a unilayer (monolayer) of chemisorbed molecules (Lima et al. 2015; Crini and Badot 2008).

2.2.3.2.1 Features of Chemisorption

The salient features of chemisorption are as follows (Grassi et al. 2012):

- **Energetics and kinetics**: In chemisorption, bond between the adsorbed molecules of the adsorbate and the adsorbent surface is characterized by higher values of enthalpy (80–450 kJ/mol). The magnitude of energy associated with chemisorption is of the same order as the energy change in a chemical reaction between a solid and a fluid. Therefore, chemisorption may be of exothermic or endothermic nature and the magnitude may vary from very small to very large. The fundamental steps in chemisorption involve activation energy (also called activated adsorption), so chemisorption possesses high activation energy and takes place steadily. The chemisorption process is practically irreversible in nature (Piccin et al. 2017).

- **Temperature**: Chemisorption involves higher kinetic energy barrier, so it does not favor lower temperature scale. Thus, resembling any other chemical reaction, chemisorption increases with increasing temperature up to a certain value and then shows downfall.
- **Specificity**: Chemical adsorption occurs only when there is any possibility of chemical interaction within the molecules of the adsorbate and the adsorbent surface for the formation of chemical bond. Consequently, chemisorption is highly specific in nature.
- **Surface area of the adsorbent**: Similar to physisorption, the extent of chemisorption also escalates with an increase in the surface area of the adsorbent.

2.3 CHEMICAL NATURE OF ADSORBENTS AND INTERACTION WITH METALS AND ANIONS

The variety of adsorbents possessing one or more unique morphological features and their characteristic properties including both physical and chemical properties that empower them to be extensively applied for the elimination of lethal organic and inorganic pollutants from water resources have been already discussed earlier. This section deals with the different types of interactions among the adsorbent and the adsorbate molecules leading to adsorption of contaminants.

Among various factors that affect the overall process of adsorption and the adsorption capacity, the presence of certain groups on the surface of the adsorbent plays a pivotal role (Cashin et al. 2018). Every adsorbent possesses its unique characteristics and functionalities such as particle size, surface area (external/internal), porosity, pore size distribution, pore volume, pore structure, bulk density including the hydrophobic and hydrophilic behavior, chemical and thermal stability, and surface morphology, in addition to attached functional groups that are quite essential for their surface chemistry and the adsorption of pollutants. In most of the cases, the interaction amid the metallic species present in the solution and the functional groups attached on adsorbents significantly improve their adsorption (Wang, Gao, et al. 2015; Bian et al. 2015). The functional groups get attached on the surface of adsorbents through the heteroatoms such as oxygen, nitrogen, sulfur, phosphorus, and halogens, which further form the basis of the classification of functional groups as oxygen-bearing functional groups, nitrogen-bearing functional groups, sulfur-bearing functional groups, etc. (Yang, Wan, et al. 2019; Cashin et al. 2018).

As a whole, the role of functional groups in the adsorption of a range of contaminants is quite complex, and the intermolecular interactions between the former and the latter are dependent upon the nature, heterogeneity, and surface chemistry of the adsorbent as well as on the ionic environment of aqueous solutions.

2.3.1 Adsorption Mechanisms of Metals and Anions

The metal-binding mechanism during adsorption of toxic metallic species from water resources onto various adsorbents takes place through either of the following processes: physical adsorption, ion exchange, electrostatic interaction, surface complexation, and precipitation (Figure 2.3); or sometimes several mechanisms work simultaneously in a particular aqueous environment (Xu and McKay 2017). Usually, chemisorption (i.e. precipitation, ion exchange, and surface complexation) significantly influences the metal adsorption than physisorption. All through the adsorption process, the mechanism that governs the whole process exclusively depends on the target metal (adsorbate), the adsorbent, ionic environment, and pH of the solution. The solution pH is an important factor because it affects the speciation of metal in the given aqueous medium, charge on the adsorbent surface, and complexation behavior of functional groups. The various underlying mechanisms of metal adsorption on the adsorbent surfaces in the course of the adsorption process are briefly discussed here (Yang, Wan, et al. 2019; Worch 2012a).

2.3.1.1 Physical Adsorption

It is the movement of metallic ions into the pores of the adsorbent and their subsequent attachment to the surface by weak forces such as H-bonding and van der walls forces. It is strongly impacted by the surface area and pore size distribution of the adsorbents along with the nature of the target metal. Further, the heterogeneity and polarity of the adsorbent surface in addition to the presence of

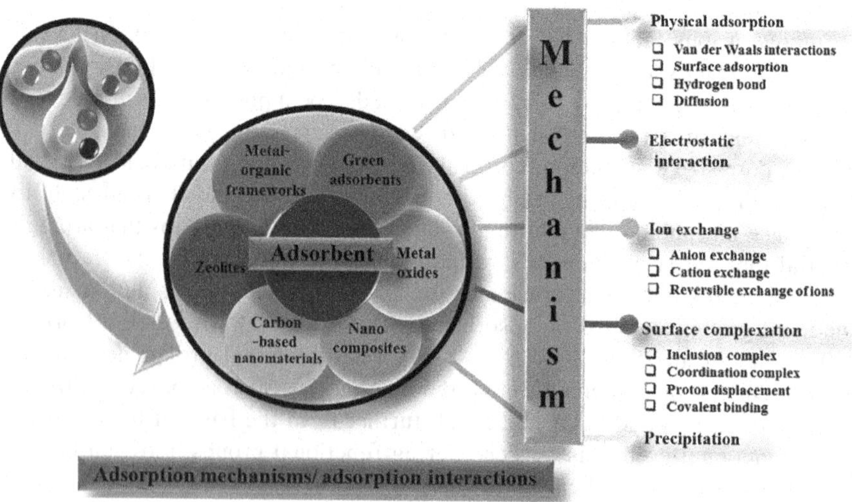

FIGURE 2.3 Illustration of mechanisms involved during adsorbate–adsorbent interaction in aqueous solutions.

functional groups facilitate the physical adsorption by helping in the movement of metal ions by means of electrostatic attraction and ion–dipole forces. It is most common and serves as a principal mechanism of adsorption of metal ions (Worch 2012a; Lu, Yu, et al. 2017; Moreno-Barbosa et al. 2013; Verma and Dutta 2015; Huynh et al. 2017).

2.3.1.2 Ion Exchange

The existence of oxygen-containing functional groups on the surface of the adsorbent leads to the adsorption of metal ions via an ion-exchange mechanism (Dong et al. 2018). Its efficiency depends on the divalent metallic species and protons of the oxygen-containing functional groups for which the ion size and surface chemistry of the adsorbent play a significant role. The solution pH is the dominating factor and cation-exchange capacity (CEC) is an imperative indicator of the adsorption of metallic species taking place via an ion-exchange mechanism (Liu, Zhu, et al. 2013; Kyzas et al. 2016; Hao et al. 2010).

2.3.1.3 Electrostatic Interaction

The electrostatic interaction mechanism is operative for removal of metallic species when there is an interaction between positively charged target metals and negatively charged adsorbents (surface containing functional groups). It is a relatively weak process and is dependent on the pH of the medium and pH_{pzc} of the adsorbent. The charged interface between the adsorbent surface and the medium is influenced by ionization of the existing functional groups on the adsorbent surface (Moreno-Barbosa et al. 2013; Afroze et al. 2016; Cui et al. 2016; Zeng et al. 2015; Yang, Tang, et al. 2014; Yuan, Zhang, et al. 2017; Cheng et al. 2012; Velazquez-Jimenez et al. 2014; Xiao et al. 2016; Lu, Yu, et al. 2017; Bayramoglu and Arica 2016).

2.3.1.4 Surface Complexation

Surface complexation leads to the formation of complexes (inner- and/or outer-sphere) during the interaction of metal ions and functional groups present on the adsorbent surface in the course of the adsorption process (Zhou et al. 2017; Kyzas et al. 2016; Shi et al. 2015; Sankararamakrishnan et al. 2014; Ma et al. 2011; Yuan, Zhang, et al. 2017; Gupta et al. 2014; Long et al. 2013; Bayramoglu and Arica 2016; Guo, Zhang, et al. 2017). The adsorption of metal onto carbon-based adsorbents is mainly governed by this mechanism (Liu, Han, et al. 2017; Li et al. 2010).

2.3.1.5 Precipitation

It includes the formation of a solid product in the solution during the adsorption of metal ions (Wang et al. 2018; Huynh et al. 2017). It serves as one of the major mechanisms for removal of metal ions and occasionally works cooperatively together with other adsorption mechanisms such as electrostatic interaction, ion exchange, and surface complexation (Inyang et al. 2016). The functional groups present on the adsorbent surfaces exhibit an insignificant direct effect on the precipitation of metal ions; however they promote precipitation indirectly by influencing other mechanisms of metal adsorption (Yang, Wan, et al. 2019).

2.4 EFFECT OF FUNCTIONALIZATION OF ADSORBENTS ON ADSORPTION OF METALS AND ANIONS

Adsorbent materials usually possess certain inherent limitations such as low adsorption capacity, small metal-binding capacities, prohibitive cost, small selectivity, and instability. Thus, in order to surmount the limitations, the surface of adsorbents is altered or modified to enable them to be well equipped with improved characteristics for desired applications. For this purpose, certain foreign functional groups are introduced into the adsorbent surface to enhance their performance for intended applications (Cashin et al. 2018; Wei et al. 2016; Yang, Wan, et al. 2019; McCarthy et al. 2012). An overview of the performance of a variety of adsorbents containing oxygen, nitrogen, and sulfur functional groups on the adsorption of hazardous metals and anions in terms of the contribution of specific functional groups and maximum adsorption capacities is given in this section and summarized in Table 2.1.

2.4.1 Oxygen-Bearing Functional Groups

Functional groups having oxygen atoms such as carboxyl and hydroxyl are the most widely used groups for surface modification of adsorbents. These groups significantly influence the surface properties, surface behavior, and surface reactions widely contributing to the adsorption of pollutants from water resources as discussed here in this section.

The introduction of a carboxylic functional group positively enhances Cr(VI) adsorption on activated carbon (AC), and on carbon nanotubes (CNTs), it showed a negligible effect on the removal of Cr(VI) (Ihsanullah, Al-Amer, et al. 2016). The acid treatment opened some of the blocked pores increasing the surface area of AC and forming more active adsorption sites, resulting in the enhancement of removal performance. For CNTs, although the acid treatment has introduced more active adsorption sites over the surface, the introduced and adhered negatively charged carboxylic groups lead to electrostatic repulsion between the CNT surface and chromate ions ($HCrO_4^-$) present in the solution. Thus, acid treatment has no major part in the adsorption capacity of CNTs, and both the adsorbents (raw and modified) have comparable adsorption capacity for chromium ions.

Zhou et al. (2017) tested magnetic gelatin-modified biochar (MG-CSB) for As(V) removal and reported that functional groups having oxygen atoms significantly contributed to the increased removal of As(V). Biochar contains various oxygen-bearing groups, e.g. carboxyl, carbonyl, and hydroxyl, and bears a net negative charge due to dissociation of these groups. Under the experimental conditions, i.e. at pH 4, As(V) exists as $HAsO_4^{2-}$ in the solution, the protonation and deprotonation of surface –OH groups make the iron oxide positively charged, and also some of the groups of biochar become protonated. All these cations and anions present in the solution consequently resulted in increased adsorption through electrostatic interaction and hydroxyl complexation between As(V) and MG-CSB.

The surface of nanosilica was modified with oxalic acid and tartaric acid and explored for Co(II) ions (Mahmoud, Yakout, et al. 2016). The modification increased the roughness of the surface of the adsorbent due to binding with

TABLE 2.1
Functionalized Adsorbents for the Adsorption of Metals and Anions

S. No.	Adsorbent	Functional Group	Contaminant	Adsorption Capacity (Unit)	References
1.	Acid-modified activated carbon (AC) and carbon nanotubes (CNTs)	-COOH	Cr(VI)	Modified CNTs = 1.314, modified AC = 18.519	Ihsanullah, Abu-Sharkh, et al. (2016)
2.	Magnetic gelatin-modified biochar	-OH	As(V)	45.8 mg/g	Zhou et al. (2017)
3.	Carboxylic acid-functionalized nanosilica	-COOH, -OH, -CO, -Si-O-H	Co(II)	NSi-Ox = 58.82 mg/g, NSi-Tar = 111.1 mg/g	Mahmoud, Yakout, et al. (2016)
4.	Acid-modified activated carbon prepared from potato peels	-COOH,-OH,-CO,-C-O-C-	Co(II)	PoP400 = 373 mg/g, PoP600 = 405 mg/g	Kyzas et al. (2016)
5.	Carboxylate-functionalized sugarcane bagasse	-COOH, -CO	Co(II), Cu(II), Ni(II)	Co(II) = 0.561 mmol/g, Cu(II) = 0.935 mmol/g, Ni(II) = 0.932 mmol/g	Ramos et al. (2016)
6.	Crown ether-modified activated carbon cloth (ACC)	H-bonding	Cr(III), Co(II), Ni(II)	Cr(III) = 0.2202 mmol/g, Co(-II) = 0.1302 mmol/g, Ni(II) = 0.1524 mmol/g	Duman and Ayranci (2010)
7.	Activated carbons of watermelon (GACW) and walnut shell (GACN) activated by phosphoric acid	-OH, -CO	Pb(II), Zn(II)	GACW: Pb(II) = 40.984 mg/g and Zn(-II) = 11.312 mg/g; GACN: Pb(II) = 32.362 mg/g and Zn(II) = 6.079 mg/g	Moreno-Barbosa et al. (2013)
8.	Base-modified *Eucalyptus sheathiana* bark	-OH	Zn(II)	250 mg/g	Afroze et al. (2016)
9.	Catechol-functionalized nanosilica	-OH	Ge(IV)	0.048 mg/m^2	Cui et al. (2016)
10.	Carboxyl functional magnetite nanoparticles	-COOH	Pb(II), Cd(II), Cu(II)	Pb(II) = 74.63 mg/g, Cd(II) = 45.66 mg/g, Cu(II) = 44.84 mg/g	Shi et al. (2015)
11.	α-Ketoglutaric acid-modified magnetic chitosan	-COOH, -OH	Cd(II)	255.77 mg/g	Yang, Tang, et al. (2014)

(continued)

TABLE 2.1 (*Continued*)
Functionalized Adsorbents for the Adsorption of Metals and Anions

S. No.	Adsorbent	Functional Group	Contaminant	Adsorption Capacity (Unit)	References
12.	Polyacrylic acid-modified magnetic mesoporous carbon	-COOH	Cd(II)	406.6 mg/g	Zeng et al. (2015)
13.	Functionalized multiwalled carbon nanotubes	-COOH, -Fe-O	As(III), As(V)	As(III) = 111.1 ± 4.8 mg/g, As(-V) = 166.66 ± 5.8 mg/g	Sankararamakrishnan et al. (2014)
14.	Catechol-functionalized activated carbon	-OH	Ge(IV)	-	Marco-Lozar et al. (2007)
15.	Phosphate-modified montmorillonite (PMM)	P-OH	Co(II), Cs(I), Sr(II)	Co(II) = 0.2143 mmol/g, Cs(I) = 0.4292 mmol/g, Sr(II) = 0.1513 mmol/g	Ma et al. (2011)
16.	Triazole-4-carboxylic acid-functionalized poly(glycidyl methacrylate) microspheres	-COOH, -OH, -NH-, -N≡	Pb(II)	69.41 mg/g	Yuan, Zhang, et al. (2017)
17.	Activated carbon from scrap tires	-OH, -CO	Ni(II)	25 mg/g	Gupta et al. (2014)
18.	$ZnCl_2$ activated *Glycyrrhiza glabra* residue carbon	-COOH, -OH, -CO	Pb(II), Ni(II)	Pb(II) = 200 mg/g, Ni(II) = 166.7 mg/g	Mohammadi et al. (2014)
19.	$ZnCl_2$ activated coir pith carbon	-COOH, -OH, -CO	V(V)	24.9 mg/g	Namasivayam and Sangeetha (2006)
20.	H_2O_2-modified attapulgite	-OH, -Si-O	Sr(II)	85.54 μg/g	Liu and Zheng (2017)
21.	KOH-modified activated carbon from textile sewage sludges	-OH, -CO	Sr(II)	12.11 mg/g	Kaçan and Kütahyalı (2012)
22.	Fe_3O_4 particle-modified sawdust	-OH, -CO, -Fe-O	Sr(II)	12.59 mg/g	Cheng et al. (2012)
23.	Zirconium–carbon hybrid sorbent	-COOH, -OH, -CO, -Zr-O	F^-	17.70 mg/g	Velazquez-Jimenez et al. (2014)
24.	Acrylamide-functionalized electrospun nanofibrous polystyrene	-NH, -NCO	Cd(II), Ni(II)	Cd(II) = 10 mg/g, Ni(II) = 4.9 mg/g	Bahramzadeh et al. (2016)
25.	Polyacrylamide-functionalized Fe_3O_4	$-NH_2$	Pb(II)	Pb(II) = 158.73 mg/g	Moradi et al. (2017)

(*continued*)

TABLE 2.1 (*Continued*)
Functionalized Adsorbents for the Adsorption of Metals and Anions

S. No.	Adsorbent	Functional Group	Contaminant	Adsorption Capacity (Unit)	References
26	Amino-functionalized metal–organic frameworks MIL-101(Cr)	-NH	Pb(II)	Pb(II) = 81.09 mg/g	Luo, Ding, et al. (2015)
27.	Amine-modified poly(glycidyl methacrylate)-grafted cellulose	$-NH^+-(CH_3)_2Cl^-$	V(V)	V(V) = 197.75 mg/g	Anirudhan et al. (2009)
28.	Dendrimer-functionalized graphene oxide	$-NH_2$	Se(IV), Se(VI)	Se(IV) = 60.9 mg/g, Se(VI) = 77.9 mg/g	Xiao et al. (2016)
29.	γ-Aminopropyltriethoxysilane-modified chrysotile	$-NH_2$	Cu(II)	Cu(II) = 1.574 mmol/g	Liu, Zhu, et al. (2013)
30.	Ammonia-modified graphene oxide	-NH	U(VI)	U(VI) = 80.13 mg/g	Verma and Dutta (2015)
31	Cetyltrimethylammonium bromide-modified graphene	$-NH_2$, -COOH, -OH	Cr(VI)	Cr(VI) = 21.57 mg/g	Wu et al. (2013)
32.	Ethylamine-modified montmorillonite	$-NH_2$, -OH	Cs(I)	Cs(I) = 80.27 mg/g	Long et al. (2013)
33.	Ethylenediamine-modified β-zeolite	$-NH_2$	Ni(II)	Ni(II) = 6.67×10^{-4}–1.44×10^{-4} mol/g	Liu, Yuan, et al. (2017)
34.	(3-Aminopropyl)triethoxysilane-functionalized coal fly ash modified with mesoporous siliceous material	$-NH_2$	Cu(II)	Cu(II) = 1.41 mg/g	Pizarro et al. (2015)
35.	Glycine-modified chitosan resin	$-NH_2$, -OH, -COOH	Mn(II)	Mn(II) = 71.4 mg/g	Al-Wakeel et al. (2015)
36.	3-(2-Aminoethylamino)-propyldimethoxymethylsilane-functionalized MCM-41	$-NH_2$	Hg(II)	Hg(II) = 231.92 mg/g	Zhu et al. (2012)
37.	Amino-functionalized magnetic nanoparticles	$-NH_2$	Cu(II)	Cu(II) = 25.77 mg/g	Hao et al. (2010)
38.	Amino-functionalized mesoporous silica SBA-15	$-NH_2$	U(VI)	U(VI) = 573 mg/g	Huynh et al. (2017)

(*continued*)

TABLE 2.1 (*Continued*)
Functionalized Adsorbents for the Adsorption of Metals and Anions

S. No.	Adsorbent	Functional Group	Contaminant	Adsorption Capacity (Unit)	References
39.	Xanthate-modified cross-linked magnetic chitosan/poly(vinyl alcohol) particles	$-NH_2$, -SH, -C=S	Pb(II), Cu(II)	Pb(II) = 59.85 mg/g, Cu(II) = 139.79 mg/g	Lv et al. (2017)
40.	Poly(allylamine)-modified magnetic graphene oxide	-OH, -NH	Se(IV), Se(VI)	Se(IV) = 120.1 mg/g, Se(VI) = 83.7 mg/g	Lu, Yu, et al. (2017)
41.	Thiol-functionalized ionic liquid-based mesoporous organosilica	-SH	Hg(II), Pb(II)	Hg(II) = 105.26 mg/g, Pb(II) = 11.40 mg/g	Elhamifar et al. (2016)
42.	Sodium dodecyl sulfate-modified chitosan beads	-S=O, -COS	Cd(II)	125 mg/g	Pal and Pal (2017)
43.	Sulfur-functionalized ordered mesoporous carbons	-C-S, -C=S, -COS, -SOx	Hg(II), Pb(II), Cd(II), Ni(II)	Hg(II) = 70.75 mg/g, Pb(II) = 29.98 mg/g, Cd(II) = 4.96 mg/g, Ni(II) = 1.2 mg/g	Saha et al. (2016)
44.	Sulfur-functionalized silica microspheres	-C-S, -SH, -SR	Hg(II)	62.3 mg/g	Saman et al. (2013)
45.	Thiol-functionalized hollow mesoporous silica microspheres	-SH	Hg(II)	118.6 mg/g	Zhang, Wu, et al. (2015)
46.	1,2-Ethanedithiol-functionalized vinyl-functionalized covalent organic frameworks	-SH, -SR	Hg(II), Hg(0)	Hg(II) = 1350 mg/g, Hg(0) = 863 mg/g	Sun et al. (2017)
47.	Sulfurized activated carbon from bagasse pith	$-SO_3H$	Zn(II)	147 mg/g	Anoop Krishnan et al. (2016)
48.	Acrylamide- and hydroxyl-functionalized metal–organic framework	-OH, -C=O	Hg(II)	278 mg/g	Luo, Chen, et al. (2015)

(continued)

TABLE 2.1 (*Continued*)
Functionalized Adsorbents for the Adsorption of Metals and Anions

S. No.	Adsorbent	Functional Group	Contaminant	Adsorption Capacity (Unit)	References
49.	Amidoxime- and carboxy-modified MCM-41 silica particles	-N=, -NH-, -OH, -COOH	U(VI)	442.3 mg/g	Bayramoglu and Arica (2016)
50.	Nitrilotriacetic acid anhydride-modified cornstalk	-COOH, $-NH_2$	Cd(II), Pb(II)	Cd(II) = 143.4 mg/g, Pb(II) = 303.5 mg/g	Huang, Yang, et al. (2015)
51.	Urea phosphate activated carbon derived from *Phragmites australis*	-COOH, -CN, -OH, -NH, -C=O	Cd(II)	40.65 mg/g	Guo, Zhang, et al. (2017)
52.	Chitosan-modified vermiculite	-OH, $-NH_2$	As(III)	72.2 mg/g	A. Saleh et al. (2016)
53.	1,2,4-Triazole-modified lignin-based adsorbent	–CH=CH–S–, –N=CH–, –NH	Cd(II)	87.4 mg/g	Jin et al. (2017)
54.	Modified chitosan microspheres	$-NH_2$, -OH	NO_3^-, PO_4^{2-}	NO_3^- = 32.15 mg/g, PO_4^{2} = 33.90 mg/g	Zhao and Feng (2016)

carboxylic acids. The roughness further increased the surface area of the sorbent due to the increased number of active sites. Further, the modification provided mechanical stability to the adsorbent to retain the spherical shape; however, a slight increase in the particle size was visualized that confirmed the complete coating of nanosilica with carboxylic acids. Metal binding to the adsorbent proceeds through a cation-exchange mechanism where the active silanol groups assisted the adsorbent in solid-phase extraction. The modification introduced additional functional groups such as -COOH, -CO, and –OH that intensely participate in metal binding along with the silanol groups that further advocated the mechanism similar to weak acid cation exchangers. In addition, a complex between Co(II) ions and the attached oxygen-donating surface functional groups also formed during the removal of target metal ions, and the selectivity of the adsorbents was governed on the basis of the soft–hard acid–base rule. Similarly, acidic modification of activated carbon prepared from potato peel was also tested for Co(II) ions where activation with H_3PO_4 introduced oxygen-containing functional groups such as carboxylate, hydroxyl, phenols, and epoxy that helped in the efficient removal of metal ions (Kyzas et al. 2016). The mechanism of adsorption was explained on the basis of FTIR where the peaks corresponding to carboxylate, hydroxyl, phenol, and epoxy groups got eliminated after adsorption, showing their involvement in adsorption. Metal binding proceeds mutually through the cation-exchange, electrostatic interaction, and surface complexation mechanisms. Ramos et al. (2016) investigated carboxylate-functionalized sugarcane bagasse for adsorption of Co(II), Cu(II), and Ni(II), and FTIR spectra revealed the underlying adsorption mechanism. The FT-IR spectrum of metal ion-loaded sugarcane bagasse exhibited splitting of the band in the region of $1685\,cm^{-1}$, which can be attributed to the stretching of C=O bond in the carboxylic acid groups. This depicted that the carboxylic groups deprotonated during the course of adsorption and involved in adsorption of metal ions. The metal ion adsorption on carboxylate-functionalized sugarcane bagasse involves an ion-exchange mechanism amid hydronium ions and metal ions with subsequent complex formation between metal ions and carboxylate groups in the order Cu(II) > Ni(II) > Co(II). Similarly, crown ether-modified activated carbon cloth (ACC) was investigated for possible enhancement in the adsorption capacity of Cr(III), Co(II), and Ni(II) ions (Duman and Ayranci 2010). Among the monobenzo and dibenzo derivatives of crown ethers, the former are more effective in enhancing the adsorption of metal ions under study than the latter irrespective of their cavity sizes. The probable reason is their attachment onto ACC through the benzene ring that provides more space on the surface, more attachment sites, and flexible cavities for adherence of metal ions. On the contrary, both the benzene rings are involved in the attachment of dibenzo derivatives to ACC, resulting in less space and less flexible cavities, thereby leading to less enhancement in the adsorption of metal ions. The H-bonding among functional groups of ACC and metal ions along with the normal ionic interactions is involved in the adsorption where Cr(III) ions showed greater removal than Co(II) and Ni(II) ions.

The adsorption of Pb(II) and Zn(II) ions on activated carbons of watermelon (GACW) and walnut shell (GACN) activated by phosphoric acid followed physisorption due to the heterogeneous distribution of pores on GACW which could avoid pore blocking and electrostatic interaction between the negatively charged GACW surface and positively charged test metal ions (Moreno-Barbosa et al. 2013).

Eucalyptus sheathiana bark was modified by NaOH for adsorption of Zn(-II) metal ions where electrostatic interaction between zinc cations and negatively charged surface increased the adsorption performance (Afroze et al. 2016). Further, cations with surface hydroxyl groups formed a surface complex with subsequent formation of other complexes after the adsorption of metal ions. A comparable remark was for the result exhibited by the removal of Ge(IV) ions on catechol-functionalized nanosilica, where the adsorption mechanism also involved electrostatic interaction and the complex formation amid germanium ions and hydroxyl groups of catechol-functionalized nanosilica (Cui et al. 2016) and catechol-functionalized activated carbon (Marco-Lozar et al. 2007). The surface complexation mechanism of adsorption was adopted by carboxyl-functionalized magnetite nanoparticles (CMNPs) for removal of Pb(II), Cd(II), and Cu(II) where negatively charged carboxylate ions capture the metal ions by chelating and forming complexes with them (Shi et al. 2015). The electrostatic interaction between Cd(II) ions and carboxyl groups containing polyacrylic acid-modified magnetic mesoporous carbon (Zeng et al. 2015) and α-ketoglutaric acid-modified magnetic chitosan (Yang, Tang, et al. 2014) is found responsible for the acceleration in the monolayer adsorption of cadmium ions.

The adsorption of As(III) and As(V) on zerovalent iron nanoparticles immobilized on oxidized multiwalled carbon nanotubes (MWCNTs) using EDTA as chelating agent (ZCNT) involved solution–oxidation–adsorption and surface complexation as the underlying mechanisms. Due to aerial oxidation, ZVI generates Fenton's reagent that oxidizes As(III) to As(V) and hydrated ferric oxides are formed on the surface with which As(V) forms complex. EDTA plays a bifunctional role as a chelating agent for arsenic and helps in the preservation of zerovalent state of iron. ZVI and EDTA conjointly aid in the enhancement of adsorption capacity (Sankararamakrishnan et al. 2014).

A combination of adsorption mechanisms were reported for the removal of Co(II), Cs(I), and Sr(II) on phosphate-modified montmorillonite (PMM) (Ma et al. 2011). Surface modification by phosphate increases the surface area and pore volume of the adsorbent due to the formation of inner-sphere complex with sorbed phosphate ions and Al-O-P-OH surface precipitates. Among the target metal ions, Cs(I) ions were preferentially adsorbed due to stronger electrostatic attraction. In addition, the phosphate modification provided extra sorption sites for Cs(I) ions. Further, high selectivity of Cs^+ as compared to K^+ resulted in its exchange with K^+ into the interlayer of PMM, by the ion-exchange method. Co(II) ions formed $COOH^+$ that interacts with negatively charged adsorbent surface, thus assisting in adsorption. In addition, the hydroxyl groups of phyllosilicate sheets bound to Co(II). Adsorption was found to be strongly pH-dependent, which advocates surface complexation as the main mechanism of Co(II) adherence onto PMM. The adsorption of Sr(II) ions involved ion exchange ($pH < 8$) and surface complexation ($pH > 8$) as the operating mechanisms.

Similarly, multiple mechanisms such as electrostatic interactions, ion exchange, and complexation were accounted for the adsorption of Pb(II) ions on triazole-4-carboxylic acid-functionalized poly(glycidyl methacrylate) microspheres (Yuan, Zhang, et al. 2017). The terminal carboxylic groups are deprotonated to form negatively charged COO^- that binds lead ions via electrostatic interactions and ion-exchange methods, and imine (-NH-) and tertiary amino groups (-N=) of triazole moieties donate their lone pair of electrons to empty atomic orbital of lead ions forming bidentate and tetradentate triazole-Pb(II) chelates or complexes. The OH^- groups also interact with Pb(II) ions through chelation or complexation.

The activated carbon prepared from scrap tires was activated by H_2O_2 and tested for adsorption of Ni(II) ions (Gupta et al. 2014). The functional groups such as hydroxyl and carbonyl on the surface of adsorbent undergo protonation at lower pH values and deprotonation at higher pH values. At $pH < pH_{pzc}$, electrostatic repulsion between positively charged species at the adsorbent surface and nickel ions lowers the adsorption, whereas at $pH > pH_{pzc}$, the adsorbent surface bears negatively charged species that undergo electrostatic attraction with nickel ions. This interaction results in the formation of complexes, thereby enhancing the adsorption process. Thus, surface complexation and ion-exchange mechanisms govern the adsorption of Ni(II) ions on activated carbon prepared from scrap tires, whereas on $ZnCl_2$ activated carbon from *Glycyrrhiza glabra* residue, Ni(II) and Pb(II) ions are adsorbed via electrostatic attractions.

For the removal of V(V) ions, $ZnCl_2$ activated coir pith carbon is used where the surface functional groups introduced by activation actively participate in the sorption. The reactive OH^- groups at the adsorbent surface enhance the removal performance via the ion-exchange mechanism (Namasivayam and Sangeetha 2006).

H_2O_2-modified attapulgite is tested for Sr(II) ions where modification with H_2O_2 brings about morphological changes in the adsorbent. It exfoliates the surface and forms more of the silicate layers that increase the surface area from 96.34 to 174.14 m^2/g with smaller pore size and larger pore volume. This leads to increased adsorption sites of clay particles resulting in the enhancement of the adsorption abilities of the modified attapulgite for Sr(II) ions. The same effect was reported for adsorption of Sr(II) ions on KOH-modified activated carbon from textile sewage sludges where treatment with KOH significantly increased the surface area from 69.13 to 135 m^2/g. Although the structure of activated carbon was less complex than that of the precursor material, the remaining functional groups contributed positively to the adsorption by reducing the equilibrium time to 5 min (Kaçan and Kütahyalı 2012). During Sr(II) removal with Fe_3O_4 particles modified sawdust, the functionalization of the adsorbent assisted in its easy separation as well as in the enhancement of adsorption performance by electrostatic interactions (Cheng et al. 2012).

Surface modification can also influence the adsorption of various anions such as F^- ions from aqueous solutions. The potential of zirconium–carbon hybrid sorbent was explored for F^- ions (Velazquez-Jimenez et al. 2014). The adsorption mechanism involved three main steps, where first of all, Zr(IV) species get adsorbed on the surface of activated carbon by COOH groups via electrostatic interactions to form C-O-Zr bonds. The next reaction involved the formation of zirconium oxalate complex by interaction of Zr(IV) with the –OH groups of oxalic acid. It regulates the nucleation and limits the growth of ZrO_2 particles that positively contribute to increased fluoride adsorption. The final step involved the fluoride attack on Zr-oxalate complexes and the displacement of hydroxyl groups by fluoride ions and the adsorption of latter on the ZrOx-AC surface.

2.4.2 Nitrogen-Bearing Functional Groups

The functional groups containing nitrogen when incorporated into adsorbents increase the basic properties of the adsorbents and hence increase their uptake capacity (Yang, Wan, et al. 2019). Usually, amine groups are most commonly used for functionalization or modification of any surface where due to the availability of lone pair of electrons and readiness in the bondformation with cationic species results in enhanced adsorption. For instance, acrylamide-functionalized electrospun nanofibrous polystyrene showed enhanced adsorption of Cd(II) and Ni(II) ions where functionalization introduced amide (-NCO) and amine (-NH-) groups onto the surface of polystyrene that lead to an increased interaction between the adsorbent and metal ions. However, the exact underlying mechanism of adsorption was unspecified (Bahramzadeh et al. 2016). Similarly, amino-functionalized metal–organic frameworks MIL-101(Cr) (Luo, Ding, et al. 2015) and polyacrylamide-functionalized Fe_3O_4 nanoparticles (Moradi et al. 2017) exhibited increased removal efficiency for Pb(II) ions, and the experimental findings suggested the adsorption mechanism to be governed by chemisorption.

The adsorption of V(V) ions on amine-modified poly(glycidyl methacrylate)-grafted cellulose (Anirudhan et al. 2009) and Se(IV) and Se(VI) ions on dendrimer-functionalized graphene oxide (Xiao et al. 2016) exhibited enhanced adsorption of these ions after functionalization. The underlying mechanism of adsorption was electrostatic attraction between positive adsorbent surface and negative vanadate ions at specified pH at which the tertiary amine moiety (-NH^+-$(CH_3)_2Cl^-$) exchanged its Cl^- ions with vanadate species. For selenium oxyanions, electrostatic attraction between charged amine groups and selenite ions resulted in increased adsorption of Se(IV). The combination of electrostatic interaction between charged amine groups and selenite ions and the formation of an inner-sphere complex on the surface of GO via ligand exchanging with hydroxyl groups occurred during the adsorption of Se(VI) ions.

Ethylenediamine-modified β-zeolite (β-zeolite-EDA) involved an inner-sphere complex formation between Ni(II) ions and amine groups of β-zeolite. The amino group (-NH_2) forms a protonated cationic (-NH_3^+) form that facilitates more availability and adsorption of nickel(II) ions in acidic solution. The formation of inner-sphere complex was explained through XPS data that evidenced that the electron density on oxygen and nitrogen atoms decreased as they easily lost electrons that migrated to nickel ions. Therefore, the complex was formed via the formation of coordination bond between oxygen and nitrogen atoms of β-zeolite-EDA and Ni(II) ions (Liu, Yuan, et al. 2017). Similarly, Hg(II) ions form coordination bonds with –NH_2 ligands that have one or more less electronegative donor atoms and have a strong coordination role with Hg^{2+} ions. This results in significant enhancement in the adsorption capacity for test metal ions. The presence of two amino groups in the structural framework further contributes to the increased adsorption. The introduction of the

–NH_2 groups in the adsorbent provides adsorption sites for Hg(II) ions on the surface of MCM-41 (Zhu et al. 2012).

The adsorption of copper(II) ions on amino-functionalized magnetic nanoparticles (Hao et al. 2010) and γ-aminopropyltriethoxysilane-modified chrysotile (Liu, Zhu, et al. 2013) followed the ion-exchange process. The latter one involved three mechanisms during the course of adsorption. The first one is the ion-exchange process that proceeds by exchange of protons of residual hydroxyl groups with the Cu(II) ions in solutions. Later, the amino groups attached to the surface of adsorbent using nitrogen atoms to complex Cu(II) ions. Finally, the protonated amino group adsorbed copper(II) ions by chelating or the ion-exchange process.

The amidation of graphene oxide significantly improved the adsorption of uranyl ions from aqueous solutions where ammonia was selected for modification of the –COOH groups of GO into –$CONH_2$ groups to accomplish higher specific binding with uranyl ions (Verma and Dutta 2015). The adsorption indicated a dual mechanism of physisorption and chemisorption due to the availability of heterogeneous binding sites in ammonia-modified graphene oxide (NH_3-GO). Amide functionalization of graphene oxide exhibited better selectivity toward uranyl ions at experimental pH. In contrast, amino-functionalized mesoporous silica SBA-15 involves a first adsorption step followed by precipitation of uranium ions when its concentration increases in the bulk phase (Huynh et al. 2017). The same dual interaction was observed for the removal of Se(IV) and Se(VI) ions on poly(-allylamine)-modified magnetic graphene oxide (PAA-MGO) due to the heterogeneous surface of PAA-MGO. However, the mechanism was similar to that reported by Xiao et al. (2016). Both physical and chemical interactions occurred during the adsorption of selenium oxyanions where ligand exchange between –OH groups present on the surface of iron oxides and selenium oxyanions and electrostatic interaction between the amine groups and selenium oxyanions take place (Lu, Yu, et al. 2017).

The functionalization of graphene with CTAB significantly increases the adsorption capacity of graphene for chromium(VI) ions (Wu et al. 2013). The amino group forms amide bond and, along with the –COOH and –OH groups on the surface of graphene, enhances the adsorption of Cr(VI) ions from water resources.

Ethylamine-modified montmorillonite showed an increase in the adsorption of Cs(I) ions, and the operating mechanism was ion exchange together with the complex formation via coordination of –NH_2 groups and surface complexation of –OH groups (Long et al. 2013). In a similar manner, the 3-(aminopropyl)triethoxysilane-functionalized coal fly ash modified with mesoporous siliceous material displayed an increased adsorption capacity of the adsorbent for Cu(II) ions by the formation of amino-copper complex through the surface amino group (Pizarro et al. 2015).

The ionizable functional groups such as –NH_2, -COOH, and –OH undergo deprotonation at high pH, making the surface negatively charged, and the adsorption of Mn(II) involves chelate interactions between the adsorbent surface and Mn(II) ions (Al-Wakeel et al. 2015).

The modification of cross-linked magnetic chitosan/poly(vinyl alcohol) particles with xanthate introduces N- and S-containing functional groups on the surface of XCMP that assists in the adsorption of Cu(II) and Pb(II) ions. The mechanism of adsorption of these metal ions involves an interaction of N and S atoms of functional groups with these ions, and the metal ions are then transformed into crystals and get adsorbed on the XCMP surface (Lv et al. 2017).

2.4.3 Sulfur-Bearing Functional Groups

The sulfur-containing functional groups positively contribute to the adsorption of organic and inorganic contaminants from water resources on a variety of adsorbents either by increasing/decreasing specific surface area and pore volume of the adsorbents or by their extraordinary binding abilities with the soft metal ions. Thiol-functionalized ionic liquid-based mesoporous organosilica exhibited soft–soft interactions for Hg(II) and Pb(II) removal where these metal ions having low charge densities are considered as soft acids that form strong covalent bonds with sulfur-containing soft bases. The adsorption mechanism involved chemisorption where covalent bonds were formed by sharing or exchanging electrons between these metal ions and –SH groups of the adsorbent (Elhamifar et al. 2016).

The modification of chitosan beads with sodium dodecyl sulfate introduces sulfur-containing –S=O and -CSO functional groups by forming a bilayer of the surfactant on the surface of chitosan. The Cd(II) ions are adsolubilized on the surfactant bilayer formed by means of electrostatic attraction (Pal and Pal 2017).

The sulfur-functionalized ordered mesoporous carbons experienced an increase in their specific surface area (837–2865 m^2/g) and pore volume (0.71–2.3 cm^3/g) after the modification and exhibited affinities for heavy metals in the following order: Hg(II) > Pb(II) > Cd(II) > Ni(II) (Saha et al. 2016). Sulfur-functionalized silica microspheres, thiol-functionalized hollow mesoporous silica microspheres, and 1,2-ethanedithiol- and vinyl-functionalized covalent organic frameworks exhibited increased Hg(II) and Hg(0) adsorption by forming a complex between the sulfur-containing functional groups (i.e. –SH, -SR) and the Hg(II) and Hg(0) metal ions, respectively (Saman et al. 2013; Zhang, Wu, et al. 2015; Sun et al. 2017).

Sulfurized activated carbon (SAC) from bagasse pith exhibited an ion-exchange mechanism for adsorption of Zn(II) ions. The surface of sulfurized activated carbon contains SAC-OH, SAC-O moieties, and $-SO_3H$ functional group that are collectively responsible for greater adsorption of Zn(II) ions. The adsorption mechanism involves the exchange of H^+ ions on the surface of SAC with Zn^{2+} ions in the aqueous medium, thus forming an ion-exchange complex (Anoop Krishnan et al. 2016).

2.4.4 Adsorbent Containing Multiple Functional Groups

The functional groups discussed so far under different subheadings in this section revealed their significant contribution to the adsorption of a variety of metal ions by enhancing the adsorption capacity and selectivity of the adsorbents. It follows that the presence of more than one of these functional groups can also bring out certain positive changes in the adsorbents influencing the overall adsorption process. The multifunctional groups on the adsorbent surface exhibit a cooperative effect on the mechanism of adsorption. This section provides an insight into adsorbents containing more than one functional group at the surface for significant removal of metallic ions from aqueous solutions.

For instance, acrylamide- and hydroxyl-functionalized metal–organic framework exhibits increased removal of Hg(II) ions where hydroxyl and acrylamide groups play a pivotal role. On functionalization, both these groups are introduced into the pore wall of MOF and provide strong sites for coordination and chemisorption of Hg(II) ions. The experimental findings suggested that the adsorption involved both physisorption and chemisorption mechanisms. However, the formation of Hg-O from both these groups advocated chemisorption as the dominating mechanism (Luo, Chen, et al. 2015).

For the removal of U(VI) ions, silica particles were modified with amidoxime (AMD) and carboxyl (CA) groups where amine and hydroxyl groups on AMD ligand act as chelation sites and on CA ligand act as ion-exchange groups for U(VI) ions (Bayramoglu and Arica 2016). The presence of amino, imino, hydroxyl, and carboxyl groups on the surface of adsorbent significantly increases the removal of test metal ions by the process of electrostatic interaction. The complexation and electrostatic interactions between AMD ligands and U(VI) at pH 5 and the deprotonation of negatively charged carboxyl groups for ion-exchange reaction with U(VI) ions after pH 3.0 are the main reasons for increased adsorption. Moreover, the affinity of uranyl ions for nitrogen and hydroxyl groups is higher than that of the carboxyl-containing counterparts.

The introduction of three carboxylic groups and amine groups into the cornstalk through the functionalization with nitrilotriacetic acid resulted in excellent adsorption capacity for Cd(II) and Pb(II) ions (Huang, Yang, et al. 2015). The mechanism involved surface chelation between carboxyl and amine groups present on the surface and Cd(II) and Pb(II) ions, and ion exchange between Na^+ on the surface and Cd^{2+} and Pb^{2+} ions in the solution. It was found that some of the metal ions undergo coordination with two carboxyl groups and one amine group and the other metal ions were adsorbed by ion exchange process through carboxyl groups.

During the activation of carbon prepared from *Phragmites australis* with urea phosphate, a number of oxygen-containing and nitrogen-containing functional groups are introduced into the prepared activated carbon (Guo, Zhang, et al. 2017). It also increases the porous structure of the carbon. The adsorption involved various mechanisms, namely microporous filtration due to the small size of Cd(II) ions, ion exchange between the protons present in the functional groups attached on the adsorbent surface and Cd(II) ions in the solutions, electrostatic attraction between the deprotonated phenolic and carboxyl groups and Cd(II) species, and surface complexation. The surface complexation occurred via covalent bonding with Cd(II) ions and O atoms in deprotonated carboxyl, ketone, ether phenol, and ester groups, and N atoms in $C\text{-}NH_2$ and O=C-NH groups and P atoms in pyrophosphate. These atoms donated their electrons to Cd(II) ions to form coordination complexes.

Chitosan modification of vermiculite introduced amine- and hydroxyl-containing functional groups that facilitate the adsorption of As(III) ions. The probable mechanism of adsorption involved chelation or complex formation between the As(III) and NH_2 groups of chitosan or ion exchange between H^+ ions of attached NH_2 groups and As(III) ions. Further, the oxalic groups adhered on the surface of vermiculite act as a bridge between the adsorbent surface and chitosan chains that aid in the enhancement of adsorption capacity. The mean adsorption energy indicated ion exchange as the dominant mechanism of removal together with the adsorption of As(III) ions into the pores present on surface or among the inner layers of adsorbent (A. Saleh et al. 2016).

Due to the introduction of multiple thio-triazole units as binding sites in the structural framework, lignin-based adsorbents exhibited an enhanced adsorption capacity for Cd(II) ions. The adsorption involved chemical interactions where the Pearson hard–soft acid–base (HSAB) principle governed the overall adsorption mechanism. Cd(II) ions acted as Lewis soft acid that reacts readily with imino and thioether groups attached to the surface considered as Lewis soft bases to form covalent bonds, and thus, the adsorbent showed promising adsorption capacity (Jin et al. 2017).

The functionalization or modification of adsorbents not only contributes to the adsorption of metallic species but also influences the removal of certain anions from aqueous solutions. For instance, chitosan microspheres were modified by formaldehyde, PEG_{2000}, and epichlorohydrin (ECH). This results in the introduction of amino and hydroxyl groups onto the surface of the adsorbent. At lower pH (i.e. 5), protonation of amino groups of chitosan occurred that enhanced the electrostatic attraction of nitrate and phosphate anions and resulted in the increased adsorption of these anions (Zhao and Feng 2016).

3 Impact of Factors on Remediation of Major Toxic Elements (Vanadium, Chromium, Nickel, Arsenic, Strontium, Cadmium, Mercury, Lead, Uranium) Via Batch Adsorption Process

Deepak Gusain
Durban University of Technology

Shikha Dubey and Yogesh Chandra Sharma
IIT(BHU), Varanasi

Faizal Bux
Durban University of Technology

CONTENTS

This chapter includes a discussion on factors affecting the adsorption of major toxic elements. The majority of these elements find their application in industrial uses. However, these elements are nonessential for human body and do not find any metabolic role in the body. Hexavalent chromium, vanadium pentoxide, radioactive strontium, cadmium (oral exposure), inorganic lead, mercury chloride and methyl mercury, and radioactive strontium are known to be carcinogens. Nickel causes allergies and bronchitis and has an adverse effect on the kidney, blood, and liver. Cadmium is associated with lung and kidney damage, and uranium (inhalation) causes lung damage. The effect of various parameters such as pH and coexisting ions on the adsorption of respective elements, and the mechanism of adsorption are discussed in this chapter.

3.1 VANADIUM

Vanadium was included in the contaminant candidate list provided by the United States Environmental Protection Agency (USEPA) (USEPA 2017). Vanadium has a harmful effect on the circulatory system. It also causes disruption in the metabolism (Treviño and Diaz 2020; Zagulski et al. 1980). Vanadium can be found out in three forms in aqueous media: cationic species (VO^{2+}, HVO^{2+}, V^{3+}), neutral species ($VO(OH)_3$), and anionic species (i.e., $V_{10}O_{26}(OH)_2^{4-}$, $V_{10}O_{27}(OH)^{5-}$, $V_{10}O_{28}^{6-}$, $VO_2(OH)^{2-}$, $VO_3(OH)^{2-}$, VO_4^{3-}, $V_2O_6(OH)^{3-}$, $V_2O_7^{4-}$, $V_3O_9^{3-}$, and $V_4O_{12}^{4-}$; Naeem et al. 2007; Padilla-Rodríguez et al. 2015). The concentration and pH affect the form of vanadium species. The cationic form of vanadium is prevalent below pH 3, whereas it exists in anionic forms between pH 4 and 11 (Peacock and Sherman 2004). An overview of the experimental parameters and optimized conditions from batch adsorption experiments for vanadium is presented in Table 3.1.

3.1.1 Effect of pH

The adsorption capacity of vanadium declined in the highly acidic conditions preferably at pH less than 4 (Padilla-Rodríguez et al. 2015; Anirudhan et al. 2009). The process can be explained when the adsorption is mainly controlled by electrostatic attraction, and it relies on speciation of vanadium and surface properties of the adsorbent. Vanadium exists as a positively charged species in the highly acidic conditions, and this leads to increased repulsion between the positively charged adsorbent's surface and the vanadium species. With an increase in pH above 4, the negatively charged species of vanadium started to predominate and led to increased attraction between the adsorbate and adsorbent below the pHzpc. Two case studies for the effect of pH on vanadium are as follows:

TABLE 3.1
Summary of Parameters and Optimized Conditions for Batch Adsorption of Vanadium

Adsorbent	Adsorbate	Surface Area (m^2/g), Pore Volume (cm^3/g), Pore Size (nm)	pHzpc	Experimental Conditions	Adsorption Capacity (mg/g)	Thermodynamic Parameters	Kinetic Model and Curve Fitting	Isotherm Model and Curve Fitting	Maximum Adsorption Conditions	References
Granular ferric hydroxide (GFH) (batch and column study)	V(V)	Surface area = 231	8	pH = 2–11 Dose = 1–5 g/l Concentration = 1–250 mg/l (isotherm study) Contact time = 24 h					pH = 3–3.5 Dose = 2.5 g/l Contact time = 24 h	Naeem et al. (2007)
Chitosan–zirconium(IV) composite	V(V)			pH = 2–9 Dose = 0.2 g/l Concentration = 10–100 mg/l (isotherm study) Contact time = up to 8 h Temperature = 277–308 K	208	ΔH = 16.7 kJ/mol ΔS = 86 J/mol K ΔG = negative	Pseudo-second-order model	Langmuir model	pH = 4 Contact time = 4 h Temperature = 308 K	Zhang et al. (2014)
Zerovalent iron	V(V)	Surface area = 31.1 Pore volume = 0.027 Pore size = 35.9 Å	8.1	pH = 4–10 Dose = 0.5–20 g/l Concentration = 0.01–100 mg/l Contact time = up to 90 min Temperature = 298–333 K	358.2	ΔH = −12.67 kJ/mol ΔS = −12.2 J/mol K ΔG = negative	Pseudo-second-order model, linear	Langmuir model, linear	pH = 4–8 Dose = 0.1 g/l Contact time = 10 min Temperature = 298 K	Erdem Yayayürük and Yayayürük (2017)
Calcined magnesium/aluminum hydrotalcite	V(V)			pH = 2–9 Dose = 0.2 g/l Agitation speed = 130 rpm Concentration = 10–50 mg/l (isotherm study) Contact time = 12 h Temperature = 25°C	513.88		Pseudo-second-order model	Sips model, non-linear	pH = 2.5	Wang, Cheng, et al. (2012)

(continued)

TABLE 3.1 (*Continued*)
Summary of Parameters and Optimized Conditions for Batch Adsorption of Vanadium

Adsorbent	Adsorbate	Surface Area (m^2/g), Pore Volume (cm^3/g), Pore Size (nm)	pHzpc	Experimental Conditions	Adsorption Capacity (mg/g)	Thermodynamic Parameters	Kinetic Model and Curve Fitting	Isotherm Model and Curve Fitting	Maximum Adsorption Conditions	References
Calcined magnesium/vanadate aluminum hydrotalcite	V(V)			do	323.78		Pseudo-second-order model	Sips model, non-linear	do	Wang, Cheng, et al. (2012)
Protonated chitin	V(III)	Surface area = 1.13		pH = 2–12 Dose = 5 g/l Agitation speed = 200 rpm Concentration = 0.5 mg/l Contact time = up to 26 h Temperature = 25°C	12.22				pH = 6 Contact time = 2 h	Padilla-Rodríguez et al. (2015)
Protonated chitin	V(IV)	Surface area = 1.13		do	6.50				do	Padilla-Rodríguez et al. (2015)
Protonated chitin	V(V)	Surface area = 1.13		do	2.58				do	Padilla-Rodríguez et al. (2015)
Aluminum-pillared clay	V(IV)	Surface area = 179	4.2	pH = 2–8 Dose = up to 10 g/l Agitation speed = 200 rpm Concentration = 5–50 mg/l Contact time = up to 30 h Temperature = 30°C	24.16			Freundlich model	pH = 4.5–6 Contact time = 24 h	Manohar et al. (2005)

(*continued*)

TABLE 3.1 (*Continued*)
Summary of Parameters and Optimized Conditions for Batch Adsorption of Vanadium

Adsorbent	Adsorbate	Surface Area (m^2/g), Pore Volume (cm^3/g), Pore Size (nm)	pHzpc	Experimental Conditions	Adsorption Capacity (mg/g)	Thermodynamic Parameters	Kinetic Model and Curve Fitting	Isotherm Model and Curve Fitting	Maximum Adsorption Conditions	References
Ti-doped chitosan bead	V(V)			pH = 2–10 Dose = 0.2 g/l Agitation speed = 200 rpm Concentration = 5–400 mg/l (isotherm study) Contact time = up to 72 h Temperature = 277–308 K	210	ΔH = 21.73 kJ/mol ΔS = 0.096 kJ/mol K ΔG = negative	Pseudo-second-order model, linear	Langmuir model, linear	pH = 4 Temperature = 308 K	Liu and Zhang (2015)
Amine-modified POLY(glycidyl methacrylate)-grafted cellulose	V(V)	Surface area = 33.2	7	pH = 2–9 Dose = 0.5–6 g/l Agitation speed = 200 rpm Concentration = 25–600 mg/l (isotherm study) Contact time = up to 120 min Temperature = 30°C	197.75		Pseudo-second-order model (only single model fitted)	Langmuir model	pH = 4–6 Contact time = 60 min	Anirudhan et al. (2009)
Zinc chloride activated carbon	V(V)	Surface area = 910 Pore volume = 0.363 Pore size = 1.60	3.20	pH = 2–11 Dose = 1–8 g/l Agitation speed = 200 rpm Concentration = 40–120 mg/dm^3 Contact time = up to 80 min Temperature = 35°C	24.9	ΔH = 29.7 kJ/mol ΔS = 164.9 J/mol K ΔG = negative	Pseudo-second-order model, linear	Freundlich isotherm, linear	pH = 4–9 Contact time = 40 min	Namasivayam and Sangeetha (2006)
Zr(IV)-impregnated collagen fiber	V(V)	Surface area = 2.07–3.87	10.3–10.7 (solid-state method)	pH = 3–11.5 Dose = 1 g/l Concentration = 2–6 mmol/l (isotherm study) Contact time = up to 24 h Temperature = 303–323 K	3.18 mmol/g				pH = 5–8 Temperature = 323 K	Liao et al. (2008)

The reduction of adsorption capacity of vanadium below pH 4 on protonated chitosan was due to the formation of VO^{2+}, HVO^{2+}, and V^{3+}, which led to generation of repulsion with the positively charged chitosan matrix (Padilla-Rodríguez et al. 2015). However, the decline in percentage removal above pH 6 (pHzpc) was due to the electrostatic repulsion of the adsorbate ($H_2VO_4^{2-}$).

In the case of vanadium removal with "amine-modified poly(glycidyl methacrylate)-grafted cellulose," percentage removal increased with a rise in pH from 2 to 4, then increased slightly (almost constant) up to pH 6, and declined thereafter (Anirudhan et al. 2009). The effect of the pH is interpreted on the basis of adsorbent's surface charge and adsorbate's speciation. The pHzpc of the adsorbent was 7, below which the adsorbent is positively charged. The cationic species of the vanadium dominated at pH lower than 2, and the increase in removal up to pH 6 was attributed to the increase in vanadium's negatively charged species. The maximum adsorption at pH 6 was attributed to preferential adsorption of the anionic species, i.e. $V_3O_9^{3-}$ of V(V) adsorbed. The anionic V(V) species undergoes ion exchange with chloride ions from the adsorbent. The decline in removal at higher pH was attributable to repulsion between negatively charged adsorbent's surface and vanadium's anionic species.

However, in the case of vanadium adsorption with calcined hydrotalcite, the adsorption capacity increased with an increase in pH from 2.0 to 2.5, with maximum adsorption capacity peaking at pH 2.5. The further increase in pH from pH 2.5 to 9.0 led to a decline in the adsorption capacity (Wang, Cheng, et al. 2012). The lower removal at pH 2 was also due to the presence of vanadium in +5 oxidation state at pH 2, dissolution of a part of the adsorbent in those conditions, and the preference for anion by hydrotalcite due to anion vacancy. The decline in the adsorption capacity after pH 2.5 was attributed to the increased competition from hydroxyl ion concentration.

The adsorption of vanadium with zinc chloride activated carbon was governed by the ion exchange mechanism in addition to the electrostatic nature of adsorption. The protonation of the adsorbent surface played a role in the alteration of behavior of adsorption. The vanadate removal with zinc chloride activated carbon declined at pH 9–11 (Namasivayam and Sangeetha 2006). This was attributed to the decrease in protonation of the surface, which led to increased electrostatic repulsion. In addition, the increase in the competitive adsorption from hydroxide ions also played a major role. The variation in optimum pH values for the batch adsorption process of vanadium along with other elements is presented in Figure 3.1.

3.1.2 Effect of Coexisting Ions

The adsorption capacity for vanadium declined in the presence of ions (Wang, Cheng, et al. 2012). The declination in the adsorption capacity was attributed to the competitive adsorption (Wang, Cheng, et al. 2012). The competitive effect of ions is also the reason for the declined adsorption capacity in groundwater as compared to synthetic aqueous solution (Namasivayam and Sangeetha 2006). However, the removal efficiency of vanadium with protonated chitosan did not significantly decrease in

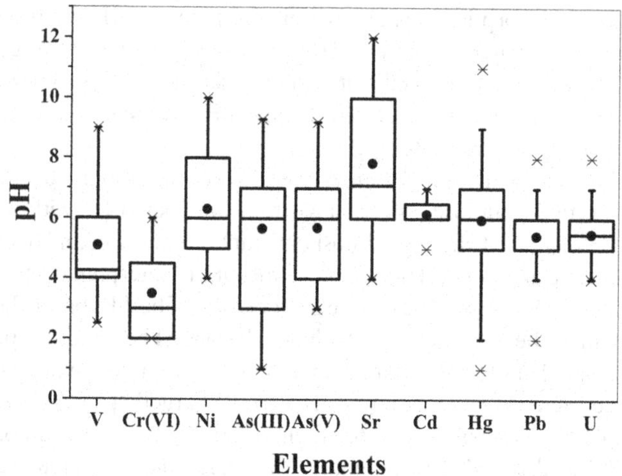

FIGURE 3.1 Box plot of optimum pH for adsorption vs toxic metal graph of the literature surveyed (the literature surveyed is in Tables 3.1–3.10). In places where the pH range is specified for optimum pH conditions, both the upper and lower limits of pH are taken into consideration. In the box plot, the point and horizontal line inside the box are mean and median values, respectively. Box is set up at the 25th and 75th percentiles and whiskers at the 5th and 95th percentiles, and the cross outside the range of whiskers indicates outliers.

the presence of chloride, sulfate, carbonate, and phosphate, and percentage removal remained more than 90% (Padilla-Rodríguez et al. 2015).

The effect of the bivalent ion, i.e. sulfate, is more than that of the monovalent ion, i.e. nitrate and chloride with chitosan–Zr(IV) composite (Zhang et al. 2014) and Ti-doped chitosan bead (Liu and Zhang 2015). The high valency of the sulfate ion was held responsible for this behavior. However, the percentage removal of vanadium with zinc chloride activated carbon was more affected by monovalent ions, i.e. perchlorate and chloride ions, as compared to bivalent sulfate and phosphate ions (Namasivayam and Sangeetha 2006).

However, chloride ions increased the percentage removal of vanadium on Zr(IV)-impregnated collagen fiber (Liao et al. 2008). Chloride ions were hypothesized to form an intermediate complex with Zr(IV) and along with this led to the reduction of activation energy for chelation reaction between vanadium(V) and Zr(IV).

3.1.3 Effect of Surface Modification

The protonated chitosan (Padilla-Rodríguez et al. 2015) and amine-modified poly(glycidyl methacrylate)-grafted cellulose (Anirudhan et al. 2009) led to increased adsorption capacity as compared to the unmodified material. Similarly, pillaring of sodium bentonite led to the formation of aluminum-pillared clay, which was 2.5 times more effective in the removal of vanadium(IV) (Manohar et al. 2005). In addition, the desorption efficiency also increased for amine-modified poly(glycidyl methacrylate)-grafted cellulose than for cellulose (Anirudhan et al. 2009).

3.1.4 Mechanism

The time-period analysis of the XRD pattern was used to predict the presence of the alternate mechanism governing the adsorption mechanism (Wang, Cheng, et al. 2012). The adsorption of vanadium on hydrotalcite was governed by memory effect and adsorption on external surface. The XRD analysis at 10, 30, and 300 min was estimated. The adsorbent recovered the original structure within 10 min. However, the adsorption capacity was not saturated, which suggests the presence of another mechanism apart from memory effect during the adsorption process, e.g. adsorption on the external surface layers.

The XPS data can be used to estimate the change in the oxidation of vanadium adsorbed on the surface of adsorbent (Wang, Cheng, et al. 2012). A new peak emerged in the O 1s XPS spectra after vanadium adsorption on titanium-doped chitosan bead at 531.5 eV corresponding to the carboxyl oxygen atoms (Liu and Zhang 2015). In addition, vanadium $2p_{3/2}$ XPS spectrum with peak position at 516.3 and 517.4 eV corresponded to vanadium(IV) and vanadium(V). This suggested the probable donation of the lone pair of electrons from oxygen atom of the hydroxyl groups to vanadium(V), leading to their reduction to vanadium(IV).

The FT-IR spectra used to estimate the chelation, the absence of any new peak, or shifting of the peak during vanadium adsorption on protonated chitin suggest the absence of any chelation and indicate the electrostatic attraction as the dominant mechanism (Padilla-Rodríguez et al. 2015). The FT-IR analysis in the adsorption of vanadium with titanium-doped chitosan bead (Liu and Zhang 2015) led to a shift of the FT-IR peaks from 1635 to 1621 cm^{-1} after adsorption. This phenomenon suggested the involvement of the hydroxyl groups in adsorption.

The mechanism of adsorption is also estimated on the basis of the activation energy (Namasivayam and Sangeetha 2006). The activation energy for film diffusion and pore diffusion is 17–21 and 21–42 kJ/mol, respectively. The activation energy for the adsorption of vanadium on zinc chloride activated carbon is in the range of 21.6–52.3 kJ/mol. Hence, the process is estimated to be pore diffusion-controlled. The value of standard change in enthalpy is also used to estimate the mechanism of adsorption, and the value of less than 40 kJ/mol is suggested to dismiss the chemisorption process (Namasivayam and Sangeetha 2006).

3.1.5 Desorption

The desorption of vanadium can be achieved with sodium hydroxide (Zhang et al. 2014; Anirudhan et al. 2009; Liu and Zhang 2015), ammonium hydroxide (Liao et al. 2008), and hydrochloric acid (Awual et al. 2019; Namasivayam and Sangeetha 2006). In the case of desorption of vanadium from protonated chitin loaded with vanadium(V), both acidic and basic solutions were used for desorption of vanadium. However, hydrochloric acid acted as the best eluent for recovery of protonated chitosan (Padilla-Rodríguez et al. 2015). Moreover, the desorption of vanadium from zinc chloride activated carbon was more at pH 12 (75%–90%) as compared to 30%–50% at pH 2 (Namasivayam and Sangeetha 2006).

In the case of vanadium desorption from Zr(IV)-impregnated collagen fiber, the desorption was preferred by ammonium hydroxide as compared to hydrochloric acid (Liao et al. 2008). This was due to the reduction in the adsorption capacity on desorption with hydrochloric acid, which was attributed to leaching of Zr(IV). The percentage removal declined with subsequent desorption cycles for vanadium(V) (Anirudhan et al. 2009; Liu and Zhang 2015) and vanadium(IV). The small amount of metal which is not recoverable upon desorption is attributed to be bound with strong interaction; it leads to a decline in adsorption capacity of adsorbent in the successive adsorption-desorption cycle (Manohar et al. 2005).

3.2 CHROMIUM

The maximum permissible level of chromium in drinking water recommended by the World Health Organization and Bureau of Indian Standards is 0.05 mg/l, whereas the USEPA has suggested the maximum level of chromium in drinking water at 0.015 mg/l (Kumari et al. 2015).

Chromium(VI) stays in the solution in the form of dichromate ($Cr_2O_7^{2-}$), hydrogen chromate ($HCrO_4^-$), or chromate (CrO_4^{2-}) ions (Li et al. 2009). They vary in solution with a change in pH and concentration of the solution (Li et al. 2009). The equations for their interconversion are as follows:

$$H_2CrO_4 \rightleftarrows H^+ + HCrO_4^- \tag{3.1}$$

$$HCrO_4^- \rightleftarrows H^+ + CrO_4^{2-} \tag{3.2}$$

$$2HCrO_4^- \rightleftarrows H_2O + Cr_2O_7^{2-} \tag{3.3}$$

$HCrO_4^-$ is the most dominant species at pH lower than 6.8 between 0.05 and 300 mg/l, and above 6.8, CrO_4^{2-} is the only dominant species. However, above 300 mg/l, $Cr_2O_7^{2-}$ is the dominant species between pH ca. 1 and 6.8 (Li et al. 2009). An overview of the experimental parameters and optimized conditions from batch adsorption experiments for chromium is presented in Table 3.2.

3.2.1 Effect of pH

Metal's adsorption on pH is largely dependent on three factors, i.e. the type of the functional group, ionic state of the functional group, and metal chemistry in the solution (Matheickal et al. 1999; Adeli et al. 2017).

The percentage removal of Cr(VI) declined with an increase in pH from acidic to alkaline region (Qiu et al. 2014; Sun et al. 2014; Sakulthaew et al. 2017). However, the adsorption of Cr(III) increased with an increase in pH of the solution with aminated iron oxide/mesoporous silica nanocomposite, with maximum adsorption at pH 5.4; on further enhancing the pH, Cr(III) precipitated as insoluble $Cr(OH)_3$ (Egodawatte et al. 2015).

The large amount of adsorption of Cr(VI) at low or acidic pH can be elucidated by the species of chromium and the nature of adsorbent's surface (Mor et al. 2007). The high protonation of the amine groups in polyaniline-coated ethyl cellulose aids in the high adsorption of Cr(VI), and at higher pH values, the hydroxide ions on the surface of adsorbent lead to repulsion between the adsorbate and adsorbent (Qiu et al. 2014). The decline in percentage removal in the case of hexavalent chromium with protonated titanate nanotubes at pH larger than 5 was attributed to the change in the species of chromium. The changed species (CrO_4^{2-}) carried divalent charge as compared to monovalent charge on species ($HCrO_4^-$) at lower than pH 5 (Wang, Liu, et al. 2013). The chromium species having divalent charge consumes two amino groups, which leads to a decline in the adsorption capacity of chromium.

The protonation helped in the adsorption of chromium(VI) at lower pH in the case of poly(glycidyl methacrylate)-grafted copolymer (Duranoğlu et al. 2012) and amino-functionalized magnetic cellulose nanocomposite (Sun et al. 2014). However, at a very low pH (lower than 2 or 2.79), the decline in percentage removal is attributed to the formation of the neutral species H_2CrO_4, which does not offer any electrostatic attraction (Wang, Liu, et al. 2013; Sun et al. 2014). However, with carbon nano-onions the decline in percentage removal from pH 3.4 to 2 was attributed to blockage of the active sites of adsorption (Sakulthaew et al. 2017).

The point of zero charge also played a major role in the effect of pH on adsorption of chromium (Duranoğlu et al. 2012). There is very little difference in the percentage removal of Cr(VI) on decreasing the pH from 7 to 2 using mesoporous magnetite (Kumari et al. 2015). The reason for this is the low pHzpc of the adsorbate (2.11). At pH higher than 2.11, the adsorbate surface is negative; thus, attraction is less between the adsorbate and adsorbent. Hence, electrostatic force of attraction for adsorption is inhibited (Kumari et al. 2015). So, the role of electrostatic interaction in the adsorption of chromium can be ascertained by adsorption capacity below and above the adsorbent's point of zero charge.

3.2.2 Effect of Coexisting Ions

The adsorption capacity of Cr(VI) decreased in the presence of coexisting ions (Wang, Liu, et al. 2013; Huang et al. 2013). The adsorption of chromium on amino-functionalized titanate nanotubes and protonated titanate nanotubes declined after the addition of chloride, nitrate, and phosphate (Wang, Liu, et al. 2013). The declination effect of anions was more for sulfate as compared to chloride, nitrate, and phosphate. The higher surface charge on sulfate was attributed to this. Phosphate exists as monohydrogen phosphate under experimental conditions, i.e., at pH 5.4. Hence, sulfate has a higher negative effect. The cations (Fe^{3+}, Cu^{2+}, Zn^{2+}, K^+, Na^+, Ca^{2+}, Mg^{2+}) did not show any significant effect on the adsorption of chromium(VI) with titanium-cross-linked carbon nanocomposites (Zhang, Xia, et al. 2015), and among anions, sulfate had a more pronounced effect than monovalent ions.

The presence of Cr(VI) in the solution had a promotional effect on the adsorption of fluoride with mesoporous alumina (Li, Xie, et al. 2016). The Cr(VI) adsorbed on the surface of mesoporous alumina led to the formation of a new surface hydroxyl group ($\equiv$CrOH or $\equiv CrOH_2^+$), which acted as a new adsorption site for fluoride (F^-). However, the addition of fluoride ions in the solution decreased the adsorption of

TABLE 3.2
Summary of Parameters and Optimized Conditions for Batch Adsorption of Chromium

Adsorbent	Adsorbate	Surface Area (m^2/g), Pore Volume (cm^3/g), Pore Size (nm)	pHzpc	Experimental Conditions	Adsorption Capacity (mg/g)	Thermodynamic Parameters	Kinetic Model and Curve Fitting	Isotherm Model and Curve Fitting	Maximum Adsorption Conditions	References
MWCNT/iron oxide composite (batch and column study)	Cr(III)	92	4	pH = 3–7 Dose = 0.1–2 g/l Agitation speed = 0–150 rpm Concentration = 20 mg/l Contact time = 10–60 min Temperature = 25°C	15.24				pH = 6 Dose = 0.4 g/l Agitation speed = 150 rpm Contact time = 60 min	Gupta et al. (2011)
Treated and activated carbon slurry waste (batch and column study)	Cr(VI)			pH = 1–7 Dose = 0.5–5.5 g/l Concentration = 10–100 mg/l (isotherm study) Contact time = 10–120 min Temperature = 303–313 K			Second-order model, linear	High R^2 for both Langmuir and Freundlich models, linear	pH = 2 Dose = 4 g/l Concentration = 100 mg/l Contact time=70 min Temperature = 303 K	Gupta et al. (2011)
Lignin	Cr(III)	Surface area = 21.7 Pore size = 14.7	2.3	pH = 1.5–6 Dose = 1–7 g/l Concentration = 0.1–2.5 mmol/l (isotherm study) Contact time = 24 h Temperature = 20°C	17.97		Pseudo-second-order model, linear	Langmuir two-surface model	pH = 5 Dose = 5 g/l	Wu et al. (2008)
Aluminum–magnesium mixed hydroxide	Cr(VI)		Zeta potential (+48 mV)	pH= 1–10 Dose = 2 g/l Agitation speed = 200 rpm Concentration = 20–200 mg/l Contact time = 5–180 min Temperature = 20°C–40°C	105.3–112.0	ΔH = 17.74 kJ/mol ΔS = 85.41 J/mol K ΔG = negative	Pseudo-second-order model, linear	Langmuir model, linear	pH = 5 Contact time = 150 min Temperature = 40°C	Li et al. (2009)

(continued)

TABLE 3.2 (*Continued*)
Summary of Parameters and Optimized Conditions for Batch Adsorption of Chromium

Adsorbent	Adsorbate	Surface Area (m^2/g), Pore Volume (cm^3/g), Pore Size (nm)	pHzpc	Experimental Conditions	Adsorption Capacity (mg/g)	Thermodynamic Parameters	Kinetic Model and Curve Fitting	Isotherm Model and Curve Fitting	Maximum Adsorption Conditions	References
Graphene modified with CTAB	Cr(VI)			pH= 1–12 Dose = 2–20 g/l Agitation speed = 150 rpm Concentration = 20–100 mg/l (isotherm study) Contact time = 2–120 min Temperature = 293–313 K	21.57	ΔH = −11.737 kJ/mol ΔS= 59.92–60.33 J/K mol ΔG = negative	Pseudo-second-order model, linear	Langmuir, linear	pH = 2 Dose = 8 g/l Time = 1 h Temperature = 293 K	Wu et al. (2013)
Fe_3O_4 nanospheres	Cr(VI)	Surface area = 11 Pore size = 24.1	2.11	pH= 2–7 Dose = 1–3 g/l Concentration = 10–100 mg/l (isotherm study) Contact time = up to 72 h Temperature = 25°C–45°C	6.64–8.90	Endothermic	Pseudo-second-order model, non-linear	Redlich–Peterson model, non-linear	pH = 4 Dose = 2 g/l Contact time = 48 h Temperature = 45°C	Kumari et al. (2015)
Dolomite	Cr(VI)	Surface area = 4.63 Pore volume = 0.0064 Pore size =15.97 Å		pH = 2–12 Agitation speed = 100–200 rpm Concentration = 5–50 mg/l (isotherm study) Contact time = up to 90 h Temperature = 20°C–60°C	10.01	ΔH = −13.21 kJ/mol ΔS = −22.47 J/mol K ΔG = negative	Pseudo-first-order model, non-linear	Freundlich, non-linear	pH = 2 Agitation speed = 200 rpm Contact time = 70 h Temperature = 20°C	Albadarin et al. (2012)
Amino-functionalized MCM-41	Cr(VI)	Surface area = 774 and 517 Pore volume = 0.4 and 0.3 Pore size = 2.5 and 2.4		pH = 2 Dose = 1 g/l Concentration = 130 mg/l Contact time = 24 h Temperature = 25°C	86.4 and 63.3					Fellenz et al. (2017)

(*continued*)

TABLE 3.2 (*Continued*)
Summary of Parameters and Optimized Conditions for Batch Adsorption of Chromium

Adsorbent	Adsorbate	Surface Area (m^2/g), Pore Volume (cm^3/g), Pore Size (nm)	pHzpc	Experimental Conditions	Adsorption Capacity (mg/g)	Thermodynamic Parameters	Kinetic Model and Curve Fitting	Isotherm Model and Curve Fitting	Maximum Adsorption Conditions	References
Mesoporous carbon mesospheres		Surface area = 1121 Pore volume = 2.7 Pore size = 9.5		pH = 1–9 Dose = 0.2 g/l Agitation speed = 100 rpm Concentration = 5–100 mg/l Contact time = 0–24 h Temperature = 15°C–45°C	156.3	ΔH = 42.5 kJ/mol ΔS= 206.3 J/mol K ΔG = negative	Pseudo-second-order model, linear	Langmuir, linear	pH = 3 Contact time = 24 h Temperature = 45°C	Zhou et al. (2016)
Activated carbon	Cr(VI)	Surface area = 1126		pH = 2–8 Dose = 25–250 mg Agitation speed = 50–250 rpm Concentration = 1 mg/l Contact time = 0–24 h Temperature = ambient temperature	1.8		Pseudo-second-order model, linear	Langmuir, linear	pH = 3 Dose = 75 mg Agitation speed = 200 rpm Contact time = 4 h	Ihsanullah, Abu-Sharkh, et al. (2016)
Acid-modified activated carbon	Cr(VI)	Surface area = 1420		do	2.02		Pseudo-second-order model, linear	Langmuir, in spite of low R^2	do	Ihsanullah, Abu-Sharkh, et al. (2016)
CNT	Cr(VI)	Surface area = 156		do	1.02		Pseudo-second-order model, linear	Langmuir model, linear	pH = 3 Dose = 200 mg Agitation speed = 150 rpm Contact time = 4 h	Ihsanullah, Abu-Sharkh, et al. (2016)
Acid-modified CNT	Cr(VI)	Surface area = 170		do	0.96		Pseudo-second-order model, linear	Langmuir model, linear	do	Ihsanullah, Abu-Sharkh, et al. (2016)

(continued)

TABLE 3.2 (*Continued*)
Summary of Parameters and Optimized Conditions for Batch Adsorption of Chromium

Adsorbent	Adsorbate	Surface Area (m^2/g), Pore Volume (cm^3/g), Pore Size (nm)	pHzpc	Experimental Conditions	Adsorption Capacity (mg/g)	Thermodynamic Parameters	Kinetic Model and Curve Fitting	Isotherm Model and Curve Fitting	Maximum Adsorption Conditions	References
Mesoporous alumina (simultaneous adsorption with fluoride)	Cr(VI)	Surface area = 211.4 Pore volume = 0.54 Pore size = 8.4		pH = 6 Dose = 1 g/l Agitation speed = 120 rpm Concentration = 100 mg/l Contact time = up to 24 h Temperature = 25°C–40°C	9.19	ΔH = −32.96 kJ/mol ΔS = −115 J/mol K ΔG = positive	Pseudo-second-order	Freundlich	Contact time = 12 h Temperature = 25°C	Li, Xie, et al. (2016)
Chitosan powder	Cr(VI)	Surface area = 6.69		pH = 2–6 Dose = 1 g/l Agitation speed = 150 rpm Concentration = 100 mg/l Contact time = up to 24 h Temperature = 25°C	60.24		Pseudo-second-order model, linear	Langmuir, linear	pH = 4.5 Contact time = 20 min	Li, Li, et al. (2015)
Chitosan fibers	Cr(VI)	Surface area = 0.489		do	131.58		Pseudo-second-order model, linear	Freundlich, linear	pH = 4.5 Contact time = 8 h	Li, Li, et al. (2015)
Polypyrrole/ Fe_3O_4 composite (column study)	Cr(VI)			pH = 2–6 Dose = 2–6 g Concentration = 50–150 mg/l Bed depth = 30 cm Bed diameter = 2 cm Flow rate = 3 ml/min Temperature = ambient temperature Breakthrough capacity = 112–125 mg/g					Column Yoon–Nelson and Thomas models	Bhaumik et al. (2013)

chromium on the adsorbent (mesoporous alumina). The reason for the decline was attributed to the competition for active sites and the competitive effect of chromium with fluoride's internal diffusion.

3.2.3 Effect of Surface Modification

The amino-functionalization increased the adsorption capacity of iron oxide/mesoporous composite (Egodawatte et al. 2015). The increase in the shell thickness of poly(m-phenylenediamine) around Fe_3O_4 led to an increase in the amino groups, which was attributed to the increased adsorption capacity (Wang, Zhang, et al. 2015).

The increase in polyaniline coating on ethyl cellulose increased the removal capacity and the rate of adsorption (Qiu et al. 2014). The increase in polyaniline coating increased the amount of protonated amine groups and led to increased Cr(VI) adsorption via electrostatic attraction. In addition, the increase in the polyaniline loading enhanced the hydrophilic property of polyaniline/ethyl cellulose composite; hence, a shorter time period was enough as compared with ethyl cellulose for chromium(VI) removal by polyaniline-coated ethyl cellulose.

3.2.4 Effect of Material

The adsorption capacity of chitosan nanofibers was more than that of chitosan powders, in spite of the reduction in the number of functional groups in chitosan nanofibers. The reason for this is the instability of the virgin fibers (powder) in the aqueous system (Li, Li, et al. 2015).

The adsorption of Cr(VI) varied with the Mg/Al ratio used for the synthesis of aluminum–magnesium mixed hydroxide (Li et al. 2009). The adsorption of Cr(VI) with the aluminum–magnesium mixed hydroxide increased with an increase in the Mg/Al ratio up to 3. On increasing the ratio from 1 to 2, the percentage removal increased from 90.2 to 97.0. However, there was only a slight increase in the percentage removal from 97.0 to 98.4 (molar ratio 3); on further increasing the ratio to 4, there was a diminutive decrease in percentage removal to 98.2. The particle size decreased and zeta potential increased with an increase in the Mg/Al ratio up to 3. The high zeta potential and lower size were attributed to the high percentage removal. The lower size was expected to have larger surface area, and the high zeta potential was estimated to enhance the interaction between the adsorbate and adsorbent. However, the surface area was not reported in the study.

3.2.5 Mechanism

The van der Waals force was postulated to be the chief reason for the adsorption of Cr(VI) on activated carbon. The small pore size and low pore volume led to slow mass transport rate (Zhou et al. 2016). However, the high adsorption capacity of mesoporous carbon was attributed to considerable surface area, high pore volume, and pore diameter. The mesopores offer an unobstructed path for Cr(VI), due to which Cr(VI) diffuse into the activated sites without any hindrance, which leads to high adsorption capacity.

The confirmation of adsorption along with the uniform distribution of chromium(VI) on adsorbent can be ascertained by energy-dispersive X-ray spectroscopy (EDS) spectrum analysis (Cao, Qu, Yan, et al. 2012). The FT-IR analysis was used to depict the mechanism of adsorption by emergence or disappearance and change in the intensity of the peak. The FT-IR analysis of poly(m-phenylenediamine)-coated iron oxide after chromium adsorption suggested oxidation of benzenoid amine units to quinoid imine units (Wang, Zhang, et al. 2015), which can be predicted by the change in the intensity of peak corresponding to quinoid imine (increase) and benzenoid amine (decrease). The results of XPS also supported the FT-IR results. In addition to oxidation or conversion of the group, FT-IR was also used to predict the involvement of functional group participating in the adsorption process. The FT-IR analysis of polyaniline-coated ethyl cellulose after adsorption suggests the participation of –OH groups via declination in the intensity at 1053 cm^{-1} along with the oxidation of half-oxidized emeraldine base to fully oxidized pernigraniline base form (Qiu et al. 2014). The declination in the intensity along with the oxidation process suggests the participation of both polyaniline and ethyl cellulose in the Cr(VI) reduction.

The XPS and XANES analyses were also used to predict the mechanism of adsorption. The ion exchange behavior on adsorption of chromium on hematite can be ascertained by XPS and XANES spectrum analyses (Cao, Qu, Yan, et al. 2012). XPS indicates the reduction in the peak of the O-H than that of hematite, which suggests the hydroxyl group exchange with chromate. The O K-edge XANES spectrum analysis supports this observation by depicting the declination in the t_{2g}-to-e_g ratio. The e_g are sensitive to the ligands linked to O atom, and their intensity (e_g) increases when hydroxide ions are replaced with chromate ions.

The XPS analysis also helped in the determination of the oxidation state of the chromium adsorbed on the surface of the adsorbent (Egodawatte et al. 2015). The hexavalent chromium adsorbed may undergo reduction to trivalent chromium (Liu et al. 2012) or partial reduction (Fan et al. 2012; Zhang, Xia, et al. 2015). The XPS spectra of the sample after the adsorption of hexavalent chromium on iron oxide hollow microspheres depict XPS peaks corresponding to trivalent chromium. This indicates that after the adsorption of Cr(VI), it has been reduced to trivalent chromium (Liu et al. 2012). The increase in pH after the adsorption of Cr(VI) with polyaniline-coated ethyl cellulose was attributed to the consumption of protons (Qiu et al. 2014) and was suggested as supporting evidence for the reduction of chromium. The adsorption on the basis of XPS, FT-IR, and change in pH was suggested to follow protonation of amino groups, followed by complexation, then reduction of hexavalent chromium along with consumption of protons, and finally adsorption of reduced form via electrostatic interaction.

Similarly, the chromium(VI) adsorption on magnetite nanospheres was postulated to follow multiple steps, i.e. complexation and hydrolysis followed by adsorption. The magnetite surface in water has ≡ Fe-OH surface sites. The pH of the solution was postulated to alter the charge on the ≡ Fe-OH group by protonation and deprotonation. At pH lower than pHzpc, protonation occurs, which leads to positive charge on the group. However, when the pH is more than pHzpc, deprotonation occurs, leading to negative charge on the surface. Upon deprotonation, the Fe-O^- site acts as a Lewis base and can coordinate with metal ion (Kumari et al. 2015).

Along with the change in the oxidation state in chromium during adsorption, the change in complexation is also postulated during the adsorption of hexavalent chromium with mesoporous alumina (Li, Xie, et al. 2016). The adsorption of Cr(VI) is postulated to be started with the formation of outer-sphere complex, which upon dehydration leads to the formation of more stable inner-sphere complex.

RAMAN spectra can also be used as supporting evidence with XPS to ascertain the mechanism of hexavalent chromium (Fellenz et al. 2017). The surface of the amino-functionalized MCM-41 after Cr(VI) adsorption contains corresponding peaks of $-NH_2$ (400 eV) and $-NH_3^+$ (402 eV) in the sample. The quantity of $-NH_3^+$ is reported to be more than that of the $-NH_2$ group, and the electrostatic attraction between $-NH_3^+$ and negatively charged species of chromium at pH 2 leads to generation of $-NH_2$ group. The Cr(VI) reduction into Cr(III) is reported with transfer of proton from $-NH_3^+$ on the surface of adsorbent to the aqueous solution. So, surface functional groups on adsorbent's surface play a role in the reduction of Cr(VI) to Cr(III). The Raman spectra also confirmed the presence of both Cr(VI) (860 cm^{-1}) and Cr(III) species (500 cm^{-1}) on the adsorbent after adsorption.

The physical mixture of iron oxide and mesoporous silica nanocomposite was also tested to nullify the hypothesis that the adsorption of trivalent chromium on aminated iron oxide/mesoporous silica nanocomposite was not due to the physical mixture of iron oxide and mesoporous silica. The physical mixture exhibited a lower adsorption capacity (0.37 mmol/g) as compared to the composite (2.08 mmol/g) (Egodawatte et al. 2015). To ascertain that both silanol and amino groups aid in the adsorption, the experiment was conducted with adsorption of copper, chromate, and arsenate. The adsorption of copper was much more as compared to chromate and arsenate, in spite of chromate and arsenate ions having electrostatic advantage over copper ions with protonated amine surface (Egodawatte et al. 2015). This suggests that both silanol and amine functional groups contribute to the adsorption.

3.2.6 Desorption

Desorption of the chromium species was conducted with sodium hydroxide (Wang, Zhang, et al. 2015; Sun et al. 2014; Mohamed et al. 2017; Huang et al. 2013; Zhang, Xia, et al. 2015; Fan et al. 2012; Daneshvar et al. 2019). The adsorption efficiency declined after desorption (Huang et al. 2013; Zhang, Xia, et al. 2015). However, during desorption of chromium from amino-functionalized magnetic cellulose nanocomposite, there was not any significant decline in the adsorption capacity of the adsorbent after successive desorption (Sun et al. 2014).

3.3 NICKEL

Nickel resists corrosion and is used as an anticorrosive material, and it resists corrosion even at higher temperatures. Hence, it finds its use in electric ovens and toasters. Nickel is also used in batteries, coins, and desalination plants (RSC 2020g). In addition, nickel finds its application in, for example, electroplating, silver refinery, and zinc-based coating (Katal, Hasani, et al. 2012). The toxicity of nickel causes dermatitis, headache, nausea, and cardiovascular and kidney diseases (Liu, Yuan, et al. 2017). An overview of the experimental parameters and optimized conditions from batch adsorption experiments for nickel is presented in Table 3.3.

3.3.1 Effect of pH

Faur-Brasquet et al. (2002) suggested the existence of nickel (30 mg/l) as Ni^{2+} up to pH ca. 8, and afterward, on increasing the pH, it transformed into $NiOH^+$, $Ni(OH)_2$, $Ni(OH)_3^-$, $Ni_2(OH)^{3+}$, and $Ni_4(OH)_4^{4+}$ at various pH values. The percentage removal of nickel increased with an increase in the pH (Liu, Yuan, et al. 2017a; Guo, Su, et al. 2017; Tirtom et al. 2012; Chen and Wang 2006; Jain et al. 2014). The high concentration of the hydronium ion in the solution at lower pH led to an increase in the competition with nickel for adsorption sites (Liu, Yuan, et al. 2017; Kara et al. 2017; Tirtom et al. 2012), and in the case of aminated groups, the high protonation at low pH was attributed to the decline in percentage removal (Guo, Su, et al. 2017). In the case of use of poly(chitosan-acrylamide) as an adsorbent, the protonation led to coiling of polymer chains via imidization of amide groups (Saleh, Ibrahim, et al. 2017). This led to the reduced interaction with the metal ion and was considered as one of the reasons for the reduced uptake.

The pH_{zpc} of the substance acts as a significant factor in the adsorption of nickel ions. At pH above the pH_{zpc}, the surface of the adsorbent is negative, which leads to an increase in attraction between the nickel ions and adsorbent's surface (Mohammadi et al. 2014; Katal, Hasani, et al. 2012). In addition, at low pH, the competition between the hydronium ions and nickel ions leads to a decline in the adsorption capacity (Kandah and Meunier 2007).

The change in pH also led to a change in the mechanism of adsorption (Fan et al. 2009); e.g. adsorption of nickel was dominated by outer-sphere complex at low pH ($pH < 8$), whereas at high pH ($pH > 8$) the adsorption mechanism followed inner-sphere complex.

The pH range studied was different in different studies. The authors have employed different pH values for the precipitation of nickel, such as 6 (Kandah and Meunier 2007; Jain et al. 2014; Guo, Su, et al. 2017; Tirtom et al. 2012), 5.5 (Mangaleshwaran et al. 2015), and 8 (Liu, Yuan, et al. 2017; Saleh, Ibrahim, et al. 2017). The adsorption of nickel adsorption on Fe_3O_4-modified tea waste was optimized at pH 4; after pH 4, the increase in the removal of nickel was attributed to precipitation (Panneerselvam et al. 2011). The concentration of nickel(II) in the solution studied by Jain et al. (2014) declined after pH 6 in the absence of adsorbent.

3.3.2 Effect of Coexisting Ions and Surface Modification

Cobalt and manganese declined the adsorption efficiency of Ni-imprinted chitosan foam adsorbents and non-imprinted chitosan foam adsorbents, and the decline was evaluated in terms of the distribution ratio and selectivity coefficient (Guo, Su, et al. 2017).

TABLE 3.3
Summary of Parameters and Optimized Conditions for Batch Adsorption of Nickel

Adsorbent	Adsorbate	Surface Area (m^2/g), Pore Volume (cm^3/g), Pore Size (nm)	pHzpc	Experimental Conditions	Adsorption Capacity (mg/g)	Thermodynamic Parameters	Kinetic Model and Curve Fitting	Isotherm Model and Curve Fitting	Maximum Adsorption Conditions	References
Activated carbon from scrap tire	Ni		6.5	pH = 1–9 Dose = 0.05–4 g/l Agitation speed = 100 rpm Concentration = 0.1–40 ppm Contact time = 5–120 min Temperature = 35°C–45°C	25	14.768 kJ/mol ΔS = 58 J/mol K ΔG = negative	Pseudo-second-order model and intra-particle diffusion model, linear	Langmuir model, linear	pH = 7 Dose = 0.5 g/l *Concentration* = 20 ppm Contact time 50 min Temperature = 55°C	Gupta et al. (2014)
Iron oxide–tea waste composite (Fe_3O_4-TW)	Ni	Surface area = 27.5	6.5	pH = 2–7 Dose = 8–12 g/l Agitation speed = 100 rpm Concentration = 50–100 mg/l Contact time = 30–180 min Temperature = 303–323 K	38.3	ΔH = 33.41 kJ/mol ΔS = 579.7 J/mol K ΔG = negative	Pseudo-first-order model, linear	No clear distinction between Langmuir and Freundlich, linear	pH = 4 Dose = 12 g/l Concentration = 100 mg/l Contact time = 120 min Temperature = 323 K	Panneerselvam et al. (2011)
Dolomite	Ni	Surface area = 1.85		pH = 2–7.5 Dose = 1 g/l Agitation speed = 200 rpm Concentration = 10 mg/l Contact time = 5–360 min Temperature = 293–323 K	1.70	−19.49 kJ/mol ΔS = 2.57 J/mol K ΔG = negative	Pseudo-second-order model	Freundlich and Langmuir isotherm models	pH = 5.5 Contact time = 105 min Temperature = 293 K	Mohammadi et al. (2015)
Coconut copra meal	Ni			pH = 2–7 Dose = 20 g/l Agitation speed = 150 rpm Concentration = 60–120 mg/l Contact time = 4 h Temperature = 288–318 K	3.37	16.67 kJ/mol ΔS = positive ΔG = negative	Pseudo-second-order model, linear	Langmuir model, linear	pH = 5 Contact time = 2 h Temperature = 288 K	Saleem et al. (2016)

(continued)

TABLE 3.3 (*Continued*)
Summary of Parameters and Optimized Conditions for Batch Adsorption of Nickel

Adsorbent	Adsorbate	Surface Area (m^2/g), Pore Volume (cm^3/g), Pore Size (nm)	pHzpc	Experimental Conditions	Adsorption Capacity (mg/g)	Thermodynamic Parameters	Kinetic Model and Curve Fitting	Isotherm Model and Curve Fitting	Maximum Adsorption Conditions	References
Fe_2O_3-carbon foam	Ni in real wastewater	Surface area = 545.5 Pore volume = 0.423 Pore size = 3.09		pH = 6.6 Agitation speed = 15 rpm Concentration = 15.5 mg/l	6.4					Lee et al. (2017)
Graphene oxide	Ni			pH = 6 Dose = 1 gm/l Concentration = 10–25 mg/l Contact time = 10–70 min Temperature = 283–308 K				Langmuir, linear	Concentration = 25 mg/l (on the basis of adsorption capacity) Contact time = 50 min Temperature = 283 K	Najafi et al. (2015)
Glycine-functionalized graphene oxide	Ni			do				Freundlich, linear	do	Najafi et al. (2015)
Serbian natural clinoptilolite (binary solution with zinc)	Ni			pH = 4–7 Dose = 10 g/l Agitation speed = 105 rpm Concentration = 50–300 mg/l Contact time = 24 h Temperature = 308–318 K		−45.85 to −13.81 kJ/mol ΔG = negative ΔS = 85.4–204.9 J/K mol	Pseudo-second-order model, linear; the result of other models is not given in the article		pH = 6 Temperature = 318 K	Stojakovic et al. (2016)

(*continued*)

TABLE 3.3 (*Continued*)
Summary of Parameters and Optimized Conditions for Batch Adsorption of Nickel

Adsorbent	Adsorbate	Surface Area (m^2/g), Pore Volume (cm^3/g), Pore Size (nm)	pHzpc	Experimental Conditions	Adsorption Capacity (mg/g)	Thermodynamic Parameters	Kinetic Model and Curve Fitting	Isotherm Model and Curve Fitting	Maximum Adsorption Conditions	References
Nano-alumina	Ni	Surface area = 78.79	7.9	Dose = 5 g/l Agitation speed = 50, 75, and 100 rpm Concentration = 25–75 mg/l Contact time = 120 min Temperature = 25°C	30.82		Pseudo-second-order model, linear	No clear distinction between various models	Concentration = 25 mg/l Agitation speed = 100 rpm	Srivastava et al. (2011)
Oleate-modified iron oxide nanoparticle	Ni		Zeta potential = 1.6×10^{-3} C/m^2	pH = 6.8 and 7.2 Dose = 10 ml nanoparticle suspension (0.1 g/l concentration of iron oxide) in 10 ml of adsorbate Agitation speed = 125 rpm Concentration = 0.06–11.7 mg/l (isotherm study) Contact time = 45 h	44 and 80 for pH 6.8 and 7.2, respectively			Langmuir model, non-linear No comparison is shown with other isotherms	pH = 7.2	Magnet et al. (2017)
Polyacrylonitrile nanofibers (PAN) modified with 2-(20-pyridyl)-imidazole	Ni			pH = 1–6 Dose = 1 g/l Concentration = 150 mg/l (pH isotherm study) Contact time = 20 h Temperature = 22°C					pH = 6	Ndayambaje et al. (2016)
RAFT-IIP (Ni ion-imprinted polymer)				pH = 2–9 Dose = 0.4 g/l Contact time = up to 240 min Temperature = 25°C	81.73				pH = 8	Liu, Meng, Liu, et al. (2015)

(*continued*)

TABLE 3.3 (*Continued*)
Summary of Parameters and Optimized Conditions for Batch Adsorption of Nickel

Adsorbent	Adsorbate	Surface Area (m^2/g), Pore Volume (cm^3/g), Pore Size (nm)	pHzpc	Experimental Conditions	Adsorption Capacity (mg/g)	Thermodynamic Parameters	Kinetic Model and Curve Fitting	Isotherm Model and Curve Fitting	Maximum Adsorption Conditions	References
Polyurethane foam cube (batch and column study)	Ni			pH = 4–8 Dose = 2–16 g/l Agitation speed = 180 rpm Concentration = 10–60 mg/l Contact time = 30–210 min Temperature = 25°C–40°C	24.39		Pseudo-second-order model, linear	Freundlich model, linear	pH = 5.5 Dose = 8 g/l Concentration = 50 mg/l Contact time = 120 min Temperature = 30°C	Mangaleshwaran et al. (2015)
Activated carbon activated by zinc chloride	Ni	Surface area = 1483	3.4	pH = 2–10 Dose = 3.33 g/l Agitation speed = 240 rpm Concentration = 25–700 mg/l Contact time = up to 30 min Temperature = room temperature	166.7		Pseudo-second-order model, linear	Langmuir model, linear	pH = 8–10 Concentration = 25 mg/l Contact time = 20 min	Mohammadi et al. (2014)
Beta zeolite	Ni			pH = 5–10 Concentration = 1.13×10^{-5} to 1.13×10^{-3} mol/l (isotherm study) Contact time = 3 days Temperature = 25°C	4.97×10^{-5} to 1.35×10^{-4} mol/g on raising the pH from 6.5 to 8	$\Delta H = 9.85$ and 11.82 at pH 6.7 and 7.5, respectively ΔG = negative $\Delta S = 95.4$ and 108.41 J/mol K at pH 6.7 and 7.5, respectively	Pseudo-second-order model, linear	Langmuir model, linear	pH = 8 (at pH 9 and 10 due to precipitation)	Liu, Yuan, et al. (2017)
Beta zeolite modified with ethylenediamine	Ni			do	6.67×10^{-4} to 1.44×10^{-4} mol/g on raising the pH from 6.5 to 8	14.97 at pH 6.7 18.51 at pH 7.5 ΔG = negative $\Delta S = 119.41$ and 137.04 J/mol K at pH 6.7 and 7.5, respectively	Pseudo-second-order model	Langmuir model	do	Liu, Yuan, et al. (2017)

(*continued*)

TABLE 3.3 (*Continued*)
Summary of Parameters and Optimized Conditions for Batch Adsorption of Nickel

Adsorbent	Adsorbate	Surface Area (m^2/g), Pore Volume (cm^3/g), Pore Size (nm)	pHzpc	Experimental Conditions	Adsorption Capacity (mg/g)	Thermodynamic Parameters	Kinetic Model and Curve Fitting	Isotherm Model and Curve Fitting	Maximum Adsorption Conditions	References
Charcoal ash	Ni	Surface area = 62.1	8	pH = 2–10 Dose = 5–40 g/l Agitation speed = 400 rpm Concentration = 100–1000 mg/l Contact time = up to 140 min Temperature = 20°C–50°C		23.83 kJ/mol ΔG = negative ΔS = positive	Pseudo-second-order model	Not clear between the Freundlich and D-R isotherms	pH = 4 Dose = 30 g/l Concentration = 600 mg/l (on the basis of adsorption capacity) Contact time = up to 120 min Temperature = 50°C	Katal, Hasani, et al. (2012)
Metakaolin-based geopolymer (batch and column study)	Ni	Surface area = 39.24		pH = 2–8 Dose = 0.4–4 g/l Agitation speed = 200 rpm Concentration = 100 mg/l (kinetic study) Contact time = 50 min Temperature = 10°C–50°C	7.24×10^{-3} mol/g		Pseudo-second-order model, linear	Langmuir model, linear	pH = 4–8 Dose = 3.2 g/l Temperature = insignificant effect	Kara et al. (2017)
Ni-imprinted chitosan foam adsorbent	Ni			pH = 2–6 Dose = 0.33 g/l Agitation speed = 120 rpm Concentration = 20–200 ppm Contact time = 1440 min Temperature = 313 K	57.59		Pseudo-second-order model, linear	Langmuir model, linear	pH = 6 Contact time = ca. 480 min	Guo, Su, et al. (2017)
Non-imprinted chitosan foam adsorbent	Ni			do	20.53		Pseudo-second-order model, linear	Freundlich model, linear	pH = 5–6 Contact time = ca. 480 min	Guo, Su, et al. (2017)

(*continued*)

TABLE 3.3 (*Continued*)
Summary of Parameters and Optimized Conditions for Batch Adsorption of Nickel

Adsorbent	Adsorbate	Surface Area (m^2/g), Pore Volume (cm^3/g), Pore Size (nm)	pHzpc	Experimental Conditions	Adsorption Capacity (mg/g)	Thermodynamic Parameters	Kinetic Model and Curve Fitting	Isotherm Model and Curve Fitting	Maximum Adsorption Conditions	References
Poly(chitosan-acrylamide)	Ni			pH = 1–8 Dose = 100–6000 ppm Concentration = 50–4000 ppm Contact time = 3 h Temperature = 25°C	63.15			Langmuir model, linear	pH = 8	Saleh, Ibrahim, et al. (2017)
Polyvinyl alcohol-based chelating sponge	Ni	Surface area = 9.55		pH = 1–6 Dose = 1 g/l Agitation speed = 120 rpm Concentration = 1–5 mmol/l (isotherm study) Contact time = up to 6 h Temperature = 278–308 K	65.39	ΔH = 36.86–37.21 kJ/mol ΔS = 162.84 J/mol K ΔG = negative	Intra-particle diffusion model, linear	Freundlich model, linear	pH = 6 Contact time = 2 h Temperature = 308 K	Cheng et al. (2014)
Chitosan–methacrylic acid nanoparticle	Ni			pH = 3–5 Dose = 5–20 g/l Agitation speed = 200 rpm Concentration = 10–50 mg/l Contact time = 120 min Temperature = room temperature	0.87		Pseudo-second-order model, linear	Freundlich model, linear	pH = 5 Dose = 20 g/l	Heidari et al. (2013)
Chitosan immobilized on bentonite	Ni			pH = 4 Dose = 6.66 g/l Agitation speed = 50 rpm Concentration = 10–50 mg/l Contact time = up to 720 min Temperature = 25°C–55°C	12.35	ΔH = −10.47 kJ/mol ΔS = −0.0234 J/mol K ΔG = negative	Pseudo-second-order model, linear	Langmuir model, linear	Contact time = 1 h for 25 mg/l and 4 h for 200 mg/l Temperature = 25°C	Futalan et al. (2011)

The Ni imprinting of chitosan foam increased the affinity for nickel in the presence of other ions. The high selective coefficient of nickel for Ni-imprinted chitosan foam adsorbent suggested the higher affinity for nickel ions (Guo, Su, et al. 2017). The calcium ion declined the adsorption efficiency of the nickel ions with polyvinyl alcohol-based chelating sponge (Cheng et al. 2014). The calcium ion was proposed to form outer-sphere complex with water, and the complex covered the surface of adsorbent, making it less available to nickel. The decline in adsorption in the presence of magnesium was less than in the presence of calcium, and the reason was attributed to less affinity of the oxygen-ligating atom toward magnesium. In addition, greater steric hindrance of calcium–water complex than that of magnesium–water complex due to the larger size of calcium was also the reason.

The higher affinity for lead and copper as compared to nickel on chitosan-coated bentonite was attributed to the higher electronegativity of lead and copper (Futalan et al. 2011). The larger electronegativity difference led to preferential adsorption on the hydroxyl and amino groups of chitosan-immobilized bentonite.

The particles became aggregated after the modification of the β-zeolite with ethylenediamine. Modification was proposed to the change in particle surface charge, which led to morphological change, and this modification of the adsorbent also led to an increase in the adsorption capacity for nickel (Liu, Yuan, et al. 2017).

3.3.3 Mechanism

The chemisorption or ion exchange process is followed for nickel adsorption on oleate-modified iron oxide nanoparticles (Magnet et al. 2017). Under acidic conditions, the surface becomes protonated and there is less force of attraction between the adsorbate and adsorbent. The maximum adsorption for nickel by Magnet et al. (2017) began from 7.5, which is close to pKa of the oleate ions in the solution.

Liu, Yuan, et al. (2017) have used the XPS study to explain the mechanism. It has been concluded that there was formation of coordination bond between the adsorbate and nickel. The XPS spectra of O 1s and N 1s for β-zeolite-modified ethylenediamine are studied. The O 1s peak is increased by 0.06 at pH 6.70 and by 0.1 and 0.09 eV at pH 7.50. Similarly, the N 1s peak is increased by 0.01 and 0.01 at pH 6.70 and 0.17 and 0.05 eV at pH 7.50. The shifting of peaks is attributed to the declination of electron density around oxygen and nitrogen and shifting of the electrons toward nickel(II).

The amino group of chitosan immobilized on bentonite helped in the adsorption of nickel (Futalan et al. 2011). The follow-up of the Langmuir model by the system led to the postulation by the authors that the system follows adsorption by only one type of functional group, i.e., amino group, whereas the follow-up of the Freundlich isotherm by lead and copper adsorption on chitosan immobilized on bentonite suggested the involvement of both carboxylic and amino groups.

3.3.4 Desorption

Desorption of nickel was carried out using acidic solutions (Mohammadi et al. 2014; Mangaleshwaran et al. 2015; Jain et al. 2014; Bartczak et al. 2018; Hamza, Wei, et al.

2019), sodium hydroxide and water saturated with carbon dioxide (Katal, Hasani, et al. 2012), ethylenediaminetetraacetic acid and ammonium hydroxide (Ndayambaje et al. 2016), and water (Bartczak et al. 2018).

In the case of desorption of nickel from polyvinyl alcohol-based chelating sponge, EDTA showed better efficiency than hydrochloric acid (Cheng et al. 2014) or nitric acid (Tirtom et al. 2012). The desorption by EDTA is based on the chelation mechanism (Heidari et al. 2013). However, desorption of nickel by sodium chloride from chitosan–methacrylic acid nanoparticle was greater than EDTA. This suggests that the interaction between the adsorbate and adsorbent was electrostatic in nature. Sodium chloride affects the electrical double layer, which weakens the interaction between chitosan-based adsorbent and the metal ion.

On desorption, treatment of 2-(2′-pyridyl)imidazole-modified polyacrylonitrile nanofiber with nitric acid and ammonium hydroxide led to disappearance of C=N and N-H in FT-IR spectrum. This was due to the destruction of bond linkage between 2-(2′-pyridyl)imidazole ligand and polyacrylonitrile nanofiber. Hence, both nitric acid and ammonium hydroxide were not suitable for desorption (Ndayambaje et al. 2016).

3.4 ARSENIC

Arsenic exists in the oxidation state of –3, 0, +3, and +5. In water, it is mostly present as As(V) and under anaerobic conditions as As(III). The provisional guideline value set up by the WHO is 10 μg/l (WHO 2017). Arsenic applies as a doping agent in semiconductors. Arsenic compounds find their application in the preservation of wood (RSC 2020a). Arsenic has not been demonstrated to be essential for human beings (WHO 2017). Chronic exposure to arsenic leads to hyperpigmentation, liver injury, cardiovascular disease, gangrene of lower extremities (black foot disease), diabetes mellitus, and immunotoxic effects (Tokar et al. 2015; Ghosh and Mukhopadhyay 2019). An overview of the experimental parameters and optimized conditions from batch adsorption experiments for arsenic is presented in Table 3.4.

3.4.1 Effect of pH

Arsenite remains in the form of H_3AsO_3 below pH 9.22 and cannot undergo any repulsion from the surface of the adsorbent (Alijani and Shariatinia 2017). The reduced removal of arsenate under the same conditions of arsenite, i.e. above pH 8, was attributed to the competition between OH^- and $H_2A_SO_4^-$ ions. The optimum pH for both arsenite and arsenate was near-neutral pH, i.e., 6–7. The optimum removal of arsenite (Ren, Zhang, et al. 2011; Li, Li, et al. 2012; Mandal et al. 2013; Alijani and Shariatinia 2017; A. Saleh et al. 2016) and arsenate (Li, Li, et al. 2012; Yu, Ma, et al. 2015; Han, Pu, et al. 2013; Vu et al. 2015; Ren et al. 2011) was mostly in the acidic or near-neutral pH. However, in some cases, optimal removal of As(III) (Martinson and Reddy 2009; Zhang, Ren, et al. 2013) or As(V) (Ren et al. 2011) occurred in alkaline conditions also.

The maximum removal of As(V) in the acidic region with hydrous cerium oxide was due to the low isoelectric point (2.7) of the adsorbent and speciation of As(V)

TABLE 3.4
Summary of Parameters and Optimized Conditions for Batch Adsorption of Arsenic

Adsorbent	Adsorbate	Surface Area (m^2/g), Pore Volume (cm^3/g), Pore Size (nm)	pHzpc	Experimental Conditions	Adsorption Capacity (mg/g)	Thermodynamic Parameters	Kinetic Model and Curve Fitting	Isotherm Model and Curve Fitting	Maximum Adsorption Conditions	References
Iron–zirconium binary oxide	As(III)	Surface area = 339 Pore volume = 0.21	5.1 (zeta)	pH = 3–11 Dose = 200 mg/l Agitation speed = 140 rpm Concentration = 5–50 mg/l (isotherm study) Contact time = up to 48 h Temperature = 25°C	46.1 (at pH = 7)		Pseudo-second-order model, linear	Freundlich model	pH = 3–5 Contact time = 25 h	Ren et al. (2011)
Iron–zirconium binary oxide	As(V)	Surface area = 339 Pore volume = 0.21 cm^3/g	5.1 (zeta)	pH = 3–10 Dose = 200 mg/l Agitation speed = 140 rpm Concentration = 5–50 mg/l (isotherm study) Contact time = up to 48 h Temperature = 25°C	120 (at pH = 7)		Pseudo-second-order model, linear	Freundlich model	pH = ca. 3–9.2 Contact time = 25 h	Ren et al. (2011)
Hydrous cerium oxide	As(III)	Surface area = 198		pH = 3–11 Dose = 0.01–0.02 g/l (kinetic study) Agitation speed = 500 rpm Concentration = 1–100 mg/l (isotherm study) and 124 μg/l (kinetic study) Contact time = 12 h Temperature = 25°C	13 at pH 7		Pseudo-second-order model (no comparison with other models), linear	Redlich–Peterson model	pH = 3–7 (no significant difference over pH = 3–11)	Li, Li, et al. (2012)

(continued)

TABLE 3.4 (*Continued*)
Summary of Parameters and Optimized Conditions for Batch Adsorption of Arsenic

Adsorbent	Adsorbate	Surface Area (m^2/g), Pore Volume (cm^3/g), Pore Size (nm)	pHzpc	Experimental Conditions	Adsorption Capacity (mg/g)	Thermodynamic Parameters	Kinetic Model and Curve Fitting	Isotherm Model and Curve Fitting	Maximum Adsorption Conditions	References
Hydrous cerium oxide	As(V)	Surface area = 198		pH = 3–11 Dose = 0.01–0.02 g/l (kinetic study) Agitation speed = 500 rpm Concentration = 1–100 mg/l (isotherm study) and 124 µg/l (kinetic study) Contact time = 12 h Temperature = 25°C	40 at pH 7		Pseudo-second-order model (no comparison with others)	Langmuir model	pH = 3 Contact time = 2 h	Li, Li, et al. (2012)
Cupric oxide	As(III)	Surface area = 85		pH = 6–11 Dose = 2 g/l Agitation speed = 250 rpm Concentration = 0.11–100 mg/l (isotherm study) and 4.4 mg/l in pH study Contact time = 30 min Temperature = 21°C–25°C	26.9			Langmuir model, non-linear	pH = 9.3	Martinson and Reddy (2009)
Cupric oxide	As(V)	Surface area = 85		pH = 6–11 Dose = 2 g/l Agitation speed = 250 rpm Concentration = 0.11–100 mg/l (isotherm study) and 4.4 mg/l in pH study Contact time = 30 min Temperature = 21°C–25°C	22.6			Langmuir model, non-linear	pH-independent	Martinson and Reddy (2009)
γ-Fe_2O_3/biochar composite	As(V)			Dose = 2 g/l Agitation speed = 200 rpm Concentration = 5–200 mg/l (isotherm study) Contact time = 24 h Temperature = 22°C	3.147		Pseudo-first-order model, non-linear	Langmuir model, non-linear		Zhang, Gao, et al. (2013)

(continued)

TABLE 3.4 (*Continued*)
Summary of Parameters and Optimized Conditions for Batch Adsorption of Arsenic

Adsorbent	Adsorbate	Surface Area (m^2/g), Pore Volume (cm^3/g), Pore Size (nm)	pHzpc	Experimental Conditions	Adsorption Capacity (mg/g)	Thermodynamic Parameters	Kinetic Model and Curve Fitting	Isotherm Model and Curve Fitting	Maximum Adsorption Conditions	References
Cerium oxide–graphene composite	As(V)		6	pH = 3–10 Dose = 0.1 g/l Concentration = 1–80 mg/l (isotherm study) and 10 mg/l (kinetic study) Contact time = 24 h Temperature = 25°C	62.33		Pseudo-second-order model, non-linear	Langmuir model, non-linear	pH = 4 Contact time = 20 min	Yu, Ma, et al. (2015)
Feldspar (Finland)	As(V)	Surface area = 1.65	6.5	pH = 3–11 Dose = 6–60 g/l Agitation speed = 200 rpm Concentration = up to 12 mg/l (isotherm study) and 2–5 mg/l (kinetic study) Contact time = up to 72 h Temperature = 288–308 K	89.41 μg/g	ΔH = 50 kJ/mol ΔS = 170 J/mol K ΔG = positive at 288 K and negative at the remaining temperatures	Pseudo-second-order model, linear	Freundlich model, linear	Temperature = 308 K	Yazdani et al. (2016)
Feldspar (Italy)	As(V)	Surface area = 0.65	8.5	do	235.3 μg/g	ΔH = 19.3 kJ/mol ΔS = 65.7 J/mol K ΔG = positive at 288 K and negative at the remaining temperatures	Pseudo-second-order model, linear	Freundlich model, linear	Temperature = 308 K	Yazdani et al. (2016)
Nano-ZIF-8	As(III)	Surface area = 1063 Pore volume = 0.53		pH = 7 Dose = 2 g/l Agitation speed = 170 rpm Concentration = 5–100 mg/l (isotherm study) and 20 mg/l (kinetic study) Contact time = up to 48 h Temperature = 25°C	49.49		Pseudo-second-order model, non-linear	Freundlich model, non-linear	Contact time = 13 h	Jian et al. (2015)

(continued)

TABLE 3.4 (*Continued*)
Summary of Parameters and Optimized Conditions for Batch Adsorption of Arsenic

Adsorbent	Adsorbate	Surface Area (m^2/g), Pore Volume (cm^3/g), Pore Size (nm)	pHzpc	Experimental Conditions	Adsorption Capacity (mg/g)	Thermodynamic Parameters	Kinetic Model and Curve Fitting	Isotherm Model and Curve Fitting	Maximum Adsorption Conditions	References
Nano-ZIF-8	As(V)	Surface area = 1063 Pore volume = 0.53		do	60.03		Pseudo-second-order model, non-linear	Freundlich model, non-linear	Contact time = 7 h	Jian et al. (2015)
Iron-impregnated sugarcane carbon (batch and column study)	As(III) +As(V) (1:1 ratio)	Surface area = 803 Pore volume = 0.63	7.6	pH = 6–8 Dose = 0.5–3 g/l Agitation speed = 5–500 rpm Concentration = 50–5000 μg/l Contact time = 10–40 min Temperature = 30°C–50°C	0.148	ΔH = −7.96 kJ/mol ΔG = negative ΔS = −6.06 J/K mol	Pseudo-second-order model, linear	Langmuir model, linear	pH = 7 Agitation speed = 500 rpm Dose = 2.5–3 g/l Concentration = 100 mg/l Contact time = 30 min Temperature = 40°C	Roy et al. (2014)
Mesoporous alumina	As(V)	Surface area = 312 Pore volume = 0.5 Pore size = 5.8		pH = 2–10 Dose = 1 g/l Concentration = 11.18–130 mg/l Contact time = 30–1440 min Temperature = 15–60°C	39.06				pH = 3.9 Concentration = 130 mg/l Contact time = 720 min Temperature = 52.8°C	Han, Pu, et al. (2013)
Zirconyl polyacrylamide(43)	As(III)	Surface area = 341		pH = 1–12 Dose = 1–15 g/l Agitation speed = 20 rpm Concentration = 10–100 mg/l Contact time = 10–140 min Temperature = 20°C–80°C	41.48	ΔH = 140–149 kJ/mol ΔG = negative ΔS = 0.257–0.383 kJ/mol	Pseudo-second-order model, linear	Langmuir model, linear	pH = 1–3 Dose = 13 g/l Concentration = 10 mg/l Contact time = 120 min Temperature = 50°C	Mandal et al. (2013)

(continued)

TABLE 3.4 (*Continued*)
Summary of Parameters and Optimized Conditions for Batch Adsorption of Arsenic

Adsorbent	Adsorbate	Surface Area (m^2/g), Pore Volume (cm^3/g), Pore Size (nm)	pHzpc	Experimental Conditions	Adsorption Capacity (mg/g)	Thermodynamic Parameters	Kinetic Model and Curve Fitting	Isotherm Model and Curve Fitting	Maximum Adsorption Conditions	References
MIL-53(Fe)	As(V)	Surface area = 14 Pore volume = 0.012		pH = 3–11 Dose = 1 g/l Concentration = 5–15 mg/l Contact time = ca. 480 min Temperature = 298 K	21.27		Pseudo-second-order model, linear	Langmuir model, linear	pH = 5 Contact time = 90 min (5 mg/l) and 120 min at (10–15 mg/l) Concentration = 10 mg/l (on the basis of adsorption capacity)	Vu et al. (2015)
Magnesium ferrite ($Mg_{0.27}Fe_{2.5}O_4$)	As(III)	Surface area = 438 Pore volume = 0.643	5.2 (IEP)	pH = 1–13 Dose = 0.002–0.01 g/l (kinetic study) Agitation speed = 300 rpm Concentration = 0.2–134 mg/l (isotherm study) and 0.101 mg/l (kinetic study) Contact time = 12 h Temperature = 25°C	127.4		Pseudo-second-order model, linear	Langmuir model, non-linear	pH = 9 Dose = 0.01 g/l	Tang et al. (2013)
Magnesium ferrite ($Mg_{0.27}Fe_{2.5}O_4$)	As(V)	Surface area = 438 Pore volume = 0.643	5.2 (IEP)	pH = 1–13 Dose = 0.002–0.01 g/l (kinetic study) Agitation speed = 300 rpm Concentration = 0.1–148 mg/l (isotherm study) and 0.097 mg/l (kinetic study) Contact time = 12 h Temperature = 25°C	83.2		Pseudo-second-order model, linear	Freundlich model, non-linear	pH = 2 Dose = 0.01 g/l	Tang et al. (2013)

(*continued*)

TABLE 3.4 (*Continued*)
Summary of Parameters and Optimized Conditions for Batch Adsorption of Arsenic

Adsorbent	Adsorbate	Surface Area (m^2/g), Pore Volume (cm^3/g), Pore Size (nm)	pHzpc	Experimental Conditions	Adsorption Capacity (mg/g)	Thermodynamic Parameters	Kinetic Model and Curve Fitting	Isotherm Model and Curve Fitting	Maximum Adsorption Conditions	References
Iron–copper binary oxide	As(III)	Surface area = 282 Pore volume = 0.31	7.9 (PZC)	pH = 3–11 Dose = 0.2 g/l Agitation speed = 200 rpm Concentration = 5–60 mg/l (isotherm study) and 10 mg/l (kinetic study) Contact time = up to 28 h Temperature = 24°C	122.3		Pseudo-second-order model, non-linear	Freundlich model, non-linear	pH = 9.1 Contact time = ca. 10 h	Zhang, Ren, et al. (2013)
Iron–copper binary oxide	As(V)	Surface area = 282 Pore volume = 0.31	7.9 (PZC)	do	82.7		do	do	pH = 3–7 Contact time = ca. 15 h	Zhang, Ren, et al. (2013)
Fe–Mn binary oxide-impregnated chitosan bead (batch and column studies)	As(III)	Surface area = 248 Pore volume = 0.37		pH = 6–11.5 Dose = 1 g/l Agitation speed = 180 rpm Concentration = 5–60 mg/l (isotherm study) Contact time = 36 h Temperature = 25°C	54.2		Pseudo-second-order model, non-linear	Freundlich model, non-linear	pH = 6–8	Qi et al. (2015)
Fe–Mn binary oxide-impregnated chitosan bead	As(V)	Surface area = 248 Pore volume = 0.37		do	39.1		Pseudo-second-order model, non-linear	Freundlich model, non-linear	pH = 6–9	Qi et al. (2015)
Zerovalent iron-doped multiwalled carbon nanotubes	As(III)			pH = 1–9 Dose = 0.05–0.3 g/l Concentration = 10–80 mg/l (isotherm study) Contact time = up to 120 min Temperature = 298–318 K	200	ΔH = 27.30 kJ/mol ΔS = 91.95 J/mol K ΔG = negative	Pseudo-first-order model, linear	Freundlich, linear	pH = 6–7 Dose = 0.2 g/l Concentration = 10 mg/l Contact time = 60 min Temperature = 318 K	Alijani and Shariatinia (2017)

(*continued*)

TABLE 3.4 (*Continued*)
Summary of Parameters and Optimized Conditions for Batch Adsorption of Arsenic

Adsorbent	Adsorbate	Surface Area (m^2/g), Pore Volume (cm^3/g), Pore Size (nm)	pHzpc	Experimental Conditions	Adsorption Capacity (mg/g)	Thermodynamic Parameters	Kinetic Model and Curve Fitting	Isotherm Model and Curve Fitting	Maximum Adsorption Conditions	References
Zerovalent iron-doped multiwalled carbon nanotubes	As(V)			do	250	ΔH = 54.17 kJ/mol ΔS = positive (181.24 J/mol K) ΔG = negative	Pseudo-first-order model, linear	Freundlich model, linear	do	Alijani and Shariatinia (2017)
Air-oxidized zerovalent iron-doped multiwalled carbon nanotubes	As(III)			do	200	ΔH = 28.70 kJ/mol ΔS = 76.42 J/mol K ΔG = positive	Pseudo-second-order model, linear	Langmuir, linear	do	Alijani and Shariatinia (2017)
Air-oxidized zerovalent iron-doped multiwalled carbon nanotubes	As(V)			do	200	ΔH = 33.82 kJ/mol ΔS = positive (94.86 J/mol K) ΔG = positive	Pseudo-second-order model, linear	Freundlich, linear	do	Alijani and Shariatinia (2017)
Hematite	As(III)			pH = 1–9 Dose = 1–6 g/l Concentration = 5–60 mg/l (isotherm study) Contact time = up to 24 h Temperature = 298–318 K	4.23	ΔH = 3.60 kJ/mol ΔS = −22.82 J/K mol ΔG = positive	Pseudo-second-order model, linear	Langmuir, linear	pH = 6–7 Dose = 4 g/l Concentration = 10 mg/l Contact time = 1320 min	Alijani and Shariatinia (2017)
Hematite	As(V)			do	5.98	ΔH = 2.91 kJ/mol ΔS = −24.07 J/mol K ΔG = positive	Pseudo-second-order model, linear	Langmuir model, linear	do	Alijani and Shariatinia (2017)

(continued)

TABLE 3.4 (*Continued*)
Summary of Parameters and Optimized Conditions for Batch Adsorption of Arsenic

Adsorbent	Adsorbate	Surface Area (m^2/g), Pore Volume (cm^3/g), Pore Size (nm)	pHzpc	Experimental Conditions	Adsorption Capacity (mg/g)	Thermodynamic Parameters	Kinetic Model and Curve Fitting	Isotherm Model and Curve Fitting	Maximum Adsorption Conditions	References
Coconut fiber	As(III)	Surface area = 226.83		pH = 4 Dose = 1–20 g/l Agitation speed = 100 rpm for 10 min, followed by 80 rpm for the rest of the experiments Concentration = 0.5–2 mg/l Contact time = 1–10 h Temperature = 298–318 K	0.118	ΔH = 22,530 kJ/mol ΔS = 96.69 J/mol K ΔG = positive	Pseudo-first-order model, linear	Freundlich model, linear	Contact time = 8 h Dose = 10 g/l	Nashine and Tembhurkar (2016)
Magnetic gelatin-modified biochar	As(V)		4	pH = 2–11 Dose = 0.1 g/l Agitation speed = 150 rpm Concentration = 0.2–50 mg/l (isotherm study) Contact time = up to 24 h Temperature = 298–318 K	45.8	ΔH = 148.12 kJ/mol ΔS = 645.2 J/mol K ΔG = negative	Pseudo-second-order model, non-linear	Langmuir model, non-linear	pH = 4 Contact time = 4 h Temperature = 318 K	Zhou et al. (2017)
Functionalized multiwalled carbon nanotube	As(III)			pH = 2–10 Dose = 2.5 g/l Agitation speed = 110 rpm Concentration = 0.1–1200 mg/l (isotherm study) Contact time = up to 8 h Temperature = 298–318 K	111	ΔH = 8.05 kJ/mol ΔS = 1.09 J/mol K ΔG = negative	Weber–Morris model	Langmuir model, linear	pH = 7 Contact time = 4 h Temperature = 318 K	Sankararamakrishnan et al. (2014)
Functionalized multiwalled carbon nanotube	As(V)			do	166.7	ΔH = 2.52 kJ/mol K ΔS = 3.42 J/mol K ΔG = negative	Weber–Morris model	Langmuir model, linear	do	Sankararamakrishnan et al. (2014)
Chitosan-modified vermiculite	As(III)			pH = 2–8 Dose = up to 40 g/l Agitation speed = 200 rpm Concentration = 10–400 mg/l Contact time = up to 120 min Temperature = 20°C–50°C	72.2	ΔH = 46.7 kJ/mol K ΔS = negative ΔG = negative	Pseudo-second-order model, linear	Langmuir, non-linear	pH = 5 Dose = 20 g/l Time = 90 min Temperature = 20°C	A. Saleh et al. (2016)

(Li, Li, et al. 2012). The increase in pH led to the increase in negative charge on the surface of the adsorbent. This led to an increase in repulsion between the arsenic species ($HAsO_4^{2-}$ and $H_2AsO_4^-$) and adsorbent, thus leading to a decline in the percentage removal with an increase in pH. However, percentage removal of As(III) was not affected by the pH till pH 9. The As(III) species remained in neutral form H_3AsO_3 till pH 9.2, and the change of surface charge did not affect the electrostatic repulsion between the adsorbent and adsorbate. The increase in pH above 9.2 led to the conversion of the H_3AsO_3 to $H_3AsO_2^-$, and hence reduction of percentage removal. However, the pH-independent adsorption behavior of As(III) is not universal. The removal of As(III) with zirconyl polyacrylamide declined with an increase in pH above 3 ($pH_{zpc}=4.8$), and the reason was attributed to the more protonation at low pH and the increased competition from the hydroxide ions for adsorption sites (Mandal et al. 2013). Similarly, the percentage removal of As(III and V) with Fe–Mn binary oxide-impregnated chitosan bead and As(III) adsorption on CuO ($pH_{zpc}=9.4$) increased with an increase in pH from 6 to 12 (Qi et al. 2015) and 6 to 10 (Martinson and Reddy 2009), respectively.

The variation in the adsorption capacity also behaved non-linearly; e.g., the adsorption of As(V) with Fe-Cu binary oxide remained constant in the region of pH 3–7 and declined after pH 7 (Zhang, Ren, et al. 2013). Similarly, the bentonite–chitosan composite adsorbent undergoes dissolution under extreme acidic conditions (below pH 2) and solidification under an alkaline environment (Dehghani et al. 2016).

The stability of the adsorbent at variable pH values led to varied arsenic adsorption (Jian et al. 2015). The dissolution of ZIF-8 increased with a decline in pH and was negligible after the neutral pH. The surface was more positive ($pH_{IEP}=9.6$) in the acidic region, but the dissolution of the adsorbent declined the adsorption efficiency. So, the As(III and V) adsorption was maximum near the neutral pH. Similarly, the declined removal of As(V) in alkaline medium was attributed to the MIL-53(Fe) (adsorbent) instability (Vu et al. 2015).

The adsorption of As(V) with mesoporous alumina led to variation in the pH of the solution. The pH of the solution increased after adsorption in acidic conditions (till pH 7.6) and declined under alkaline conditions (pH 8.6 and 10). The increase in pH was due to the protonation of the adsorbent (Han, Pu, et al. 2013), and the decline in pH was due to the competition between hydroxide ions and anionic species of arsenic. The change in pH during adsorption was also time-dependent at near-neutral pH, i.e., 6.6. The decline in pH during the first hour of the adsorption process was due to the protonation of the surface and adsorption of hydroxide ions. Afterward, the pH of the solution got increased due to the ion exchange between hydroxide and arsenic species.

3.4.2 Effect of Coexisting Ions

As(III and V) adsorption with ZIF-8 was not significantly affected by the presence of sulfate and nitrate ions at neutral pH, i.e., 7, although their amount (sulfate and nitrate) was present 100 times than arsenic (Jian et al. 2015). However, CO_3^{2-} and PO_4^{3-} ions hindered the adsorption of As(III and V), and this was due to a similar adsorption behavior of arsenic with PO_4^{3-} ions (Table 3.5).

TABLE 3.5
Effect of Coexisting Ions on the Adsorption of Arsenate and Arsenite

		As(V)		As(III)		
Adsorbent	**Effect**	**Yes**	**No**	**Yes**	**No**	**References**
ZIF-8		Carbonate and phosphate	Sulfate and nitrate			Jian et al. (2015)
Hydrous cerium oxide		Humic acid	Chloride and nitrate		Chloride and nitrate	Li, Li, et al. (2012)
Chitosan fibers		Sulfate phosphate, humic acid	Carbonate and fluoride			Min et al. (2015)
Feldspar		Fluoride and phosphate	Chloride, nitrate, and sulfate			Yazdani et al. (2016).

Fluoride and phosphate caused a decline in percentage removal of As(V) on feldspar, whereas chloride, nitrate, and sulfate did not affect the percentage removal of As(V) (Yazdani et al. 2016). This was attributed to the categorization of fluoride and phosphate ions as inner-spherical adsorbing ions, similar structure of phosphate ions (triprotic acid form) to arsenate ions, and even similar ionization potential of phosphate (Yu, Ma, et al. 2015).

The charge on ionic species in addition to a similar structure of phosphate to arsenate (As(V)) was reported to be the reason for the decline in the percentage removal of As(V) on electrospun chitosan fiber (Min et al. 2015). It was postulated on the basis of negligible effect of carbonate and fluoride ion and the significant effect by the presence of sulfate and phosphate. In addition to inorganic species, humic acid also had a negative effect on the adsorption of As(V) (Min et al. 2015; Li, Li, et al. 2012).

The presence of ferrous ions minimally affected the adsorption of arsenate on goethite and hematite (Catalano et al. 2011). However, the near-neutral pH and high concentration of ferrous ions and arsenate led to precipitation. The precipitation can be evidenced by XRD. The minimum effect of ferrous ions on arsenate adsorption was attributed to the incompatible structure of ferrous ion species with iron oxide.

The ferrous ion enhanced the adsorption of arsenate under anoxic conditions without any change in its mechanism of adsorption on gamma-Al_2O_3 (Zhu and Elzinga 2015). This was attributed to the lowering of electrostatic barrier. The surface of the aluminum oxide became positively charged due to the inner-sphere complexation of the ferrous ions, which led to the increased adsorption of the anionic species, i.e., $HAsO_4^{2-}$. However, ferrous ions minimally affect the adsorption of the arsenite ion in a ternary sample (Fe(II), As(III), and As(V)) under anoxic conditions. The arsenite ion forms the inner-sphere complex confirmed by the XAS results, whereas Fe(II) forms the positively charged Fe(II)-Al(III)-LDH precipitate. The nonsignificant effect was attributed to the small amount of the ferrous ion in contrast to aluminum, which limits the prospective of Fe(II)-Al(III)-LDH to act as the quantitatively important adsorbent relative to gamma-alumina. The arsenic species was neutral and not affected by positively charged Fe(II)-Al(III)-LDH precipitate (H_3AsO_3).

3.4.3 Effect of Temperature and Agitation

The adsorbed amount increases with temperature for As(III) (Alijani and Shariatinia 2017; Liu, Chuang, et al. 2015; A. Saleh et al. 2016) and As(V) (Alijani and Shariatinia 2017; Liu, Chuang, et al. 2015; Yazdani et al. 2016). The increase in the adsorption of arsenic with adsorbent temperature was attributed to the reduction in adsorbent's boundary layer, so that the mass transfer resistance decreased (Nashine and Tembhurkar 2016). The enhancement of adsorption by an increase in temperature was also attributed to variation in diffusion rate across the boundary layer by a decrease in solution viscosity (Yazdani et al. 2016). The adsorption of arsenic was also increased with increasing agitation speed (Roy et al. 2014).

3.4.4 Mechanism

The iron in "zerovalent iron-doped multiwalled carbon nanotubes" was supposed to be oxidized to Fe^{2+} and Fe^{3+}. Fe^{2+} and Fe^{3+} further form hydroxide species. The adsorption of arsenic on these hydroxide and oxide surfaces was attributed to three phenomena, i.e., electrostatic, ion exchange, and coagulation (Alijani and Shariatinia 2017). The non-effectiveness of pH on arsenic adsorption between pH 4 and 8 ruled out the ion exchange and electrostatic phenomena. Hence, coagulation through ligand exchange–surface complexation remained as the principle phenomenon of adsorption. Whenever iron oxide interacted with water, the coordination shell of the metal ion on the oxide surface filled with -OH groups. During the period of ligand exchange, $H_2AsO_4^-$ replaced the two -OH groups on the surface of the adsorbent. The high adsorption capacity of iron-doped multiwalled carbon nanotubes as compared to hematite was attributed to the large quantity of functional groups. The decreased efficiency of air-oxidized iron-doped multiwalled carbon nanotubes as compared to iron-doped multiwalled carbon nanotubes was attributed to oxidation of Fe to hematite. The high efficiency of zerovalent iron-doped multiwalled carbon as compared to hematite for arsenite was attributed to the requirement of a step of pre-oxidation of arsenite to arsenate.

The adsorption of As(v) on the surface of hematite was confirmed by EDS spectra (Cao, Qu, Yan, et al. 2012). The elemental mapping using EDS suggested that the arsenic and chromium were evenly distributed on the surface of the adsorbent. To further confirm that adsorption occurred on the surface, XPS analysis of the adsorbent after and before the adsorption was conducted (Cao, Qu, Yan, et al. 2012). The analysis depicted the corresponding peaks of arsenic(V) at 45.4 eV (As-O bond) and chromium(VI) at 579.5 eV ($2p_{3/2}$-O) and 588.7 eV ($2p_{1/2}$-O). The O 1s XPS peak analysis was used to decipher the mechanism of adsorption. The deconvolution of O 1s spectra to two peaks at 530.0 and 531.5 eV was attributed to the oxygen bonded to metal atom in the lattice (Fe-O) of the adsorbent and oxygen atom present at the surface of the adsorbent(O-H), respectively. The peak intensity corresponding to O1s was higher for oxygen at surface (O-H) than for oxygen of the lattice (Fe-O). The intensity of the O1s peak (O-H) decreased after the adsorption of arsenic and chromium. This suggests the exchange of hydroxyl groups with $H_2AsO_4^-$ or $Cr_2O_7^{2-}/HCrO^{4-}$.

The XPS spectra were also helpful to estimate the oxidation/reduction behavior of adsorbed species (Liu, Chuang, et al. 2015). Magnetite contains both Fe^{2+} and Fe^{3+} ionic species. The initial Fe^{3+}/Fe^{2+} ratio in the sample without adsorption was 2, which increased to 3.2 on adsorption under anoxic conditions (oven-dried) of As(V) with a simultaneous increase in As(III) species by reduction. The ratio of Fe^{3+}/Fe^{2+} after the adsorption of As(III) changed to 2.9, indicating the simultaneous oxidation of Fe^{2+} and As(III). In addition to XPS, the XANES spectrum study is also applied for determining the mechanism of adsorption (Cao, Qu, Yan, et al. 2012). The O K-edge spectra are sensitive to the environment surrounding oxygen. The e_g spectra intensity increased after adsorption causing a change in the ratio of t_{2g} to e_g. This indicates the replacement of O-H groups with $H_2AsO_4^-$ or $Cr_2O_7^{2-}/HCrO^{4-}$. This indicates the ion exchange mechanism working in the respective study.

The XANES spectra are sensitive to the oven-drying process, if adsorption experiments are performed under anoxic conditions (Liu, Chuang, et al. 2015). The analysis of wet paste sample yielded As(III) and As(V) fraction as 2% and 4%, respectively. However, on oven-drying, the amount of As(III) and As(V) raised to 34% and 12%, respectively. The EXAFS analysis is helpful in determining the structure of arsenic species with coordination number and interatomic distance (Liu, Chuang, et al. 2015). The changes in interatomic distance and the coordination shell number suggest the reduction and oxidation of the material on the surface of adsorbent.

In addition to the reduction of peak size (Jian et al. 2015), the shift and emergence of new XPS peaks can be used as evidence for the adsorption phenomenon (Jian et al. 2015). The shifting of Zn 2p peaks (ZIF-8) to higher binding energy after adsorption indicates the involvement of Zn in the adsorption process. The emergence of new peaks at 400.6, 530.8, and 531.8 eV was assigned to the protonated nitrogen atom, Zn-O-As, and N-O-As, respectively.

The XPS study for the increase in peak area corresponds to metal-oxygen with a simultaneous decrease in metal-OH and H_2O concentration. The decrease in –OH groups suggests the role played by them in the adsorption of As(V) via ion exchange (Yu, Ma, et al. 2015). The adsorption of As(III and V) on the surface of Fe-Cu binary oxide did not occur with a change in the oxidation state of the arsenic species (Zhang, Ren, et al. 2013). However, curve fitting analysis of the sample is not the part of the manuscript.

The presence of any redox reaction during the adsorption process is also determined by XPS (Li, Li, et al. 2012). The XPS study suggested oxidation of As(III) with adsorption on CuO nanoparticles (Martinson and Reddy 2009). It has been hypothesized that As(III) undergoes oxidation prior to adsorption on the adsorbent. However, the study by Liu, Chuang, et al. (2015) suggested the oxidation of As(III) during drying of the sample. The mechanism of adsorption of As(V) on MIL-53(Fe) is proposed to occur on the phenomenon of Lewis acid–base interaction and the electrostatic interaction between central metal ion and adsorbate anionic species (Vu et al. 2015).

The XANES and XPS results differ in the percentage of species, e.g., content of oxidized and reduced species (Liu, Chuang, et al. 2015). The reason is attributed to the nature of analysis. The XPS analysis gives the idea about the surface of the adsorbent (< 10 nm), but the XANES analysis detects signals from the interior and exterior surfaces of the adsorbent. The material adsorbed inside the aggregates remains protected from the redox environment.

In the study of FT-IR spectra, emergence, shift and flattening of peaks are used as the evidence for the interaction between the adsorbate and arsenate species (Jian et al. 2015). In FT-IR spectrum, the peaks of C-H, N-H, and O-H in the region of 2500–3500 cm^{-1} which weakened after adsorption suggested the interaction of protonated groups (formed by water cluster acting as the Lewis acid site) with the adsorbate ionic species. The FT-IR peaks at 1591 cm^{-1} (N-H) and 1849 cm^{-1} (N-H) underwent red (−12 cm^{-1})- and blueshift (+225 cm^{-1}) after the adsorption of arsenate species, indicating the involvement of –NH groups. In addition, a strong band appeared at around 478 cm^{-1} (Zn-O), indicating the interaction of zinc with adsorbate species. Han, Pu, et al. (2013) suggested the band shift in FT-IR spectra as the interaction between the arsenic(V) species and the adsorbent -OH groups.

The peak in FT-IR spectra of As(III)-O occurred at 795 cm^{-1}; however, in the adsorption of As(III) with hydrous cerium oxide, the peak occurred at 773 cm^{-1}. The redshift was attributed to the weakening of As-O bond's strength. The decrease in the strength in As-O bond was due to the formation Ce-As-O bond (Li, Li, et al. 2012). This led to the conclusion by the authors for the formation of inner-sphere complex. Similar results were observed for the adsorption of As(V) species (805 cm^{-1}), indicating inner-sphere complex formation was also followed by the As(V) adsorption (Li, Li, et al. 2012).

Mineql version 4.6 computer models have also been used to predict the mechanism of adsorption (Han, Pu, et al. 2013; Yu, Ma, et al. 2015). The software suggested the change of arsenic(V) species on changing the pH along with dominant phase at a particular pH. The software suggested the existence of $H_2AsO_4^-$ and $HAsO_4^{2-}$ in the pH ranges of 3.5–5.5 and 8–10 (Han, Pu, et al. 2013). The optimized adsorption results at different pH values along with the results from Mineql version 4.6 computer model suggested preferred adsorption of arsenic species. The preferred species of arsenic for adsorption on alumina were suggested to be in the following order: $H_2AsO_4^- > H_3AsO_4 > HAsO_4^{2-} > AsO_4^{3-}$.

The adsorption mechanism was also hypothesized on the basis of arsenic(V) species distribution over different pH values, FT-IR, and change in pH after adsorption (Han, Pu, et al. 2013). At acidic pH (pH=2), the arsenic species ($H_2AsO_4^-$) was adsorbed by the acidic sites and protonated groups (OH). In addition, hydrogen bonding also played a role in the adsorption of uncharged H_3AsO_4 with unprotonated -OH groups. At near-neutral pH (6.6), the adsorption was proposed to be governed by electrostatic interaction between acidic centers and protonated hydroxyl groups with $H_2AsO_4^-$ species and with $HAsO_4^{2-}$ under alkaline conditions (pH= 10). In addition, the $HAsO_4^{2-}$ species was adsorbed on weakly acidic centers and protonated hydroxyl groups (pH= 10). Under near-neutral conditions (pH=6.6), the alkalized acid centers were also hypothesized to exchange -OH groups with $H_2AsO_4^-$ than $HAsO_4^{2-}$ (Han, Pu, et al. 2013).

3.4.5 Desorption

Both acids and bases were applied for desorption and reuse of the adsorbent (Alijani and Shariatinia 2017). However, in most of the cases, alkali were used (Mandal et al. 2013; Min et al. 2015; Zhang, Ren, et al. 2013; Qi et al. 2015; Tang et al. 2013) with different molar strengths ranging from 0.001 M (Min et al. 2015) to 2 M (Tang et al. 2013).

3.5 STRONTIUM

Strontium is found in nature in the +2 oxidation state only (Watts and Howe 2010). Strontium can find its way into environment by natural sources such as weathering of rocks. Anthropogenic sources are milling, use of phosphate fertilizers, and pyrotechnic devices. In addition, strontium-90 is an abundant radionuclide in nuclear fission and can be accidentally released into the environment (RSC 2020i). The high ingestion of the stable strontium led to a decline in the serum levels of calcitriol, which led to an adverse effect on calcium absorption (Armbrecht et al. 1998). An overview of the experimental parameters and optimized conditions from batch adsorption experiments for strontium is presented in Table 3.6.

3.5.1 Effect of pH

The adsorption of strontium increases on increasing the pH from acidic toward basic region (Zhao et al. 2014; Chen and Wang 2012; Zhang, Liu, Jiang, et al. 2015). The removal of strontium with graphene oxide–hydroxyapatite composite weakly depends on the pH (Wen et al. 2014). However, percentage removal increases with a rise in pH from 2 to 11. The protonation at a low pH was held responsible for low percentage removal.

In some cases, the increase in percentage removal was not uniform (Chen and Wang 2012; Zhao et al. 2014). The percentage removal in the case of Na-montmorillonite increased with a rise in pH (Zhang, Liu, Jiang, et al. 2015). Further increasing the pH of strontium solution after the attainment of maximum removal did not lead to any significant change in adsorption capacity (Yu, Mei, et al. 2015). Similarly, the adsorption increased very fast from pH 1 to 6 on SBA-15, and afterward, the rate of increase was not significant till pH 10 (Zhang, Liu, Jiang, et al. 2015). The protonation of the adsorbent at lower pH and deprotonation at higher pH caused the change in electrostatic force of attraction, leading to a change in the adsorption behavior with pH.

The pH dependency of strontium adsorption on Na-rectorite and Na-montmorillonite was attributed to three factors, i.e., speciation of Sr(II), surface property of the adsorbent, and functional group dissociation (Zhao et al. 2014; Yu, Mei, et al. 2015). The strontium hydrolysis constants were log $\beta 1 = -13.29$ and log $\beta 2 = -28.51$. Strontium was present as Sr^{2+} between 3 and 12. In addition, $Sr(OH)^-$ was present in negligible amount at lower pH. The increase in pH led to a decrease in $\equiv SOH^{2+}$ and an increase in $\equiv SO^-$ sites. This led to an increase in adsorption due to the deprotonation.

The percentage removal also declined in some cases on increasing the pH after maximum removal at a certain pH (Liu, Meng, Luo, et al. 2015). The adsorption

TABLE 3.6
Summary of Parameters and Optimized Conditions for Batch Adsorption of Strontium

Adsorbent	Adsorbate	Surface Area (m^2/g), Pore Volume (cm^3/g), Pore Size (nm)	pHzpc	Experimental Conditions	Adsorption Capacity (mg/g)	Thermodynamic Parameters	Kinetic Model and Curve Fitting	Isotherm Model and Curve Fitting	Maximum Adsorption Conditions	References
Graphene oxide–hydroxyapatite nanocomposite	Sr	Surface area = 91.85		pH = 2–11 Dose = 0.5 g/l Concentration = 20, 60, and 100 mg/l (kinetic study) Contact time = up to 48 h Temperature = 25°C	702.18		Pseudo-second-order model, linear, high r^2 and qe close No comparison with first order	Langmuir	Maximum at pH 11, but isotherm at pH 7 Contact time = 2 h	Wen et al. (2014)
Hydrophilic ion-imprinted polymer	Sr			pH = 2–8 Dose = 0.4 g/l Concentration = 3–8 mg/l (kinetic study) Contact time = up to 240 min Temperature = 25°C–45°C	135.28		Pseudo-second-order model	Langmuir	pH = 6 Contact time = 60 min Temperature = 45°C	Liu, Meng, Luo, et al. (2015)
Niobium-doped tungsten oxide	Sr	Surface area = 153		pH = 0–7 Dose = 2 g/l Concentration = 60–180 mg/l (isotherm study) and 90 mg/l (kinetic study) Contact time = up to 180 min Temperature = 298 K			Pseudo-second-order model	Freundlich	pH = 4 Contact time = 60 min	Liu, Mu, et al. (2015)

(*continued*)

TABLE 3.6 (*Continued*)
Summary of Parameters and Optimized Conditions for Batch Adsorption of Strontium

Adsorbent	Adsorbate	Surface Area (m^2/g), Pore Volume (cm^3/g), Pore Size (nm)	pHzpc	Experimental Conditions	Adsorption Capacity (mg/g)	Thermodynamic Parameters	Kinetic Model and Curve Fitting	Isotherm Model and Curve Fitting	Maximum Adsorption Conditions	References
Na-rectorite	Sr	Surface area = 11.9		pH = 2–12 Dose = 0.1–1 g/l Concentration = 10 mg/l (most studies) Contact time = up to 24 h Temperature = 293–333 K	10.78–14.28	ΔH = 1.31–132 kJ/mol ΔS = 90.30 ΔG = negative	Pseudo-second-order model, linear	Langmuir model, linear	pH = 10 Dose = 0.6 g/l (not maximum) Equilibrium achieved in 5 h, but using the 24-h contact period Temperature = 333 K	Zhao et al. (2014)
Na-montmorillonite	Sr			pH = 2–12 Dose = 0.1–1.2 g/l Concentration = 10 mg/l (most studies) Contact time = up to 12 h Temperature = 293–333 K	10.93	ΔH = 15.08–15.25 kJ/mol ΔS = 119.33 ΔG = negative	Pseudo-second-order model, linear (no comparison)	Freundlich, linear	pH > 9 Dose = 1.2 g/l, but 0.5 g/l used in most studies Contact time = 12 h, but equilibrium achieved in 6 h Temperature = 333 K	Yu, Mei, et al. (2015)
Dolomite	Sr			pH = 2.5–8.5 Dose = 10 g/l Agitation speed = 200 rpm Concentration = 10–50 mg/l (isotherm study) Contact time = up to 360 min Temperature = 293–333 K		ΔH = −16.68 kJ/mol ΔS = 18.60 ΔG = negative	Pseudo-second-order kinetic model, linear	Langmuir model, non-linear	pH = 5.5 Contact time = 120 min Temperature = 293 K	Ghaemi et al. (2011)

(continued)

TABLE 3.6 (*Continued*)

Summary of Parameters and Optimized Conditions for Batch Adsorption of Strontium

Adsorbent	Adsorbate	Surface Area (m^2/g), Pore Volume (cm^3/g), Pore Size (nm)	pHzpc	Experimental Conditions	Adsorption Capacity (mg/g)	Thermodynamic Parameters	Kinetic Model and Curve Fitting	Isotherm Model and Curve Fitting	Maximum Adsorption Conditions	References
Fe_3O_4 particle-modified sawdust	Sr			pH = 4–9 Dose = 0.4–5 g/l Concentration = 5–50 mg/l (isotherm study) Contact time = up to 120 min Temperature = 20°C	12.59			Langmuir model	pH = 9 Dose = 0.25 and 0.4 g/l in terms of adsorption capacity Contact time = 30 min	Cheng et al. (2012)
Magnetic chitosan bead	Sr			pH = 3.3–8.2 Dose = 0.67–10 g/l Agitation speed = 150 rpm Concentration = 5–300 mg/l Contact time = up to 10 h Temperature = 30°C	2.28			Langmuir model, linear	pH = 8.2 Contact time = 6 h	Chen and Wang (2012)
Activated carbon from textile sewage sludge	Sr	Surface area = 135		pH = 4–8 Dose = 0.33–13.66 g/l Concentration = 10–90 mg/l Contact time = up to 360 min Temperature = 20°C–60°C	12.2	ΔH = 44.70 kJ/mol ΔS = 195.84 J/mol K ΔG = negative			pH = 6 Dose = 13.33 g/l Initial concentration = 10 mg/l Contact time = 5 min	Kaçan and Kütahyalı (2012)

TABLE 3.6 (*Continued*)
Summary of Parameters and Optimized Conditions for Batch Adsorption of Strontium

Adsorbent	Adsorbate	Surface Area (m^2/g), Pore Volume (cm^3/g), Pore Size (nm)	pHzpc	Experimental Conditions	Adsorption Capacity (mg/g)	Thermodynamic Parameters	Kinetic Model and Curve Fitting	Isotherm Model and Curve Fitting	Maximum Adsorption Conditions	References
Sodium trititanate	Sr	Surface area = 29.37 Pore volume = 0.0012 Pore size = 9.89		pH = 2–6 Dose = 0.1–2 g/l (isotherm and pH studies) Concentration = 10–500 mg/l (isotherm study) and 10–30 mg/l (kinetic study) Contact time = up to 15 h Temperature = 298–318 K	79.37–108.7	ΔH = 14.23 kJ/mol ΔS = 30.79 J/mol K ΔG = positive (may be due to value of Kl taken as qe/ce)	Pseudo-second-order model, linear	Langmuir model, linear	pH = 6 Temperature = 318 K	Guan et al. (2011)
Potassium tetratitanate	Sr	Surface area = 38.35 Pore volume = 0.044 Pore size = 4.59		do	88.50–111.1	ΔH = 11.97 kJ/mol ΔS = 24.57 J/mol K ΔG = positive (may be due to value of Kl taken as qe/ce)	Pseudo-second-order model, linear	Langmuir model, linear	pH = 6 Temperature = 318 K	Guan et al. (2011)
SBA-15 (Santa Barbara Amorphous-15)	Sr			pH = 1–10 Concentration = 5–80 mg/l Contact time = 5–260 min Temperature = ambient temperature	17.67		Pseudo-second-order model, linear	Langmuir, linear	pH = 6–10 Contact time = 100 min	Zhang, Liu, Jiang, et al. (2015)
H_2O_2-modified attapulgite	Sr	Surface area = 174.4 Pore volume = 0.2556		pH = up to pH 11 Dose = 60 g/l (isotherm study) Agitation speed = 145 rpm Concentration = 10^{-5} to 8×10^{-5} mg/l Contact time = up to 24 h Temperature = 25°C–50°C	86.8 μg/g		Pseudo-second-order model, linear	Langmuir, linear	Contact time = 8 h Temperature = 40°C	Liu and Zheng (2017)

of hydrophilic ion-imprinted polymer based on graphene oxide, i.e., RAFT-IIP, increased slightly from pH 2 to 3, and afterward, it increased up to pH 6. The adsorption capacity decreased after pH 6. The low percentage removal at low pH was attributed to proton competition.

3.5.2 Zeta Potential

The change of percentage removal with pH is akin to change in the zeta potential (Liu, Mu, et al. 2015). The decrease in zeta potential occurred with the simultaneous increase in adsorption. In addition, the stagnant behavior of zeta potential led to no further increase in percentage removal after pH 4. The simultaneous variation in adsorption with respect to the change in zeta potential suggests that the adsorption mechanism was electrostatic in nature.

The zeta potential varies from case to case. The zeta potential of graphene oxide–hydroxyapatite composite was negative and became more negative with an increase in pH (Wen et al. 2014). In the case of niobium-doped tungsten oxide, the zeta potential became more negative with a rise in pH up to 4, and afterward, it became stagnant (Liu, Mu, et al. 2015).

3.5.3 Effect of Coexisting Ions

Cadmium and lead slightly influence the removal of strontium, whereas magnesium, aluminum, and sodium have no effect on strontium adsorption (Wen et al. 2014). The reason for this phenomenon is attributed to the difference in radii by less than 15% for lead and cadmium.

Humic acid increased the adsorption of strontium at pH lower than ca. 8.3, but at higher pH, it caused a decline in percentage removal (Zhao et al. 2014). Similarly, the adsorption of strontium with Na-montmorillonite increased in the presence of humic acid below pH 7, and an antagonistic effect was noticed at pH higher than 7 (Yu, Mei, et al. 2015).

3.5.4 Effect of Adsorbent's Modification

The treatment of adsorbent increased the removal efficiency (Wen et al. 2014; Liu and Zheng 2017). The strontium adsorption on graphene oxide by Wen et al. (2014) was more than that previously reported on graphene oxide. This is attributed to the pre-oxidization process and the residing functional groups of graphene oxide. The modification of attapulgite with H_2O_2 also increased its adsorption capacity (Liu and Zheng 2017). The H_2O_2 modification after treatment led to a decrease in agglomeration and an increase in surface area (Liu and Zheng 2017). The FT-IR spectra also showed a blueshift after H_2O_2 treatment. This was taken as the evidence of formation of smaller particles (Liu and Zheng 2017).

In addition, composite formation and doping led to an increase in percentage removal. The composite of graphene oxide and hydroxyapatite had adsorption capacity two times that of hydroxyapatite and nine times that of graphene oxide (Wen et al.

2014). The doping of niobium in tungsten oxide increased the adsorption capacity of strontium (Liu, Mu, et al. 2015). This was attributed to the degradation of crystallinity of tungsten oxide on addition of niobium.

3.5.5 Mechanism

The ion exchange mechanism of strontium adsorption on graphene oxide–hydroxyapatite nanocomposite was estimated by XPS analysis (Wen et al. 2014). The Ca 2p peak's intensity (347.4 and 350.9 eV) declined after strontium adsorption on graphene oxide–hydroxyapatite nanocomposite. This was attributed to the replacement of calcium ion with cadmium by the ion exchange phenomenon. The ionic radius of strontium (0.125 nm) is comparable to that of calcium (0.103 nm). Firstly, H^+ ion liberated from the solid surface into the solution; afterward, exchange of ions occurred; and finally, the complex was formed.

The ion exchange mechanism for the adsorption of Sr^{2+} on Na-rectorite was estimated by the determination of exchanged Na^+ ion in the solution (Zhao et al. 2014). The amount of Na^+ ion present in the solution was 1.25 times that of Sr^{2+} adsorbed. However, at pH 10.5 and above, Na^+ had no change with the amount of Sr^{2+} adsorbed. This was attributed to the formation of outer-sphere complex at low pH and inner-sphere complex at higher pH. The same case was reported in strontium adsorption with Na-montmorillonite (Yu, Mei, et al. 2015). The theoretical diffuse-layer model (DLM) model via Visual Minteq 3.0 was also used to explain the experimental data (Yu, Mei, et al. 2015). The FT-IR analysis was also used to ascertain the functional groups involved in the adsorption process (Chen and Wang 2012).

3.5.6 Desorption

The strontium regeneration was achieved by means of lowering the pH, e.g., the use of hydrochloric acid (Liu, Meng, Luo, et al. 2015; Zuo et al. 2019). The adsorbent, i.e., RAFT-IIP (hydrophilic ion-imprinted polymer based on graphene oxide), underwent five adsorption–desorption cycles without any significant loss of capacity (Liu, Meng, Luo, et al. 2015).

3.6 CADMIUM

Cadmium finds its application in batteries, galvanization of alloys, pigment, plasticizer, barrier to control nuclear fission, and plastic stabilizer (Tokar et al. 2015). Cadmium can be released into the environment by natural means such as volcanic eruption, forest fire, and sea-salt aerosol (ATSDR 2012). Cadmium is known to cause birth defects and cancer (RSC 2020b). The IARC has also classified it as a carcinogen. Food is a major source of cadmium to human exposure, and kidney is the target organ showing adverse effects of its toxicity (WHO 2017). An overview of the experimental parameters and optimized conditions from batch adsorption experiments for cadmium is presented in Table 3.7.

TABLE 3.7
Summary of Parameters and Optimized Conditions for Batch Adsorption of Cadmium

Adsorbent	Adsorbate	Surface Area (m^2/g), Pore Volume (cm^3/g), Pore Size (nm)	pHzpc	Experimental Conditions	Adsorption Capacity (mg/g)	Thermodynamic Parameters	Kinetic Model and Curve Fitting	Isotherm Model and Curve Fitting	Maximum Adsorption Conditions	References
Thiol-functionalized polyacrylonitrile fiber (mixed solution)	Cd			pH = 7 Dose = 1 g/l Agitation speed = 100 rpm Concentration = 30–500 mg/l Contact time = up to 180 min Temperature = 298 K	350.6		Pseudo-second-order model	Langmuir		Deng et al. (2017)
Chemically modified cellulose	Cd			pH = 2–6 Agitation speed = 180 rpm Concentration = 002 mol/l Contact time = up to 150 min Temperature = 15°C–25°C	279.33		Pseudo-second-order model, linear	Langmuir model, linear	pH = 5.2 Contact time = 90 min Temperature = 25°C	Zhou et al. (2013)
1,2,4-Triazole-3-thiol-modified lignin-based adsorbent	Cd			pH = 2–6 Dose = 1 g/l Agitation speed = 100 rpm Concentration = 25–300 mg/l (isotherm study) and 200 mg/l (kinetic study) Contact time = up to 180 min Temperature = 15°C–65°C	72.4	ΔH = 23.55 kJ/mol ΔS = 73.57 J/mol K ΔG = 288–318 K and negative at 328 and 338 K	Pseudo-second-order model, linear	Langmuir model, linear	pH = 6 Contact time = 180 min Temperature = 65°C	Jin et al. (2017)

(continued)

TABLE 3.7 (*Continued*)
Summary of Parameters and Optimized Conditions for Batch Adsorption of Cadmium

Adsorbent	Adsorbate	Surface Area (m^2/g), Pore Volume (cm^3/g), Pore Size (nm)	pHzpc	Experimental Conditions	Adsorption Capacity (mg/g)	Thermodynamic Parameters	Kinetic Model and Curve Fitting	Isotherm Model and Curve Fitting	Maximum Adsorption Conditions	References
Urea phosphate activated carbon from Phragmites australis	Cd			pH = 2–9.75 Dose = 1.6 g/l Agitation speed = 120 rpm Contact time = up to 24 h Temperature = 30°C			Pseudo-second-order model, linear	Langmuir, linear	pH = 5.6–6.5	Guo, Zhang, et al. (2017)
Carboxyl-functionalized magnetite	Cd		ISP = 4.8	pH = 2–9 Dose = 1 g/l Concentration = 10 mg/l in most studies Contact time = 10–150 min Temperature = 15°C–35°C	45.66	ΔH = −19.19 kJ/mol ΔS = −31.85 J/mol K ΔG = negative	Pseudo-second-order model, linear	Freundlich model, linear	pH = 6 Contact time = 120 min	Shi et al. (2015)
Carboxyl-terminated superparamagnetic iron oxide nanoparticles	Cd				20 µg/mg					Feng et al. (2012)
Acrylamide monomer-functionalized electrospun polystyrene	Cd			pH = 2–6 Dose = 0.5–5 g/l Concentration = 25 mg/l Contact time = 2 h Temperature = 23°C	10			Langmuir model, linear	pH = 5 Dose = 5 g/l	Bahramzadeh et al. (2016)
Tourmaline	Cd			pH = 2–8 Dose = 6 g/l Concentration = 25 mg/l Contact time = 10 min–24 h Temperature = 15°C–35°C	31.8	ΔH = 24.69 kJ/mol ΔS = 140.5–136.6 J/mol K ΔG = negative		Langmuir (the author suggested Freundlich, multiple metals)	pH = 6 Temperature = 35°C	Wang, Liu, et al. (2012)

(*continued*)

TABLE 3.7 (*Continued*)
Summary of Parameters and Optimized Conditions for Batch Adsorption of Cadmium

Adsorbent	Adsorbate	Surface Area (m^2/g), Pore Volume (cm^3/g), Pore Size (nm)	pHzpc	Experimental Conditions	Adsorption Capacity (mg/g)	Thermodynamic Parameters	Kinetic Model and Curve Fitting	Isotherm Model and Curve Fitting	Maximum Adsorption Conditions	References
Fe–Mn binary oxide	Cd	Surface area = 142	7.8	pH = 2–6 Dose = 0.5 g/l Agitation speed = 200 rpm Concentration = 50 mg/l Contact time = 24 h Temperature = 25°C	74.76	ΔH = 5.174 kJ/mol ΔG = negative ΔS = 0.078 J/mol K	Pseudo-second-order model, non-linear	Langmuir model, non-linear	pH = 6	Zhong et al. (2016)
Dandelion nanostructured TiO_2	Cd	Surface area = 226 Pore volume = 0.236 Pore size = 4.1		pH = 1–6 Concentration = 40–200 mg/l (isotherm study) Contact time = 80 min Temperature = 25°C	396		Pseudo-second-order model, linear	Langmuir model, linear	pH = 5 (maximum is 6) Contact time = 60 min	Zha et al. (2014)
Spherical flower-shaped nanostructured TiO_2	Cd	Surface area = 172 Pore size = 4.5		pH = 1–6 Concentration = 40–200 mg/l (isotherm study) Contact time = 150 min Temperature = 25°C	282		Pseudo-second-order model, linear	Langmuir, linear	pH = 5 (maximum is 6) Contact time = 150 min	Zha et al. (2014)
Diethyl-4-(4 amino-5-mercapto-4H-1,2,4-triazol-3-yl)phenyl phosphonate-coated Fe_3O_4	Cd	Surface area = 11.25 Pore volume = 0.0669		pH = 2–8 Dose = 0.02–014 g/l Concentration = 20–60 mg/l Contact time = 120 min Temperature = 25°C	49.1		Pseudo-second-order model, linear	Langmuir, linear	pH = 6 Dose = 0.1 g/l Contact time = 60 min	Venkateswarlu and Yoon (2015b)
Nitrilotriacetic acid anhydride-modified lignocellulosic material	Cd	Surface area = 1.8006 Pore size = 8.92		pH = 2–8 Dose = 0.02–014 g/l Concentration = 20–60 mg/l Contact time = 120 min Temperature = 25°C	143.4	ΔH = 10.32 kJ/mol ΔS = 35.91 J/K mol ΔG = negative	Pseudo-second-order model, linear	Langmuir, linear		Huang, Yang, et al. (2015)

(continued)

TABLE 3.7 (*Continued*)
Summary of Parameters and Optimized Conditions for Batch Adsorption of Cadmium

Adsorbent	Adsorbate	Surface Area (m^2/g), Pore Volume (cm^3/g), Pore Size (nm)	pHzpc	Experimental Conditions	Adsorption Capacity (mg/g)	Thermodynamic Parameters	Kinetic Model and Curve Fitting	Isotherm Model and Curve Fitting	Maximum Adsorption Conditions	References
Graphene oxide–Al_{13} composite	Cd		6.3	pH = 2–10 Dose = 0.1–0.4 g/l Concentration = 5–25 mg/l Contact time = 2 h Temperature = 298 K	89.74	ΔH = 17.38 kJ/mol ΔS = 73.27 J/mol K ΔG = negative	Pseudo-second-order model, linear	Langmuir model, linear		Yan, Zhao, et al. (2015)
L-arginine-modified magnetic adsorbent	Cd			pH = 1–6 Dose = 0.5 g/l Concentration = 30–210 mg/l Contact time = up to 30 min Temperature = 25°C–40°C	102	ΔH = 3.10 kJ/mol ΔS = 17.46 J/mol K ΔG = negative	Pseudo-second-order model	Langmuir model, linear	pH = 6 Concentration = 120 mg/l (on the basis of adsorption capacity) Contact time = 20 min Temperature = 40°C	Guo, Jiao, et al. (2017)
TMU-16-NH_2 metal–organic framework	Cd		4.9	pH = 3–10 Dose = 0.4–4 g/l Agitation speed = 240 rpm Concentration = 50–200 mg/l Contact time = 5–40 min Temperature = 298–318 K	126.6	ΔH = 10.39 kJ/mol ΔG = negative ΔS = 54.71 J/mol K	Pseudo-second-order model, linear	Both Langmuir and Freundlich (R^2 is 0.9766 for Langmuir and 0.9986 for Freundlich)	pH = 6 Dose = 2 g/l Concentration = 50 mg/l Contact time = 30 min Temperature = 318 K	Roushani et al. (2017)
Polylysine–resorcinol-coated alumina nanotube	Cd			pH = 2–8 Dose = 0.1–0.4 g/l Concentration = 0.5–200 mg/l (isotherm study) Contact time = 5–15 min	220			Freundlich model, linear	pH = 6.5 Contact time = 11 min	Beyki et al. (2017)

(*continued*)

TABLE 3.7 (*Continued*)
Summary of Parameters and Optimized Conditions for Batch Adsorption of Cadmium

Adsorbent	Adsorbate	Surface Area (m^2/g), Pore Volume (cm^3/g), Pore Size (nm)	pHzpc	Experimental Conditions	Adsorption Capacity (mg/g)	Thermodynamic Parameters	Kinetic Model and Curve Fitting	Isotherm Model and Curve Fitting	Maximum Adsorption Conditions	References
Iron oxide activated red mud	Cd		6.8	pH = 2–12 Dose = 1–12 g/l Concentration = 100–500 µg/l Contact time = up to 150 min Temperature = 293–308 K	117.64 µg/g	ΔH = 7.266 kJ/mol ΔG = negative ΔS = 0.06 kJ/mol K	Pseudo-second-order model, linear	Freundlich model, linear	pH = 6.5 Dose = 6 g/l Initial concentration = 400 µg/l Contact time = 90 min Temperature = 300 K (optimum conditions not maximum)	Khan et al. (2015)
Sodium dodecyl sulfate-modified chitosan	Cd			pH = 4–8 Dose = 0.09–1.35 g/l Concentration = 10–100 mg/l Contact time = up to 48 h Temperature = 25°C	125		Pseudo-second-order model, linear	Langmuir model, linear	pH = 7 Concentration = 10–30 mg/l Contact time = 28–30 h	Pal and Pal (2017)
Polyacrylic acid-modified magnetic mesoporous carbon	Cd	surface area = 689.7 Pore size = 3 and 5	ca. 4	pH = 2–9 Dose = 0.2 g/l Agitation speed = 150 rpm Concentration = 50–1200 mg/l (isotherm study) and 100 mg/l (kinetic study) Contact time = up to 300 min Temperature = 30°C	406		Pseudo-second-order model (no comparison with pseudo-first order, but high R^2 and near adsorption capacity with experimental data)	Langmuir model, linear	pH = 7 Contact time = 120 min	Zeng et al. (2015)

(*continued*)

TABLE 3.7 (*Continued*)
Summary of Parameters and Optimized Conditions for Batch Adsorption of Cadmium

Adsorbent	Adsorbate	Surface Area (m^2/g), Pore Volume (cm^3/g), Pore Size (nm)	pHzpc	Experimental Conditions	Adsorption Capacity (mg/g)	Thermodynamic Parameters	Kinetic Model and Curve Fitting	Isotherm Model and Curve Fitting	Maximum Adsorption Conditions	References
Graphene oxide	Cd	Surface area = 117.3	4.1	pH = 2–10 Dose = 1.2 g/l Concentration = 10 mg/l Contact time = up to 24 h Temperature = 293–313 K	111.11	ΔH = 11.49 kJ/mol ΔG = negative ΔS = 120.63 J/mol K	Pseudo-second-order model, linear	Langmuir model, linear	pH = 7 Contact time = 6 h Temperature = 313 K	Huang, Wu, et al. (2015)
Alumina-decorated multiwalled carbon nanotube	Cd	Surface area = 109.2 Pore volume = 0.2492 Pore size = 10.97	6.2	pH = 4–10 Dose = 1 g/l Agitation speed = 150 rpm Concentration = 1 mg/l Contact time = up to 16 h Temperature = 25°C	27.21		Pseudo-second-order model (no comparison with pseudo-first order, but high R^2 and near adsorption capacity with experimental data)	Langmuir model, non-linear	pH = 7 Contact time = 4 h	Liang, Liu, et al. (2015)
Alpha-ketoglutaric acid-modified magnetic chitosan	Cd			pH = 1–8 Dose = 1.33 g/l Agitation speed = 150 rpm Concentration = 10–1000 mg/l (isotherm study) 100–500 mg/l (kinetic study) Contact time = up to 90 min Temperature = 25°C–75°C	201.2	ΔH = 10.21 kJ/mol ΔG = negative ΔS = 58.92 J/mol K	Pseudo-second-order model, linear	Langmuir	pH = 6 Contact time = 30 min	Yang, Tang, et al. (2014)

3.6.1 Effect of pH

The percentage removal of cadmium increased with an increase in the pH from acidic pH to near-neutral pH (Zhou et al. 2013). The maximum amount of adsorption was achieved at different pH values, e.g., 5 (Bahramzadeh et al. 2016), 5.2 (Zhou et al. 2013), 6 (Jin et al. 2017; Sreenu et al. 2016), 7 (Pal and Pal 2017), and 8 (Wang, Liu, et al. 2012). The pH range studied was different in different studies. The authors have recommended different pH values for precipitation of cadmium, such as 6 (Zha et al. 2014; Sreenu et al. 2016; Wang, Liu, et al. 2012; Jin et al. 2017; Guo, Jiao, et al. 2017; Khan et al. 2015), 6.8 (Zhong et al. 2016), 7 (Zeng et al. 2015), and 8 (Chen, Shah, et al. 2017). However, most have limited their experiments at pH 6 or pH lower than 6, to avoid precipitation.

The low adsorption capacity at lower pH was due to the competition from hydronium ions (Zhou et al. 2013; Shi et al. 2015; Bahramzadeh et al. 2016; Zuo et al. 2017; Venkateswarlu and Yoon 2015b), protonation of the surface (Zha et al. 2014; Jin et al. 2017; Roushani et al. 2017), and electrostatic interaction, e.g., an increase in positive surface charge with pH change (Bahramzadeh et al. 2016). In addition, dissolution of the adsorbent, i.e., calcite dissolution at lower pH (Zuo et al. 2017); ion exchange and complex formation (Yakout et al. 2016; Zha et al. 2014); and steric hindrance (Guo, Jiao, et al. 2017) were also responsible for the variable adsorption capacity at different pH values.

The adsorption capacity declined after attaining the maximum pH in some cases (Zhou et al. 2013; Bahramzadeh et al. 2016). The decrease after maximum removal was attributed to hydroxide precipitation (Sreenu et al. 2016; Venkateswarlu and Yoon 2015b; Yan, Zhao, et al. 2015).

The increase in removal was not smooth with pH. The increase is fast in some regions and becomes slow or constant in some other regions (Zha et al. 2014; Shi et al. 2015). The adsorption of cadmium on $CaCO_3$-modified sewage sludge biochar increased with an increase in the pH from 4.5 to 8.5 (Zuo et al. 2017). The higher adsorption at higher pH was attributed to the increase in calcite ion in the solution, which led to an increased ion exchange process.

The adsorbent had a positive charge below the pH_{zpc} (point of zero charge) (Ma, Zhu, et al. 2015; Roushani et al. 2017). The treatment of montmorillonite with aluminum led to an increase in pH_{zpc} (at 6.81 and 7.56 for two adsorbents), which caused to a positive charge on the surface of the adsorbent at the adsorption experiment conditions (pH = 5). Hence, it led to reduced adsorption capacity after treatment. In addition, the tendency to form chelate with adsorbate and competition from proton also depends on the isoelectric point (Shi et al. 2015).

3.6.2 Effect of Coexisting Ions and Comparison with Other Ions

Cu, Ni, Pb, and Zn showed a competitive effect with Cd in binding to 2-mercaptobenzaldehyde SBA-15 (Sreenu et al. 2016). The adsorption capacity of cadmium declined in the presence of zinc on L-arginine-modified magnetic adsorbent due to the competition for active sites (Guo, Jiao, et al. 2017). The presence of competitive ions in seawater was held responsible for the lower adsorption capacity in comparison with tap and industrial water (Yakout et al. 2016). The presence of phosphate enhanced the adsorption efficiency of cadmium on Al_{13}-pillared montmorillonite (Ma, Zhu, et al. 2015). This was attributed to the adsorption of phosphate on the adsorbent followed by the adsorption of cadmium on phosphate.

The cadmium(II) ion removal efficiency on diethyl-4-(4 amino-5-mercapto-4H-1, 2,4-triazol-3-yl)phenyl phosphonate-coated Fe_3O_4 was more than that of Zn, Cu, Co, and Ni (Venkateswarlu and Yoon 2015b). The mercapto and amine groups on the adsorbent were held responsible for this. The soft acid (Cd) interaction with soft base (mercapto) predominates as compared to the interaction with other metal ions (borderline acid according to the HSAB theory). However, zinc has more efficiency for L-arginine-modified magnetic adsorbent as compared to cadmium (Guo, Jiao, et al. 2017). This is attributed to the Irving–Williams series in which Mn(II) < Fe(II) ~ Cd(II) < Zn(II) < Ni(II) << Cu(II). So, Zn is more approachable than Cd.

3.6.3 Effect of Surface Modification

The surface modification can be investigated by XRD, BET, and FT-IR (Shi et al. 2015; Sreenu et al. 2016; Zeng et al. 2015). The change in crystallinity can be estimated by XRD (Shi et al. 2015). The presence of functional groups on the adsorbent is estimated by FT-IR. The carboxyl functionalization on magnetite led to the presence of new peaks at 1618 and 1419 cm^{-1}. This is indicative of the formation of a bond between COO- and Fe atom (Shi et al. 2015). The zeta potential change is also taken as the macroscopic evidence for surface modification (Zeng et al. 2015).

The TEM analysis is also used to study the structure of the adsorbent post-functionalization. The TEM analysis depicts that 2D hexagonal mesostructure of magnetic mesoporous carbon was retained even after modification with polyacrylic acid (Zeng et al. 2015). The voids were present after modification, which suggests the nonsignificant impact of chelation on magnetic mesoporous carbon.

The functionalization increased the adsorption efficiency of the adsorbent (Sreenu et al. 2016; Venkateswarlu and Yoon 2015b). This can happen in spite of the decrease in surface area after surface modification (Sreenu et al. 2016). The increased number of binding sites in the form of functional groups was held responsible for the increased adsorption efficiency (Zeng et al. 2015).

The increase in surface-modifying agent concentration increased the adsorption up to a point, and this can be deduced from sodium dodecyl sulfate modification of chitosan. The sodium dodecyl sulfate concentration varied from 200 to 6000 mg/l during the preparation of sodium dodecyl sulfate-modified chitosan (Pal and Pal 2017). The percentage removal increased with a rise in sodium dodecyl sulfate concentration to 200–3000 mg/l. This was attributed to the increased negative surface charge on the surface of the adsorbent by sodium dodecyl sulfate. However on further raising the concentration, sodium dodecyl sulfate (critical micelle concentration = 2333 mg/l) remained as micelle in solution.

3.6.4 Effect of Material

Aluminum and clay ratio changed from 1 to 4 in the synthesis of Al_{13}-pillared montmorillonite (Ma, Zhu, et al. 2015). Different variants of Al_{13}-pillared montmorillonite were used for removal of cadmium. The adsorption capacity was more for the least amount of Al in comparison with the material having larger specific surface area (ratios of 4 and 2). This was attributed to the higher ion exchange capacity with high-charge species (cation in current case), and this led to an enhanced interaction with the charged sites. The intercalated cations (Al) occupying the adsorption sites were not readily available to exchange with Cd(II). So, the amount of adsorption decreased with the high amount of aluminum in Al_{13}-pillared montmorillonite. However, the incorporation of Al into graphene oxide raised its adsorption efficiency from 61% to 98.92% (Yan, Zhao, et al. 2015).

The morphology of the material also affects the adsorption of cadmium. The uptake by dandelion-structured TiO_2 was larger than that of spherical flower-shaped TiO_2 (Zha et al. 2014). The higher number of active sites was held responsible for the higher percentage removal of dandelion-structured TiO_2 as compared to that of spherical flower-structured TiO_2. The higher number of active sites was attributed to the higher surface area. In addition, larger pore volume was also attributed to the fast diffusion of Cd(II).

3.6.5 Effect of Temperature

The adsorption capacity increased with a rise in temperature in some cases (Chen, Shah, et al. 2017; Huang, Wu, et al. 2015; Singh et al. 2014). Temperature can influence the rate of diffusion and the viscosity of the solution (Guo, Jiao, et al. 2017). When the temperature increased from 25°C to 40°C for cadmium adsorption on L-arginine-modified magnetic adsorbent, the capacity raised from 102 to 109 mg/g for Cd(II). The increase in temperature led to a decrease in the viscosity of the solution, which led to an increase in the diffusion rate of metal across the boundary layer. On the contrary, an increase in temperature also decreased the adsorption in few cases (Shi et al. 2015; Khan et al. 2015).

3.6.6 Mechanism

The mechanism for adsorption of cadmium on Fe–Mn binary oxide can be estimated by XPS (Zhong et al. 2016). The Ols peak of the adsorbent metal lattice and surface hydroxyl group decreased from 530.2 to 529.9 and by 0.1 eV, respectively. The decrease in binding energy corresponds to the formation of Cd-O bond. In addition, the proportion of O in the surface hydroxyl group declined with a simultaneous increase in O in metal lattice. This suggests the participation of surface hydroxyl group with cadmium via a complexation process (Zhong et al. 2016). In FT-IR analysis, the peak corresponding to surface hydroxyl group (3425 cm^{-1}) decreased (3422 cm^{-1}) with a decrease in the intensity. This was attributed to the interaction between hydroxyl groups and cadmium.

Multiple mechanisms were held responsible for the remediation of cadmium using urea phosphate activated carbon from *Phragmites australis*. XPS was used to understand the mechanism of participation groups in adsorption (Guo, Zhang, et al. 2017). The mechanisms considered were microporous filtration, ion exchange, electrostatic attraction, and surface complexation.

The FT-IR analysis is used to determine the role of functional groups in adsorption. The FT-IR of diethyl-4-(4 amino-5-mercapto-4H-1,2,4-triazol-3-yl)phenyl phosphonate-coated Fe_3O_4 after adsorption (Venkateswarlu and Yoon 2015b) depicts a shift of IR peaks at 3332 cm^{-1} (amine) and 2522 cm^{-1} (mercapto group) to 3312 and 2508 cm^{-1}, respectively. This suggests the removal by the coordination process. The FT-IR is also used to suggest the presence of heavy metals after adsorption. The change in broadband that occurred at 1650, 1200–955, 805–695, and 608–509 cm^{-1} after adsorption indicated the presence of heavy metals (Wang, Liu, et al. 2012).

The explanation of the mechanism of adsorption was also done with the theoretical model. The remediation of cadmium on graphene oxide was explained using a diffusion layer model via FITEQL software (Huang, Wu, et al. 2015). The adsorption data at different pH values were fitted with theoretical prediction by software. The model predicted different cadmium species at different pH values. The results indicate the formation of inner-sphere complex at near-neutral pH.

The adsorption of cadmium on tourmaline occurred by ion exchange and electrostatic mechanisms (Wang, Liu, et al. 2012). Tourmaline is predicted to maintain the pH during adsorption. Tourmaline polarizes the water molecules. The hydrogen ion from polar water molecule pulled the cations from the adsorbent (tourmaline). This led to increased negative charge on the surface due to the loss of cations, which caused the enhancement in adsorption of cadmium, and it was also predicted to maintain the pH (Wang, Liu, et al. 2012). The value of zeta potential raised from −24 to −13 mV, −23 to −8 mV, and −19 to −4 mV at 15°C, 25°C, and 35°C, respectively, at all concentrations. The trend of change coincides with that of adsorption isotherm. This can suggest that adsorption occurred through electrostatic mechanisms (Wang, Liu, et al. 2012). However, above pH 6, hydroxide precipitation along with the ion exchange was held responsible for cadmium removal.

The measurement of the final pH also took into account whether precipitation occurred or not. The final pH in the adsorption of cadmium with tourmaline was higher in studies with initial pH 2–6, but lower in the case of initial pH 7 and 8 (Wang, Liu, et al. 2012). The final pH in the case of initial pH 5–8 tended to be close to 6. As the final pH was adjusted below the precipitation pH, precipitation was not the only mechanism for the removal of cadmium.

3.6.7 Desorption

Cadmium was removed by using different desorbing agents, e.g., nitric acid (Khan et al. 2015; Zhong et al. 2016; Sreenu et al. 2016; Huang, Yang, et al. 2015) and hydrochloric acid (Zhou et al. 2013; Jin et al. 2017; Shi et al. 2015; Zha et al. 2014; Guo, Jiao, et al. 2017; Awual 2019), by lowering the pH, or by using an acidic medium (Venkateswarlu and Yoon 2015b; Feng et al. 2012). The desorption of cadmium from synthetic mineral adsorbent was performed with nitric acid at various pH values, i.e. 3, 2, 1, and 0.5 (Chen, Shah, et al. 2017). Desorption was nonsignificant at pH 3, and it raised to 100% by lowering the pH to 1. This phenomenon was taken as the macroscopic evidence of the electrostatic binding of the adsorbate to adsorbent. In addition, XPS analysis depicted a decrease in the Ca 2s/C 1s and Ca2p/C1s intensity ratios and an increase in the Cd3d/C1s intensity ratio. This was taken as the evidence for the exchange of calcium ion with cadmium as the major mechanism of adsorption (Chen, Shah, et al. 2017). The adsorption efficiency decreased after desorption (Zhong et al. 2016; Jin et al. 2017; Shi et al. 2015; Huang, Yang, et al. 2015). In addition to acidic medium in a majority of studies, sodium hydroxide was also used for desorption of cadmium from alpha-ketoglutaric acid-modified magnetic chitosan up to six cycles (Yang, Tang, et al. 2014).

3.7 MERCURY

Mercury exists in the environment in many forms, i.e. metallic mercury, inorganic mercury, and organic mercury (ATSDR 1999). Mercury can enter into the environment by natural and anthropogenic sources. Mercury finds its application in chlorine production and electrical applications. Food is the main source of mercury intake by humans. The WHO has set up the guideline value of 0.006 mg/l for inorganic mercury (WHO 2017). The main target organs of mercury toxicity are brain and kidney (WHO 2017; ATSDR 1999). An overview of the experimental parameters and optimized conditions from batch adsorption experiments for mercury is presented in Table 3.8.

3.7.1 Effect of pH

The pH of the solution is considered to play a crucial role in adsorption as it affects the functional groups and surface charges present on the adsorbent's surface and the speciation of adsorbate in the solution (Guo et al. 2016). The adsorption of mercury increased with an increase in the pH, exhibiting the maximum at variable pH values, e.g. 5 (Saha et al. 2016; Chandra and Kim 2011; Ravi et al. 2014; Luo, Chen, et al. 2015), 6 (Zhang, Wu, et al. 2016; Liu, Ding, et al. 2016; Zhu et al. 2012; Ravi and Selvaraj 2014; Huang, He, et al. 2017; Patra and Kim 2017), 6.2 (Zhang, Wu, et al. 2015), 6.5 (Wang, Lv, et al. 2016), and 7 (Huang et al. 2014; Wang, Cheng, Chen, et al. 2013; Salamun et al. 2015). However, in some cases, the optimum removal of mercury occurred at lower pH values also, e.g. 3 (Huang, He, et al. 2015; Huang et al. 2016) and 4 (Lin and Zou 2017).

The adsorption of mercury gets hindered at pH values both lower and higher than the optimum pH after the attainment of maximum removal. The decreased adsorption at lower pH than optimum pH is due to the competition of hydronium ions with the Hg^{2+} ions for active sites on the adsorbent's surface and the formation of certain negatively charged ($Hg(OH)^{3-}$) (Saha et al. 2016) and insoluble species ($Hg(OH)_2$) (Chandra and Kim 2011) at higher pH values than the optimum pH.

The removal of Hg^{2+} and Hg^0 on 1,2-ethanedithiol-modified vinyl-functionalized sulfur derivatives of covalent organic frameworks (COF-S-SH) occurred at a wide pH range (3–10) (Sun et al. 2017), whereas the removal on Fe_3O_4 superparamagnetic nanoparticles was independent of pH and remained constant at the range of 6–9 (Qi et al. 2014). The strong interaction between –SH functional group of magnetic MOF composite (MOF stands for metal–organic framework) and mercury species makes the removal independent of pH, but to avoid hydrolysis of mercury at higher pH, the process was carried out at pH 3 (Huang, He, et al. 2015).

3.7.2 Effect of Coexisting Ions and Temperature

The presence of metal ions (Pb^{2+}, Ni^{2+}, Cu^{2+}, Co^{2+}, and Cd^{2+}) as interferences in the medium did not affect the binding affinity of reduced graphene oxide–thymine composite for Hg(II) and exhibited high selectivity due to the specific binding of Hg(II) with the thymine base pairs that form a thymine–Hg(II)–thymine complex on the surface of graphene oxide (Liu, Ding, et al. 2016). In a similar way, Hg(II) uptake capacity remained unaffected with quaternary ammonium-functionalized magnetic mesoporous silica in the presence of metallic species (K^+, Na^+, Ba^{2+}, Zn^{2+}, Cu^{2+}, and Pb^{2+}) and anions (PO_4^{3-}, CrO_4^{2-}, SO_4^{2-}, CH_3COO^-, Cl^-, and NO_3^-)except ClO_4^-, the ClO_4^- being weakly hydrated soft anion formed strong ion pairs with weakly hydrated quaternary ammonium cations (Liu and Du 2011).

The adsorption of mercury exhibited enhancement with increasing temperature, and the probable reason for this was attributed to either the increased availability of active site for occupation or the diffusion rate of ions (Chandra and Kim 2011; Elhamifar et al. 2016; Thakur et al. 2013). However, in some cases adsorption capacity decreased with a rise in temperature (Saman et al. 2013; Liu, Ding, et al. 2016; Diagboya et al. 2015).

3.7.3 Mechanism

The mechanism of adsorption of Hg^{2+} ions over poly(2-aminothiazole) involved complexation (Wang, Lv, et al. 2016), which was explained on the basis of FT-IR and XPS analysis. The disappearance of infrared peaks after adsorption corresponding to the N-H stretching vibration and C-S stretching vibration showed that nitrogen in the secondary amine and sulfur in the thiazole ring participated in the adsorption, while the retention of the peak of C=N ruled out its involvement in the process. The presence of Hg 4f signal in XPS spectrum after adsorption revealed the adherence of Hg(II) on the adsorbent. Further, the decrease in peak intensity and its shift toward higher binding energy in S 2p spectrum of poly(2-aminothiazole) after adsorption showed the formation of a complex between sulfur and mercury.

The XPS spectra of uracil-functionalized polymer poly[1-(4-vinylbenzyl uracil)] after adsorption of Hg^{2+} ions (Wang, Cheng, Chen, et al. 2013) exhibited two oxidized peaks at 101.3 and 105.4 indicating the existence of mercury in divalent state for binding with uracil groups. The FT-IR spectrum depicted decreased intensity of N-H stretching vibration on combining with Hg^{2+}, which suggested towards displacement of imido protons of poly[1-(4-vinylbenzyl uracil)] by Hg^{2+}. Further, the creation of imide–Hg–imide complex caused a shift of stretching vibration of C=O to lesser wave numbers and an immediate decrease in N-H stretching vibration. Similarly, the presence of mercury within COF-S-SH post-adsorption was established by the presence of two XPS signals corresponding to Hg^{2+}, and the strong binding interaction between Hg^{2+} and sulfur species in Hg@COF-S-SH was approved by S 2p XPS spectra, which depicted a shift in S binding energy than that of in COF-S-SH. The absence of a characteristic S-H stretching peak in FT-IR spectrum of Hg@COF-S-SH further showed the interaction between Hg^{2+} and S-H groups (Sun et al. 2017).

The adherence of Hg^{2+} on MNPC-T700-M3 (magnetic nitrogen-doped porous carbon) leads to two XPS signals in the Hg 4f spectrum having a shift of 2.1 eV toward higher binding energy side of Hg^{2+} in MNPC-T700-M3-Hg^{2+} ; which suggests towards the formation of a covalent bond between an empty orbital of Hg^{2+} and π-electrons of the N species (present in MNPC-T700-M3). The C1s spectrum of MNPC-T700-M3-Hg^{2+} depicted no remarkable change, suggesting no interaction between Hg^{2+} and carbon atoms. Upon adsorption of Hg^{2+}, the N 1s peaks of

TABLE 3.8
Summary of Parameters and Optimized Conditions for Batch Adsorption of Mercury

Adsorbent	Adsorbate	Surface Area (m^2/g), Pore Volume (cm^3/g), Pore Size (nm)	pHzpc	Experimental Conditions	Adsorption Capacity (mg/g)	Thermodynamic Parameters	Kinetic Model and Curve Fitting	Isotherm Model and Curve Fitting	Maximum Adsorption Conditions	References
Uracil-functionalized poly[1-(4-vinylbenzyl uracil)] polymer	Hg(II)			pH = 1.5–11 Dose = 0.25 g/l (kinetic study) Concentration = 5–50 mg/l (isotherm study) Contact time = up to 300 min	543.9				pH = 5–11 Contact time = 300 min	Wang, Cheng, Chen, et al. (2013)
Polypyrrole–reduced graphene oxide composite	Hg(II)	Surface area = 166		pH = 2–11 Dose = 0.2 g/l Concentration = 50–250 mg/l Contact time = 20–200 min Temperature = 5°C–60°C	980 (at pH = 5)		Pseudo-second-order model (only no comparison with other models)	Langmuir and Freundlich models	pH = 5–11 Contact time = 20 min Temperature = 40°C–60°C	Chandra and Kim (2011)
Graphene oxide (GO)–iron oxide (Fe_3O_4) magnetic nanoparticle composite	Hg(II)	Surface area = 214 Pore size = 13.1		pH = 5 Dose = 20 mg (kinetic study) Concentration = 50 mg/l (kinetic study) Contact time = 5–180 min Temperature = 20°C–40°C	16.6		Elovich model; pseudo-second-order model (at 20°C); intra-particle diffusion (at higher temperatures), linear		Contact time = 120 min (at 30°C and 40°C) Temperature = 20°C	Diagboya et al. (2015)
Thymine (T)-grafted poly(vinyl alcohol) (PVA)–polyethyleneimine (PEI) nanofibers	Hg(II)			pH = 7 Dose = 80 mg Concentration = 20 mg/l Contact time = 1 h Temperature = room temperature	40					Huang et al. (2014)

(continued)

TABLE 3.8 (*Continued*)
Summary of Parameters and Optimized Conditions for Batch Adsorption of Mercury

Adsorbent	Adsorbate	Surface Area (m^2/g), Pore Volume (cm^3/g), Pore Size (nm)	pHzpc	Experimental Conditions	Adsorption Capacity (mg/g)	Thermodynamic Parameters	Kinetic Model and Curve Fitting	Isotherm Model and Curve Fitting	Maximum Adsorption Conditions	References
Thiol-functionalized ionic liquid-based periodic mesoporous organosilica	Hg(II)			pH = 2–9 Dose = 0.4–2.8 g/l Agitation speed = 250 rpm Concentration = 30–80 mg/l Contact time = 50 min Temperature = 298–323 K	105.26	ΔH = 25.19 kJ/mol ΔS = 104.59 J/mol K ΔG = negative	Pseudo-second-order model, linear	Langmuir model, linear	pH = 8 Dose = 2.4 g/l Concentration = 30 mg/l Temperature = 323 K	Elhamifar et al. (2016)
Thymine-grafted reduced graphene oxide complexes	Hg(II)			pH = 2–6 Concentration = 250 mg/l (kinetic study and pH study) Contact time = 0–180 min Temperature = 298–318 K	128	ΔH = −84.8 kJ/mol ΔS = 29.1 J/mol K ΔG = negative	Pseudo-second-order model, linear (no comparison with other models)	Langmuir model, linear	pH = 6 Contact time = 20 min Temperature = 298 K	Liu, Ding, et al. (2016)
Magnetic graphene oxide composite	Hg(II)	Surface area = 58.6 Pore volume = 0.088	Zeta potential = pH 3	pH = 2–9 Dose = 0.25–2 g/l Agitation speed = 200 rpm Concentration = 100 mg/l Contact time = up to 6 h Temperature = 298–333 K	71.3 at pH = 8	ΔH = 13.4 kJ/mol ΔS = 46.8 J/mol K ΔG = negative	Pseudo-second-order model and intra-particle diffusion, linear	Langmuir model, linear	pH = 9 Dose = 0.5 g/l in most studies Contact time = 3 h Temperature = 333 K	Guo et al. (2016)
Magnetic MOF composite (Fe_3O_4@SiO_2@HKUST-1)	Hg(II)	Surface area = 775 Pore volume = 0.54	6	pH = 2–9 Dose = 1 g/l Concentration = 25–225 mg/l (isotherm study) and 20 mg/l (kinetic study) Contact time = up to 120 min Temperature = 25°C	264 at pH = 3		Pseudo-second-order model, linear (no comparison with other models)	Langmuir model, linear (no comparison with other models)	pH = 3 (above pH 6 Hg exists as $Hg(OH)_2$) Contact time = 20 min	Huang, He, et al. (2015)

(*continued*)

TABLE 3.8 (*Continued*)

Summary of Parameters and Optimized Conditions for Batch Adsorption of Mercury

Adsorbent	Adsorbate	Surface Area (m^2/g), Pore Volume (cm^3/g), Pore Size (nm)	pHzpc	Experimental Conditions	Adsorption Capacity (mg/g)	Thermodynamic Parameters	Kinetic Model and Curve Fitting	Isotherm Model and Curve Fitting	Maximum Adsorption Conditions	References
Magnetic N-doped porous carbon composites	Hg(II)	Surface area = 653 Pore volume = 3.97 Pore size = 0.48		pH = 2–7 Concentration = 50–500 mg/l (isotherm study) and 20 mg/l (kinetic study) Contact time = up to 30 min Temperature = room temperature	429 at pH = 6		Pseudo-second-order model, linear	Langmuir model, linear	pH = 4–7 Contact time = 10 min	Huang, He, et al. (2017)
Fe_3O_4@SiO_2/ZIF-8	Hg(II)	Surface area = 1050 Pore volume = 0.76 Pore size = 19.6		pH = 2–7 Concentration = 10 mg/l Contact time = 60 min Temperature = room temperature					pH = 6–7	Huang, He, et al. (2017)
Graphene–diatom silica aerogels	Hg(II)	Surface area = 368	Zeta potential = pH 6	pH = 2–10 Dose = 40 mg/l Concentration = 50–400 mg/l (isotherm study) and 200 mg/l (kinetic study) Contact time = 90 min Temperature = 25°C	528		Pseudo-second-order model, linear	Langmuir model, linear	pH = 6.5	Kabiri et al. (2015)
Magnetic β-cyclodextrin inclusion complex with phosphonobutane tricarboxylic acid	Hg(II)	Surface area = 13.84 Pore volume = 9.9×10^{-2} Pore size = 1.64 nm		pH = 1–7 Dose= 0.2 g/l Concentration = 5–100 mg/l (isotherm study) and 20 mg/l (kinetic study) Contact time = up to 400 min Temperature = 15°C–55°C	77.59 at pH = 4	ΔH = 24.09–56.94 kJ/mol ΔS = 97.87–181.77 J/mol K ΔG = negative	Pseudo-second-order model, linear	Langmuir model, linear	pH = 1 Contact time = 150 min Temperature = 55°C	Lin and Zou (2017)

(*continued*)

TABLE 3.8 (*Continued*)
Summary of Parameters and Optimized Conditions for Batch Adsorption of Mercury

Adsorbent	Adsorbate	Surface Area (m^2/g), Pore Volume (cm^3/g), Pore Size (nm)	pHzpc	Experimental Conditions	Adsorption Capacity (mg/g)	Thermodynamic Parameters	Kinetic Model and Curve Fitting	Isotherm Model and Curve Fitting	Maximum Adsorption Conditions	References
Quaternary ammonium-functionalized magnetic mesoporous silica	Hg(II)	Surface area = 427.25 Pore size = 2.1		pH = 4–8.5 Dose = 20 g/l Concentration = 0.75 mmol/l (pH study) Contact time = up to 18 min Temperature = 298 K	0.098 mmol/g at pH = 5.5			Langmuir model, non-linear	pH = 4–5.5 Contact time = 7 min	Liu and Du (2011)
Acrylamide- and hydroxyl-functionalized metal–organic framework	Hg(II)			pH = 1–7 Dose = 50 mg/l Concentration = 100 ppb (kinetic study) Contact time = up to 180 min Temperature = 25°C–45°C	278	ΔH = −3.93 kJ/mol ΔS = 67.33 J/mol K ΔG = negative	Pseudo-second-order model, linear	Langmuir model, linear	pH = 5 Contact time = 90 min Temperature = 25°C	Luo, Chen, et al. (2015)
Poly(2-aminothiazole) (PAT) polymer	Hg(II)			pH = 2.5–6.5 Dose = 0.4 g/l Agitation speed = 100 rpm Concentration = 200 mg/l (kinetic study) Contact time = up to 40 h Temperature = 288–308 K	325.7	ΔH = 11.6 kJ/mol ΔS = 104 J/mol K ΔG = negative	Pseudo-second-order model	Langmuir model	pH = 6.5 Contact time = 28 h Temperature = 308 K	Wang, Lv, et al. (2016)
Mesoporous α-FeOOH nanoparticles	Hg(II)	Surface area = 152.4 Pore volume = 0.089		pH = 4–10 Dose = 1 g/l Concentration = 20–200 mg/l Contact time = 240 min Temperature = room temperature	146		Pseudo-second-order model, linear	Langmuir and Freundlich models, linear	pH = 6 Concentration = 20 mg/l	Patra and Kim (2017)

(*continued*)

TABLE 3.8 (*Continued*)
Summary of Parameters and Optimized Conditions for Batch Adsorption of Mercury

Adsorbent	Adsorbate	Surface Area (m^2/g), Pore Volume (cm^3/g), Pore Size (nm)	pHzpc	Experimental Conditions	Adsorption Capacity (mg/g)	Thermodynamic Parameters	Kinetic Model and Curve Fitting	Isotherm Model and Curve Fitting	Maximum Adsorption Conditions	References
Magnetite Fe_3O_4 superparamagnetic nanoparticle–polymer composite	Hg(II)			pH = 6–9 Dose = up to 1 g/l Agitation speed = 350 rpm Concentration = 5–100 μg/l Contact time = 2–60 min Temperature = 298 K	16.9 at pH 7			Langmuir model, linear	pH-independent Dose = 0.4 g/l Concentration = independent in the studied range Contact time = 10 min	Qi et al. (2014)
Chain-like mesoporous silica	Hg(II)	Surface area = 302 Pore volume = 0.45 Pore size = 3.2		pH = 1–10 Dose = 0.4 g/l Agitation speed = 200 rpm Concentration = 10–300 mg/l Contact time = up to 24 h Temperature = 25°C	401 at pH = 5			Langmuir model, linear	pH = 5 Concentration = independent in the studied range Contact time = 12 h	Ravi and Selvaraj (2014)
Sulfur-functionalized ordered mesoporous carbons	Hg(II)	Surface area = 837 Pore volume = 0.71		pH = 2–9 Dose = 0.25 g/l Concentration = 100 mg/l (kinetic and pH studies) Contact time = up to 4 h	70.75 at pH = 5		Pseudo-second-order model, linear		pH = 5	Saha et al. (2016)
Acid-vacuo heat-treated banana stem fiber	Hg(II)		5.8 (PZC)	pH = 2–9 Dose = 0.05 g/l Concentration = 100 mg/l (kinetic study) Contact time = up to 180 min Temperature = 303–323 K	372	ΔH = −86.4 kJ/mol ΔS = −304.4 J/mol K ΔG = negative	Pseudo-second-order model	Langmuir model	pH = 7 Contact time = 90 min Temperature = 303 K	Salamun et al. (2015)
Silica microsphere	Hg(II)	Surface area = 960 Pore volume = 0.64 Pore size = 2.65 nm		pH = 3–9.8 Dose = 1 g/l Contact time = up to 48 h Temperature = 30°C–60°C	20	ΔH = −9.41 kJ/mol ΔS = 0.09 J/mol K ΔG = negative	Pseudo-second-order model	Langmuir model, linear	pH = 5.8 Time = 6 h Temperature = 40°C–50°C	Saman et al. (2013)

(*continued*)

TABLE 3.8 (*Continued*)
Summary of Parameters and Optimized Conditions for Batch Adsorption of Mercury

Adsorbent	Adsorbate	Surface Area (m^2/g), Pore Volume (cm^3/g), Pore Size (nm)	pHzpc	Experimental Conditions	Adsorption Capacity (mg/g)	Thermodynamic Parameters	Kinetic Model and Curve Fitting	Isotherm Model and Curve Fitting	Maximum Adsorption Conditions	References
Sulfur-functionalized silica microspheres	Hg(II)	Surface area = 622 Pore volume = 0.08 Pore size = 2.59 nm		pH = 2.5–9 Dose = 1 g/l Contact time = 48 h Temperature = 30°C–60°C	62.3 at pH = 7.5	ΔH = −47.77 kJ/mol ΔS = 1.77 J/mol K ΔG = negative	Pseudo-second-order model	Langmuir model, linear	pH = 2–7.5 Contact time = 24 h Temperature = 40°C	Saman et al. (2013)
1,2-Ethanedithiol-modified vinyl-functionalized sulfur derivatives of covalent organic frameworks (batch and column study)	Hg(II)	Surface area = 546		pH = 3–10 Dose = 0.45 g/l Concentration = 25–700 mg/l (isotherm study) and 5 mg/l (kinetic study) Contact time = 3 h	1350		Pseudo-second-order model, linear	Langmuir model, linear	pH = 3, but at pH 10, K_d = 1.5×10^8 ml/g, and at pH 3, K_d = 4.67×10^8 ml/g	Sun et al. (2017)
1,2-Ethanedithiol-modified vinyl-functionalized sulfur derivatives of covalent organic frameworks	Hg(0) vapor	Surface area = 546		Dose = 20 mg Concentration = 300 mg Hg (elemental) mercury as source Contact time = 72 h Temperature = 140°C	863					Sun et al. (2017)
Sulfur/reduced graphene oxide nanohybrid	Hg(II)			pH = 2–8 Contact time = up to 180 min Temperature = 301–323 K	965.55	ΔH = 128.9 kJ/mol ΔS = 500.7 J/mol K ΔG = negative	Pseudo-second-order model, linear	Langmuir model, linear	pH = 6–8 Contact time = 30 min Temperature = 323 K	Thakur et al. (2013)
Magnetic ion-imprinted polymer/Fe_3O_4@SiO_2 nanoparticle composite	Hg(II)	Surface area = 97.32		pH = 2–7 Dose = 0.25–1.5 g/l Concentration = 20–300 mg/l (isotherm study) and 200 mg/l (kinetic study) Contact time = up to 18 min Temperature = 25°C	78.3		Pseudo-second-order model, linear (no comparison with other models)	Langmuir model, linear	pH = 6 Dose = 1 g/l Contact time = 5 min	Zhang, Wu, et al. (2016)

(*continued*)

TABLE 3.8 (*Continued*)
Summary of Parameters and Optimized Conditions for Batch Adsorption of Mercury

Adsorbent	Adsorbate	Surface Area (m^2/g), Pore Volume (cm^3/g), Pore Size (nm)	pHzpc	Experimental Conditions	Adsorption Capacity (mg/g)	Thermodynamic Parameters	Kinetic Model and Curve Fitting	Isotherm Model and Curve Fitting	Maximum Adsorption Conditions	References
Hollow mesoporous silica spheres with magnetic cores modified with (3-mercaptopropyl)-trimethoxysilane	Hg(II)	Surface area = 380.75 Pore size = 1.91		pH = 2–7 Dose = 10–80 mg/l Concentration = 0.1–50 mg/l (isotherm study) Contact time = up to 600 min Temperature = room temperature	131 at pH 6.2		Pseudo-second-order model, linear (no comparison with other models at both 1 and 6 mg/l initial concentrations, high value of correlation coefficient)	Langmuir model	pH = 6.2–6.8 Dose = 80 mg/l Contact time = 130 min at the initial concentration of 1 mg/l and 300 min at the initial concentration of 6 mg/l	Zhang, Wu, et al. (2015)
3-Aminopropyltriethoxysilane-functionalized MCM-41	Hg(II)	Surface area = 648.53 Pore volume = 0.71		pH = 1–6 Dose = 2 g/l Agitation speed = 180 rpm Concentration = 150 and 250 mg/l (kinetic study) Contact time = up to 180 min Temperature = 298–318 K	125.06	ΔH = −15.96 kJ/mol ΔS = 32.16 J/mol K ΔG = negative	Pseudo-second-order model, linear	Langmuir model, linear	pH = 6 Contact time = 60 min Temperature = 298 K	Zhu et al. (2012)
3-(2-Aminoethylamino)-propyldimethoxymethylsilane-functionalized MCM-41	Hg(II)	Surface area = 508.73 Pore volume = 0.681		pH = 1–6 Dose = 2 g/l Agitation speed = 180 rpm Concentration = 4 and 50 mg/l (kinetic study) Contact time = up to 180 min Temperature = 298–318 K	231.92	ΔH = −15.62 kJ/mol ΔS = 32.16 J/mol K ΔG = negative	Pseudo-second-order model, linear	Langmuir model, linear	pH = 6 Contact time = 30 min at 50 mg/l and 10 min at 4 mg/l Temperature = 298 K	Zhu et al. (2012)

MNPC-T700-M3 exhibited a shift toward the higher binding energy side, which advocates the chelation of only nitrogen with Hg^{2+} (Huang, He, et al. 2017).

The mechanistic pathway of adsorption of Hg^{2+} on graphene–diatom silica composites involved the combination of a number of factors including physical entrapment via the electrostatic attraction between α-FeOOH nanoparticles and the graphene edges and Hg^{2+} ions, ion exchange in the solution, and the amino functional groups present on the APTES-modified diatomaceous earth. These factors collectively resulted in enhanced adsorption of Hg^{2+} ions from aqueous solutions (Kabiri et al. 2015).

The removal of mercury(II) with acrylamide- and hydroxyl-functionalized metal–organic framework (Luo, Chen, et al. 2015), magnetic ion-imprinted polymer (MIIP) with incorporated $Fe_3O_4@SiO_2$ nanoparticles (Zhang, Wu, et al. 2016), and magnetic β-cyclodextrin inclusion complex with phosphonobutane tricarboxylic acid followed chemisorption (Lin and Zou 2017), where the coordination of hydroxyl and acrylamide groups of the acrylamide- and hydroxyl-functionalized metal–organic framework with mercury led to its chemical adsorption (Luo, Chen, et al. 2015). Similarly, soft acid–soft base interaction between allylthiourea of magnetic ion-imprinted polymer and Hg(II) resulted in chemisorption of Hg(II) ions from aqueous medium (Zhang, Wu, et al. 2016). The formation of metal-chelate complexes by donation of lone pair of electrons of hydroxyl oxygen present in the inner cavity of β-cyclodextrin to the vacant orbital of Hg^{2+} was the reason for the chemical adsorption of Hg^{2+} (Lin and Zou 2017).

3.7.4 Desorption and Reuse

Regeneration of the adsorbents can be achieved with several eluents such as hydrochloric acid (Zhang, Wu, et al. 2015; Salamun et al. 2015; Huang, He, et al. 2017), nitric acid (Wang, Lv, et al. 2016), potassium iodide (Saman et al. 2013), 2-mercaptoethanol (Qi et al. 2014), and mixtures of acid and thiourea (Zhang, Wu, et al. 2016; Ravi and Selvaraj 2014; Liu, Ding, et al. 2016). 5% thiourea in 0.5 mol/l nitric acid efficiently desorbed Hg(II)-MIIP (Zhang, Wu, et al. 2016), while 0.1% thiourea in 0.01 M hydrochloric acid successfully desorbed the adsorbent mercapto-functionalized magnetic Zr-MOF (MFC-S) (Huang et al. 2016). However, the elution of Hg^{2+} from the $Fe_3O_4@SiO_2@HKUST\text{-}1$ was not possible with any eluent due to the extremely strong binding between Hg^{2+} and –SH group of the adsorbent (Huang, He, et al. 2015).

3.8 LEAD

Lead is resistant to corrosion, and it finds its application in batteries, radiation protection, pipes, solder, and ammunition (ATSDR 2007). The contact of humans with water occurred with dissolution from the natural sources and corrosion of lead in water pipes (WHO 2017). The guideline value set up by the WHO is 10 μg/l. Exposure to lead causes neurodevelopmental effects, kidney impairment, and hypertension. An overview of the experimental parameters and optimized conditions from batch adsorption experiments for lead is presented in Table 3.9.

TABLE 3.9
Summary of Parameters and Optimized Conditions for Batch Adsorption of Lead

Adsorbent	Adsorbate	Surface Area (m^2/g), Pore Volume (cm^3/g), Pore Size (nm)	pHzpc	Experimental Conditions	Adsorption Capacity (mg/g)	Thermodynamic Parameters	Kinetic Model and Curve Fitting	Isotherm Model and Curve Fitting	Maximum Adsorption Conditions	References
Cellulose-based carbogel (CeOvaS)	Pb(II)	Surface area =38 Pore volume = 0.18	Zeta = 3	pH = 3–6 Dose = 2 g/l Agitation speed = 200 rpm Concentration = 50 mg/l (pH study) and 5–1500 mg/l (kinetic study) Contact time = up to 24 h	240 (at pH = 5)		Pseudo-second-order model, non-linear	Sips model	pH = no significant difference at pH 3–6, but removal was slightly higher at pH 4 Contact time = 300 min	Alatalo et al. (2015)
Porous magnesium silicate (PMS)	Pb(II)	Surface area = 650.5 Pore volume = 1.121 Pore size = 6.89		pH = 2.8–5.8 Dose = 0.34 g/l Concentration = 0–265 mg/l (isotherm study) Contact time = 24 h Temperature = 25°C	436.7 (at pH = 5.8)	-	Pseudo-second-order model, linear	Langmuir model, linear (no comparison with other models)	pH = 5–5.8	Huang, Wu, et al. (2017)
Hollow hydroxyapatite microsphere	Pb(II)	Surface area = 164.73 Pore size = 15.9		pH = 1–6 Concentration = 50 μg/ml Contact time = 2 h Temperature = 25°C	99.79 (at pH = 5)			-	pH = 3–6 (not significant difference in the range of 3–6) Contact time = 45 min	Jiang et al. (2012)

(continued)

TABLE 3.9 (*Continued*)

Summary of Parameters and Optimized Conditions for Batch Adsorption of Lead

Adsorbent	Adsorbate	Surface Area (m^2/g), Pore Volume (cm^3/g), Pore Size (nm)	pHzpc	Experimental Conditions	Adsorption Capacity (mg/g)	Thermodynamic Parameters	Kinetic Model and Curve Fitting	Isotherm Model and Curve Fitting	Maximum Adsorption Conditions	References
Fly ash/poly(acrylic acid) composite	Pb(II)			pH = 1–6 Dose = 2 g/l Concentration = 100 mg/l (kinetic study) Contact time = up to 48 h	41–44 (at pH = 5)		Pseudo-second-order model, linear		pH = 6 Contact time = 24 h	Jiang and Liu (2014)
Graphene oxide–$MnFe_2O_4$	Pb(II)	Surface area = 196		pH = 2–6 Dose = 200 mg/l Concentration = 1–400 mg/l (isotherm study) and 20 mg/l (kinetic study) Contact time = up to 120 min Temperature = 298–333 K	673 (at pH = 5)	ΔH = 4.01 kJ/mol ΔG = negative	Pseudo-second-order model, linear	Langmuir model No comparison with other models	pH = 5–6 Temperature = 333 K	Kumar et al. (2014)
Diethylenetriaminepentaacetic acid-modified magnetic Graphene Oxide (DTPA/MGO)	Pb(II)		2.2	pH = 1.5–6 Dose = 0.4 g/l Agitation speed = 180 rpm Concentration = 10–400 mg/l (isotherm study) and 100 mg/l (kinetic study) Contact time = 6 h Temperature = 293–323 K	387.6 mg/g (at pH = 3)	ΔH = 120.57 J/mol ΔG = negative ΔS = 30.6 kJ/mol/K	Pseudo-second-order model, linear	Langmuir model, linear	pH = 3–6 Contact time = 20 min Temperature = 323 K	Li, Wang, et al. (2017)

(continued)

TABLE 3.9 (*Continued*)
Summary of Parameters and Optimized Conditions for Batch Adsorption of Lead

Adsorbent	Adsorbate	Surface Area (m^2/g), Pore Volume (cm^3/g), Pore Size (nm)	pHzpc	Experimental Conditions	Adsorption Capacity (mg/g)	Thermodynamic Parameters	Kinetic Model and Curve Fitting	Isotherm Model and Curve Fitting	Maximum Adsorption Conditions	References
Amino-functionalized Cr-based MIL-101 MOFs (ED-MIL-101)	Pb(II)	Surface area = 347 Pore volume = 0.28		pH = 2–7 Dose = 1 g/l Concentration = 500 mg/l (kinetic study) Contact time = 6 h Temperature = 298 K	81.09		Pseudo-second-order model, linear	Langmuir model, linear	pH = 6 Contact time = 90 min	Luo, Ding, et al. (2015)
EDTA–graphene oxide	Pb(II)	Surface area = 623		pH = 2–8 Dose = 0.1 or 0.125 g/l Concentration = 5–300 mg/l (isotherm study) Contact time = 24 h Temperature = 25°C	479 ± 46 mg/g at pH = 6.8			Langmuir model	pH = 8 Isotherm study at pH 6.8 (to avoid precipitation)	Madadrang et al. (2012)
Graphene oxide–zirconium phosphate (GO–Zr-P)	Pb(II)			pH = 1–8 Dose = up to 1.5 g/l Concentration = 10–400 mg/l Contact time = 20 min Temperature = room temperature	363.4 mg/g (at pH = 6)		Pseudo-second-order model, linear	Langmuir model, linear	pH = 6 Dose = 1.5 g/l Concentration = adsorption capacity saturated at 200 mg/l	Pourbeyram (2016)

(continued)

TABLE 3.9 (*Continued*)
Summary of Parameters and Optimized Conditions for Batch Adsorption of Lead

Adsorbent	Adsorbate	Surface Area (m^2/g), Pore Volume (cm^3/g), Pore Size (nm)	pHzpc	Experimental Conditions	Adsorption Capacity (mg/g)	Thermodynamic Parameters	Kinetic Model and Curve Fitting	Isotherm Model and Curve Fitting	Maximum Adsorption Conditions	References
Polydopamine microspheres (PDA-Ms)	Pb(II)		3.2	pH = 0.5–6.5 Dose = 0.5 g/l (isotherm and pH studies), 0.1 g/l (kinetic study) Concentration = 40 mg/l (kinetic study) Contact time = up to 55 min Temperature = 293–333 K	165.8 mg/g (at pH = 6.5)			Langmuir model	pH = 6.5 Contact time = 1 min Temperature = 293 K	Zhang, Yang, et al. (2017)
Biochar/ZnS composites	Pb(II)			pH = 2–6 Dose = 1.2 g/l (isotherm study) and 0.4 g/l (kinetic study) Concentration = 20–700 mg/l (isotherm study) and 100 mg/l (kinetic study) Contact time = up to 24 h Temperature = 298–308 K	367.65 mg/g (at pH = 6)	ΔH = 2.05 kJ/mol ΔS = 27.41 J/mol/K ΔG = negative	Pseudo-first-order model, linear	Langmuir model, linear	pH = 6 Contact time = 12 h Temperature = 308 K	Yan, Kong, et al. (2015)

(*continued*)

TABLE 3.9 (*Continued*)
Summary of Parameters and Optimized Conditions for Batch Adsorption of Lead

Adsorbent	Adsorbate	Surface Area (m^2/g), Pore Volume (cm^3/g), Pore Size (nm)	pHzpc	Experimental Conditions	Adsorption Capacity (mg/g)	Thermodynamic Parameters	Kinetic Model and Curve Fitting	Isotherm Model and Curve Fitting	Maximum Adsorption Conditions	References
Chitosan/graphene oxide composite	Pb(II)			pH = 2–7 Dose = 0.5 g/l Concentration = 100 mg/l (pH study) Agitation = 200 rpm Contact time = 1 h Temperature = 25°C–45°C	423.8	ΔH = 52.31 kJ/mol ΔS = 0.167 kJ/mol K ΔG = negative	Double-exponential	Redlich–Peterson model, non-linear	pH = 6 Contact time = 30 min Temperature = 45°C	Hadi Najafabadi et al. (2015)
Xanthate-modified chitosan/poly(vinyl alcohol)	Pb(II)			pH = 2–7 Dose = 3 g/l Concentration = 100 mg/l (pH study) Contact time = 12 h Temperature = 308–328 K	59.85 mg/g (at pH = 6.0)		Pseudo-second-order model	Langmuir and Freundlich models	pH = 6.88 Temperature = 328 K	Lv et al. (2017)
HAP/α-Fe_2O_3	Pb(II)	Surface area = 80.7		pH = 2–10 Dose = 10 mg/l Concentration = 20 mg/l Contact time = 120 min	754.14		Pseudo-second-order model, linear	Freundlich model, linear	pH = 4	Bharath and Ponpandian (2015)

(*continued*)

TABLE 3.9 (*Continued*)
Summary of Parameters and Optimized Conditions for Batch Adsorption of Lead

Adsorbent	Adsorbate	Surface Area (m^2/g), Pore Volume (cm^3/g), Pore Size (nm)	pHzpc	Experimental Conditions	Adsorption Capacity (mg/g)	Thermodynamic Parameters	Kinetic Model and Curve Fitting	Isotherm Model and Curve Fitting	Maximum Adsorption Conditions	References
Iodate-doped chitosan composite	Pb(II)	Surface area =0.87 Pore volume =2.13×10^{-3} Pore size =9.77	3.9	pH=3–8 Dose=0.1–0.5 g/l Agitation speed =200 rpm Concentration =35–115 mg/l (isotherm study) Contact time =up to 280 min Temperature =room temperature	22.22		Pseudo-second-order model, linear	Langmuir and Freundlich models, linear	pH=6 Dose =0.5 g/l Contact time =240 min	Gedam and Dongre (2015)
APTES-modified Fe_3O_4@ glycidyl methacrylate–acrylamide	Pb(II)			pH=2–7 Dose=0.1–0.6 g/l Agitation speed =200 rpm Concentration =1–50 mg/l Contact time =up to 30 min Temperature =10°C–45°C	158.73 mg/g (at pH=6)	ΔH=50.79 kJ/mol ΔS=213.59 J/mol K ΔG=negative	Pseudo-second-order model, linear	Langmuir model, linear	pH=6–7 Dose=0.14 g/l Concentration =10 mg/l Contact time =2 min Temperature =25°C–45°C	Moradi et al. (2017)
1,2,3-triazole-4-carboxylic acid-modified poly(glycidyl methacrylate) (batch and column studies)	Pb(II)	Surface area=9.53		pH=1–6 Dose=1 g/l Agitation=150 rpm Concentration=20–100 mg/l (isotherm study) and 25 mg/l (kinetic and pH studies) Contact time=5 h Temperature =ca. 298–318 K	69.4	ΔH=20.13 kJ/mol ΔS=88.01 J/mol K ΔG=negative	Pseudo-second-order model, non-linear	Langmuir model, non-linear	pH=5 Temperature =328 K	Yuan, Zhang, et al. (2017)

(*continued*)

TABLE 3.9 (*Continued*)
Summary of Parameters and Optimized Conditions for Batch Adsorption of Lead

Adsorbent	Adsorbate	Surface Area (m^2/g), Pore Volume (cm^3/g), Pore Size (nm)	pHzpc	Experimental Conditions	Adsorption Capacity (mg/g)	Thermodynamic Parameters	Kinetic Model and Curve Fitting	Isotherm Model and Curve Fitting	Maximum Adsorption Conditions	References
Fe_3O_4@DAPF (2,3-diaminophenol–formaldehyde)	Pb(II)	Surface area = 48.8 Pore volume = 0.159 Pore size = 17	Isoelectric point between 6 and 7	pH = 2–7 Dose = 0.02–0.12 g/l Agitation speed = 200 rpm Concentration = 10–100 mg/l Contact time = 5–90 min Temperature = 303–323 K	83.30	ΔH = −46.57 kJ/mol ΔS = 0.141 J/mol K ΔG = negative	Pseudo-second-order model, linear	Langmuir model, non-linear	pH = 5 Dose = 0.10 g/l Concentration = 10 mg/l Contact time = 60 min Temperature = 303 K	Venkateswarlu and Yoon (2015a)
Chelating polyethylene fiber (PE-MA-NN)	Pb(II)	Surface area = 175.4 Pore volume = 1.045		pH = 1–5.5 Dose = 2, 20, or 0.8 g/l in various studies Concentration = 980 mg/l (pH study) and 820 mg/l (kinetic study) Agitation = 120 or 150 rpm Contact time = 4 h Temperature = 278–308 K	526.31 (at pH = 4.5)	ΔH = 16.56–30.70 kJ/mol ΔS = 95.3–96.01 J/mol K ΔG = negative	Pseudo-second-order model	Langmuir model, non-linear	pH = 4–5.5 Contact time = 50 min Temperature = 308 K	Wang, Cheng, Yang, et al. (2013)

(continued)

TABLE 3.9 (*Continued*)
Summary of Parameters and Optimized Conditions for Batch Adsorption of Lead

Adsorbent	Adsorbate	Surface Area (m^2/g), Pore Volume (cm^3/g), Pore Size (nm)	pHzpc	Experimental Conditions	Adsorption Capacity (mg/g)	Thermodynamic Parameters	Kinetic Model and Curve Fitting	Isotherm Model and Curve Fitting	Maximum Adsorption Conditions	References
Nickel ferrite ($NiFe_2O_4$)	Pb(II)		7.22	pH = 2–8 Dose = 0.2–4 g/l (pH study) Concentration = 10–40 mg/l (kinetic study) Contact time = 150 min Temperature = 298–308 K	48.98 (at pH = 5)		Pseudo-second-order model, double-exponential non-linear	Langmuir and Sips model, non-linear	pH = 5 Dose = 1.4 g/l (pH study) Contact time = 60 min	Reddy and Lee (2013)
Albumin-functionalized magnetic nanoparticles	Pb(II)		4.7	pH = 2–8 Dose = 1.4 g/l Concentration = 10 mg/l Contact time = 8 min Temperature = 25°C–45°C	15.15			Langmuir, linear	pH = not much variation in the pH range studied Dose = 1.4 mg/l	Malekpour and Khodadadi (2016)
nZVI@Mg-$(OH)_2$ under nitrogen environment	Pb(II)	Surface area = 40.2		pH = 2–7 Dose = 0.5 g/l Agitation = 300 and 160 rpm Concentration = 250–1000 mg/l Contact time = 120 min Temperature = 25°C	1986.6 (at pH = 6.86)				Concentration = 1000 mg/l	Liu, Wang, et al. (2015)

3.8.1 Effect of pH

The adsorption process is profoundly influenced by pH by affecting the surface properties of the adsorbent and the solution chemistry of the metal at a particular pH. However, removal of Pb^{2+} over cellulose-based carbogel appears to be independent of the solution pH due to the electrostatic interaction between adsorbate and adsorbent having the same charge resulting in cationic exchange in the interlayers of the adsorbent (Alatalo et al. 2015). In some cases, the adsorption is feebly influenced by the pH of the solution, and highest adsorption is exhibited in acidic conditions (Li, Wang, et al. 2017; Rathore et al. 2017). At lower pH values, diethylenetriaminepentaacetic acid (DTPA) is present as H_2DTPA^{3-}, H_3DTPA^{2-}, H_4DTPA^-, and H_6DTPA^+ species on the surface of diethylenetriaminepentaacetic acid-modified magnetic graphene oxide ($pH_{zpc}=2.2$) in various proportions, with the least proportion of H_6DTPA^+ (3.51%), and forms stable complexes with the Pb^{2+} ions favoring adsorption in acidic condition (Li, Wang, et al. 2017). The soft basic S^{2-} ligands are preferably combined with soft acidic Pb^{2+} ions (HSAB principle), as hard protons have lower affinity for the former in acidic conditions ($pH=1.75$) (Rathore et al. 2017).

The pH_{zpc} value of the adsorbent plays a significant role in adsorption where the adsorbent bears negative charge above it on its surface and leads to an electrostatic interaction between the surface of adsorbent and lead ions. The surface hydroxyl groups were combined with the lead ions for surface adsorption on nickel ferrite ($pH_{zpc}=7.22$) (Reddy and Lee 2013).

The maximum removal of lead ions was achieved at various pH values, e.g., 5 (Jiang and Liu 2014; Kumar et al. 2014; Jiang et al. 2012), 5.5 (Lei et al. 2015), 5.8 (Huang et al. 2011; Huang, Wu, et al. 2017), 6 (Pourbeyram 2016; Yan, Kong, et al. 2015; Hadi Najafabadi et al. 2015; Gedam and Dongre 2015; Moradi et al. 2017), 6.5 (Zhang, Yang, et al. 2017), 6.8 (Madadrang et al. 2012), 6.88 (Lv et al. 2017), and 7 (Mahmoud, Abdou, et al. 2016), under different experimental conditions. Electrostatic attraction (Huang, Wu, et al. 2017;Gedam and Dongre 2015; Moradi et al. 2017; Kumar et al. 2014), surface complex formation (Pourbeyram 2016; Huang et al. 2011), ion exchange and complex formation (Madadrang et al. 2012), dissolution–precipitation (Lei et al. 2015) were held responsible for optimum adsorption at a particular pH. The hindered adsorption at lower pH values was predominantly attributed to the competition of hydronium ion with the lead ions for active sites on the adsorbent surface. In most of the cases, the decrease in adsorption capacity at $pH>6$ is attributed to the prevalence of precipitation of lead ions (Pourbeyram 2016; Yan, Kong, et al. 2015). However, some authors have reported different pH values for precipitation, e.g., 6.5 (Zhang, Yang, et al. 2017), 7 (Lv et al. 2017), and 8 (Madadrang et al. 2012).

3.8.2 Effect of the Presence of Ions

The presence of some interfering ionic species affects the adsorption efficiency of the Pb^{2+} ions to some extent due to the competition for active sites on the surface of the adsorbents. The addition of cations (Zn^{2+} and Cd^{2+}) and anions (NO^{3-}, SO_4^{2-}) in the solution has a negligible effect, whereas HCO_3^- and HPO_4^{2-} significantly decline the adsorption capacity of graphene oxide–$MnFe_2O_4$ magnetic nanohybrids for Pb^{2+} ions (Kumar et al. 2014).

The presence of calcium ions has a negligible effect on the adsorption capacity of chelating PE-MA-NN (porous chelating) fiber (Wang, Cheng, Yang, et al. 2013) and layered metal chalcophosphate (Rathore et al. 2017) due to the formation of a stable five-membered complex between [–NH –$(CH_3)_2$–NH_2] of PE-MA-NN fiber and Pb^{2+} and the strong interaction between soft acidic Pb^{2+} and soft basic S^{2-} ligands of layered metal chalcophosphate (HSAB principle), respectively.

Iron oxide@diaminophenol–formaldehyde (core-shell ferromagnetic nanorod) polymer composite (Fe_3O_4@DAPF) (Venkateswarlu and Yoon 2015a) exhibited the highest removal efficiency for Pb^{2+} than other metal ions i.e. Cd^{2+}, Zn^{2+}, Hg^{2+}, and As^{3+} . The high efficiency of Pb^{2+} ion can be explained by Pearson acid–base classification, which advocates the binding preference of borderline acids to borderline bases, and thus, the borderline acid i.e. Pb^{2+} preferentially attracted to the borderline base i.e. aminefunctional group of the DAPF resin.

The presence of Ni^{2+}, Co^{2+}, Zn^{2+}, and Cu^{2+} did not affect the metal uptake capacity of ethylenediamine-grafted MIL-101 (ED-MIL-101, chromium-based) because the –NH_2 groups of ethylenediamine that coordinated with metal ions in ED-MIL-101 did not match well with other competing ions (Luo, Ding, et al. 2015), whereas Zn^{2+}, Ni^{2+}, Cd^{2+}, and Cu^{2+} ions significantly decreased the adsorption capacity of the diethylenetriaminepentaacetic acid-modified magnetic graphene oxide composites (Li, Wang, et al. 2017).

The adsorption of Pb^{2+} ions on cross-linked poly(glycidyl methacrylate) microspheres functionalized with triazole-4-carboxylic acid was marginally affected by the presence of other interfering heavy metal (Mg^{2+}, Ca^{2+}, and Cu^{2+}) ions. The adsorption followed Pauling's electronegativity order, in addition to the opposite order of hydrate's radii ($Pb^{2+} > Cu^{2+} > Ca^{2+} > Mg^{2+}$). The cations with greater electronegativity and smaller hydrate radii tended to bind more favorably via surface complexation and electrostatic interactions (Yuan, Zhang, et al. 2017).

3.8.3 Zeta Potential

The zeta potential of EDTA–graphene oxide (EDTA-GO) showed variation with the pH values as it decreased initially when the pH increased from 3 to 6, formed a plateau between 6 and 8, and then increased at pH 12. The adsorbent surface was negatively charged at all the pH ranges due to ionization of functional groups present on the surface leading to more negative charges on the surface, resulting in stronger interaction between lead ions and EDTA-GO (Madadrang et al. 2012). However, the surface of graphene oxide–zirconium phosphate (GO-Zr-P) was slightly positively charged at pH 1 (Pourbeyram 2016). The slight negative charge present on the surface slowly increased in the pH range of 2–6, and at pH > 6, deprotonation of phosphate groups led to a quick rise in surface negative charge. At pH > 8, a gradual detachment of Zr-P nanoparticles from the surface of GO-Zr-P was observed.

During the adsorption of Pb^{2+} ions on layered metal chalcophosphate (Rathore et al. 2017), Pb^{2+} ions occupy the Mn^{2+} vacancies in the layer and pull out the intercalated hydrated K^+ ions from the interlayer, decreasing the inter-lamellar spacing.

This process generates the excess negative charge on surface S atoms promoting the adsorption of extra Pb^{2+} over the surface. The adherence of Pb^{2+} ions on the surface diminishes the Mn^{3+} concentration in Pb-MPS-1, which was evident by XPS studies. Therefore, Pb^{2+} ions were adsorbed both in the Mn^{2+} vacancies and on the surfaces. The zeta potential measurement showed that the intercalation of K^+ led to increased negative charge on MPS-1, which was diminished by the sorption of Pb^{2+} into K-MPS-1.

3.8.4 Effect of Surface Modification and Material

The modification of the graphene oxide (GO) surface with EDTA groups through silanization significantly enhances the adsorption capacity of GO for heavy metal ions (Madadrang et al. 2012). Formaldehyde- and 2,3-diaminophenol-based polymer use for surface modification of ferromagnetic Fe_3O_4 nanorods led to the protection of Fe_3O_4 surfaces and enhancement of the stability as well as surface area of the composite (Venkateswarlu and Yoon 2015a). The introduction of poly(acrylamide) into reduced graphene oxide (RGO) significantly enhanced its dispersion in solution and its metal uptake capacity (Yang, Xie, et al. 2013). Similarly, functionalization of PE-MA-NN fibers synthesized with polyethylene via with hydrophilic groups (amino/amide groups) (Wang, Cheng, Yang, et al. 2013), chitosan through α-MnO_2 with valine amino acid (Mallakpour and Madani 2016), and MIL-101 modification with ethylenediamine (Luo, Ding, et al. 2015) increased their adsorption capacities for lead ions.

XRD of the ED-MIL-101 and Pb/ED-MIL-101 (material after Pb^{2+} adsorption on ED-MIL-101) depicted the same XRD patterns, which suggested towards no change in the crystal structures of adsorbent after adsorption. However, the modification of MIL-101 with excess ED resulted in the absence of its characteristic peaks in XRD patterns that may be due to the destruction of crystalline order of framework or decomposition of crystalline MIL-101 (Luo, Ding, et al. 2015). In the XRD pattern of Fe_3O_4@DAPF, an additional broad peak appeared along with the characteristic peaks that were attributed to the scattering by DAPF resin. The intensities of peaks of Fe_3O_4@DAPF were also diminished probably by the coating of polymer resin shell on the surface of Fe_3O_4 nanorods (Venkateswarlu and Yoon 2015a). Upon modification of α-MnO_2 rods with valine, the resulting material showed well-defined diffraction lines of α-MnO_2 phases with higher crystallinity and L-valine. The XRD pattern of chitosan revealed it as hydrated crystals in which chitosan chains aligned antiparallelly forming sheet-like structure stacked together by bonding with intermolecular hydrogen leading to 3D crystals. On modification with α-MnO_2–valine, the characteristic peak of chitosan disappeared due to expansion of spacing between chitosan chains in each sheet by incorporation of functional groups in the backbone (Mallakpour and Madani 2016).

The modification was also established by FT-IR spectrum where strong adsorption bands at 3386 and 3185 cm^{-1} corresponded to the bonding between O-H in DAPF and Fe-O by covalent bonding (Venkateswarlu and Yoon 2015a). The FT-IR bands at 1581, 1051, and 882 cm^{-1} were attributed to the N-H plane stretching, C-N bond stretching, and $-NH_2$ stretching, respectively, advocating the grafting of ED on

MIL-101 (Luo, Ding, et al. 2015). A peak at 2925 and 1647 cm^{-1} corresponding to C-H asymmetric stretching and C=O stretching vibrations, respectively, proved successful grafting of poly(acrylamide) chains on RGO (Yang, Xie, et al. 2013).

3.8.5 Mechanism

The adsorption mechanism of lead ions on MgO nanostructures involved solid–liquid interfacial reaction between MgO and Pb^{2+} ions (Cao, Qu, Wei, et al. 2012) where Mg^{2+} ions got exchanged with Pb^{2+} ions during adsorption and lead ions replaced the Mg^{2+} ions in the MgO crystal lattice. There exists a nearly linear relationship between the amount of Mg^{2+} ions released in the solution and the amount of Pb^{2+} ions adsorbed on the surface of material. The XRD pattern of material after adsorption containing diffraction peaks of MgO and PbO showed the partial chemical reaction between MgO and Pb^{2+} ions. The presence of FT-IR peaks of $Mg(OH)_2$ after adsorption, but the absence of peaks of $Pb(OH)_2$ after adsorption, is another indication of chemical reaction amid MgO and Pb^{2+} ions. Further, the SEM image revealed that the flowerlike morphology of MgO after adsorption got distorted because the formation of PbO hindered and blocked the reaction of MgO with lead ions when the first few layers of the Mg^{2+} got exchanged with Pb^{2+} ions.

The adsorption of lead ions on graphene oxide–$MnFe_2O_4$ magnetic nanohybrids occurred through the complexation mechanism – i.e. the competition between protons and metal ions gets diminished at higher pH values and –OH groups get ionized to $–O^-$, leading to enhancement in adsorption; and cation exchange reaction where some of the $GO\text{-}COO^-$ and $GO\text{-}O^-$ groups present on the surface of graphene helped in the adsorption of Pb^{2+} ions (Kumar et al. 2014).

The sorption of lead ions on graphene oxide–zirconium phosphate nanocomposite involved the complexation mechanism (Pourbeyram 2016). The XPS spectrum of graphene oxide–zirconium phosphate nanocomposite after the adsorption of Pb^{2+} ions displayed five peaks assigned to the different forms of oxygen, viz. C-O, P-O-H, Zr-O-C, Zr-O-P, and P-O-Cd. The significant reduction in the peak intensity of P-O-H and appearance of a new peak of P-O-Cd after adsorption indicated that chemical interactions such as complexation with phosphate groups on the surface of graphene oxide–zirconium phosphate nanocomposite were responsible for the adsorption of Pb^{2+} ions.

3.8.6 Desorption

The desorption of lead ions was carried out by various desorbing agents such as hydrochloric acid (Pourbeyram 2016; Kumar et al. 2014; Madadrang et al. 2012; Jiang and Liu 2014; Li, Wang, et al. 2017; Gedam and Dongre 2015; Lv et al. 2017; Bayuo et al. 2020), nitric acid (Yuan, Zhang, et al. 2017), EDTA (Luo, Ding, et al. 2015), and both hydrochloric acid and EDTA (Wang, Cheng, Yang, et al. 2013).

The surface –OH and –COOH groups protonated at lower pH values resulted in desorption (Kumar et al. 2014). The decrease in pH increased the desorption with a maximum of 90% at $pH < 2$, and ~92% of the material was regenerated within 1 h. EDTA–graphene oxide exhibited retention of about 80% of its initial adsorption capacity on repeated use up to ten cycles (Madadrang et al. 2012). The higher concentration of H^+ at lower pH changed $-COO^-$ into $-COOH$, which was competitive against the $-COO^-$ and Pb^{2+} interaction, and thus accelerated the desorption of lead ions (Jiang and Liu 2014).

The adsorption capacity and the desorption rate of lead ions remained constant after PE-MA-NN regeneration up to five consecutive cycles (Wang, Cheng, Yang, et al. 2013). The desorption of lead ions from poly(glycidyl methacrylate) microspheres functionalized with triazole-4-carboxylic acid was rapid chemical desorption, and it followed pseudo-second-order kinetics (Yuan, Zhang, et al. 2017).

3.9 URANIUM

Uranium is a naturally occurring radioactive element, and its three isotopes, i.e., ^{234}U, ^{235}U, and ^{238}U, have similar chemical properties with varied radioactive properties (ATSDR 2013). Uranium is used in nuclear power plants, counterbalance on helicopter rotors, shield against ionizing radiation, and armor in military vehicles. Natural (^{238}U) and depleted uranium (^{234}U, ^{235}U) owing to chemical nature have similar health effects. Kidney is the chief target organ of uranium toxicity. An overview of the experimental parameters and optimized conditions from batch adsorption experiments for uranium is presented in Table 3.10.

3.9.1 Effect of pH

The percentage removal increased with a rise in pH from pH 3 to 6 and then declined afterward using bovine serum albumin-coated graphene oxide (Yang, Liu, et al. 2017), SBA-15 (Huynh et al. 2017), and carbon nanofiber (Sun et al. 2016). In other cases, also adsorption increased up to pH 6 and then declined, such as with Fe_3O_4@C@-Ni-Al layered double hydroxide (Zhang, Wang, et al. 2013), bisphenol A and tannic acid embedded in silica (Yamazaki et al. 2015), and cobalt ferrite/multiwalled carbon nanotube composite (Tan, Liu, et al. 2015). The maximum adsorption was also found at different pH values apart from pH 6, such as at pH 4.5 (Zhang, Zhang, et al. 2015) and 5 (Zhang, Jing, et al. 2015).

The adsorbent surface charge and speciation were held responsible for the effect of pH on removal (Huynh et al. 2017; Sun et al. 2016). The adsorption rise up to pH 6 or optimum pH was attributed to electrostatic force (Zhang, Wang, et al. 2013). The uranyl ions remained as positively charged species (UO_2^{2+}) up to pH 6 (Sun et al. 2016; Zhang, Wang, et al. 2013), and after pH 6, the uranyl ions mainly existed as negatively charged species ($UO_2(OH)^{3-}$ and $(U_2\ O)_3(OH)^{7-}$), and this led to the electrostatic repulsion above pH 6 (Huynh et al. 2017). The decreased percentage removal above pH 6 was also attributed to increased negative zeta potential with a rise in pH, in addition to speciation of U(VI) ions (Yang, Liu, et al. 2017; Huynh et al. 2017; Sun et al. 2016).

The decline in percentage removal of uranium on magnetic cobalt ferrite/multiwalled carbon nanotube composite after pH 6 was attributed to soluble hydroxide and carbonate species (Tan, Liu, et al. 2015). In addition, adsorbate–adsorbent degradation at a higher pH was attributed to the decline in percentage removal. The change in pH was insignificant in the case of addition of hydroxyapatite and tributyl phosphate-coated hydroxyapatite as adsorbent (Kim, Um, et al. 2017).

The low percentage removal at a very low pH was attributed to the competition from hydronium ions and increased protonation of the adsorbent functional groups (Zhang, Jing, et al. 2015; Yang, Liu, et al. 2017; Sun et al. 2016; Tan, Wang, Liu, Sun, et al. 2015). The stability of adsorbent with pH also affects the adsorption capacity (Li, Wang, Yuan, et al. 2015). The dissolution of Fe at a low pH, i.e., 2.1, hampered the remediation of uranium; however, on raising pH to 4–9, the percentage removal remained constant and unaffected. This was attributed to the absence of dissolution of iron.

The large K_d (distribution coefficient) value suggests higher adsorption (Ma, Huang, et al. 2015). In the case of uranyl ion adsorption on tannic acid-embedded mesoporous silica, the K_d value increased with a rise in pH (Yamazaki et al. 2015). This was in spite of the formation of different uranyl species at higher pH, i.e., $(UO_2)_{11}(CO_3)_6(OH)_{12}^{2-}$. This suggested that the adsorption was not due to the cation–anion exchange process and remediation followed the complexation process.

The adsorption of uranyl ions at two different pH values led to different species adsorbed on the surface of the adsorbent (Sun et al. 2016). The adsorption of U(VI) on carbon nanofibers at pH 4 led to the presence of $U^{VI}O_2^{2+}$, whereas at pH 7 it indicated the presence of $U^{IV}O_2$ also. This was due to the reduction of U(VI) (35.8%) to U(IV). The reduction of uranyl ions was also reported in anoxic conditions in Fe/RGO composite (Li, Wang, Yuan, et al. 2015). In the case of functionalized adsorbents, protonation–deprotonation, functional group ionization, and electrostatic interaction were held responsible for the percentage removal of U(VI) (Pan et al. 2017; Dolatyari et al. 2016).

3.9.2 Zeta Potential

The shift in zeta potential after adsorption was considered as the macroscopic evidence for the formation of inner-sphere complex (Tan, Liu, et al. 2015; Tan, Wang, Liu, Sun, et al. 2015). The zeta potential increase indicated that surface charge shifted positively after adsorption for uranyl removal with MnO_2–Fe_3O_4 reduced graphene oxide composite (Tan, Wang, Liu, Sun, et al. 2015) and cobalt ferrite/multiwalled carbon nanotube composite (Tan, Liu, et al. 2015). Hence, uranyl ions were adsorbed as cationic or neutral inner-sphere complex. The EXAFS data can be used as the evidence of inner-sphere complex; e.g., U(VI) formation as inner complex on Na-dentonite was evidenced by EXAFS data (Sheng et al. 2014). The zeta potential of graphene oxide is negative in the pH range of 2–9 (Verma and Dutta 2015). The zeta potential becomes more negative with a rise in pH up to 6, and afterward, it declines slightly with a rise in pH up to 9. However, the zeta potential of graphene oxide treated with ammonia increases toward positive, and it changes from positive at pH 2 to negative at pH 9.

3.9.3 Effect of Coexisting Ions

The presence of inorganic/organic species affects the adsorption depending on the adsorbent and concentration of interfering ion and pH of the system. The variation in

TABLE 3.10
Summary of Parameters and Optimized Conditions for Batch Adsorption of Uranium

Adsorbent	Adsorbate	Surface Area (m^2/g), Pore Volume (cm^3/g), Pore Size (nm)	pHzpc	Experimental Conditions	Adsorption Capacity (mg/g)	Thermodynamic Parameters	Kinetic Model and Curve Fitting	Isotherm Model and Curve Fitting	Maximum Adsorption Conditions	References
ZVInps	U(VI)	Surface area = 9.7		pH = 5 Dose = 0.08 or 2 g/l Concentration = 0–714 mg/l Contact time = up to 1440 min	8173 (anoxic) 1354 (oxic)				pH = 3.5–9 Contact time = 20 min at 24 ppm and 40 min at 333 ppm	Li, Wang, Yuan, et al. (2015)
Fe/RGO	U(VI)	Surface area = 218.9		do	4174 (anoxic)					Li, Wang, Yuan, et al. (2015)
PANI/GO	U(VI)			pH = ca. 1–8 Dose = 0.05 g/l Concentration = 50 mg/l Contact time = 48 h Temperature = 20°C	1960				pH = ca. 6–8	Shao et al. (2014)
$CoFe_2O_4$/MWCNTs	U(VI)	Surface area = 170.3 Pore volume = 0.441 Pore size = 10.37	4.5	pH = 2–12 Dose = 0.4 g/l Agitation speed = rpm Concentration = 50 mg/l Contact time = 360 min Temperature = 25°C–55°C	212.2	ΔH = 8.471 kJ/mol ΔS = 54.71 kJ/mol ΔG = negative	Pseudo-second-order model, linear	Langmuir model, linear	pH = 6 Temperature = 55°C	Tan, Liu, et al. (2015)
Magnetic calcium silicate hydrate	U(VI)	Surface area = 196		pH = 2–12 Dose = 1–2.4 g/l Agitation speed = 200 rpm Concentration = 50–5000 mg/l Contact time = 120 min Temperature = 298–318 K	2500 at 298 K 2778 at 318 K	ΔH = 10.38 kJ/mol ΔS = 105.92 J/mol/K ΔG = negative	Pseudo-second-order model, linear	Langmuir model, linear	pH = 4–6 Dose = 1.2 g/l Concentration = 200 mg/l Temperature = 318 K	Zhang, Liu, Wang, et al. (2015)

(continued)

TABLE 3.10 (*Continued*)
Summary of Parameters and Optimized Conditions for Batch Adsorption of Uranium

Adsorbent	Adsorbate	Surface Area (m^2/g), Pore Volume (cm^3/g), Pore Size (nm)	pHzpc	Experimental Conditions	Adsorption Capacity (mg/g)	Thermodynamic Parameters	Kinetic Model and Curve Fitting	Isotherm Model and Curve Fitting	Maximum Adsorption Conditions	References
Ammonia-modified graphene oxide	U(VI)		3.66	pH = 2–9 Dose = 0.2–0.7 g/l Agitation speed = 150 rpm Concentration = 5–100 mg/l Contact time = 4 h and 12 h Temperature = 288–313 K	80.13	ΔH = 35.087 kJ/mol ΔS = 126.83 J/mol ΔG = negative	Pseudo-second-order model, linear	Both Langmuir and Freundlich, linear	pH = 4–6 Dose = 1.2 g/l Concentration = 200 mg/l Temperature = 318 K	Verma and Dutta (2015)
Magnesium carbonate basic coating on cotton cloth	U(VI)	Surface area = 128.2 Pore size = 6–7		pH = 2–10 Dose = 0.5 g/l Agitation speed = 150 rpm Concentration = 200 mg/l Contact time = 5–70 min Temperature = 298–318 K	371.26	ΔH = 37.35 kJ/mol ΔS = 0.137–0.142 kJ/mol K ΔG = negative	Pseudo-second-order model, linear	Freundlich model, linear	pH = 5 Temperature = 318 K Contact time = 50 min	Zhang, Jing, et al. (2015)
rGO/NiAl-LDH composite	U(VI)	Surface area = 256.80 Pore volume = 0.66		pH = 2–12 Dose = 0.25–5 g/l Agitation speed = 150 rpm Concentration = 200 mg/l Contact time = 5–70 min Temperature = 298–318 K	277.80	ΔH = 22.90 kJ/mol ΔS = 117 J/mol K ΔG = negative	Pseudo-second-order model, linear	Langmuir model, linear	pH = 5 Contact time = 50 min Temperature = 318 K	Tan, Wang, Liu, Wang, et al. (2015)
Amidoxime-functionalized MCM-41	U(VI)			pH = 2–7 Dose = 1–4 g/l Agitation speed = 150 rpm Concentration = up to 400 mg/l Contact time = up to 60 min Temperature = 293–323 K	442.3	ΔH = 81.4 kJ/mol ΔS = 0.35 J/mol K ΔG = negative	Pseudo-second-order model, linear	Langmuir model, linear	pH = 5 Contact time = 30 min Temperature = 323 K	Bayramoglu and Arica (2016)

(continued)

TABLE 3.10 (*Continued*)
Summary of Parameters and Optimized Conditions for Batch Adsorption of Uranium

Adsorbent	Adsorbate	Surface Area (m^2/g), Pore Volume (cm^3/g), Pore Size (nm)	pHzpc	Experimental Conditions	Adsorption Capacity (mg/g)	Thermodynamic Parameters	Kinetic Model and Curve Fitting	Isotherm Model and Curve Fitting	Maximum Adsorption Conditions	References
Carboxyl-functionalized MCM-41	U(VI)			pH = 2–7 Dose = 1–4 g/l Agitation speed = 150 rpm Concentration = up to 600 mg/l Contact time = up to 60 min Temperature = 293–323 K	296.7	ΔH = 42.6 kJ/mol ΔS = 0.21 kJ/mol K ΔG = negative	do	do	pH = 5 Contact time = 30 min Temperature = 323 K	Bayramoglu and Arica (2016)
Bisphenol A-type resin embedded in silica beads	U(VI)			pH = 7.9 (seawater) and 2–9.1 (simulated seawater) Dose = 10–33.33 g/l Concentration = 100 ppb Contact time = 24 h Temperature = 298 K	K_d value = 0.4×10^3				pH = 6	Yamazaki et al. (2015)
Tannic acid-type resin embedded in silica beads	U(VI)			do	K_d value = 1.1×10^3					Yamazaki et al. (2015)
Chitosan@attapulgite	U(VI)		4.3	pH = 3–10 Dose = 0.3 g/l Concentration = 2.8×10^{-5} mol/l Contact time = up to 48 h Temperature = 25°C	53.5			Freundlich model, non-linear	pH = 7	Pan et al. (2017)
SBA-15 bearing N-propylsalicylaldimine	U(VI)			pH = 2–8 Dose = 0.05–3 g/l Concentration = 10 mg/l Contact time = 1–240 min Temperature = 288–328 K	54	ΔH = 24.1 kJ/mol ΔS = 94.5 J/mol K ΔG = negative	Pseudo-second-order model, linear	Langmuir, linear	pH = 4 Dose = 1 g/l Contact time = 15 min Temperature = 328 K	Dolatyari et al. (2016)

(continued)

TABLE 3.10 (*Continued*)
Summary of Parameters and Optimized Conditions for Batch Adsorption of Uranium

Adsorbent	Adsorbate	Surface Area (m^2/g), Pore Volume (cm^3/g), Pore Size (nm)	pHzpc	Experimental Conditions	Adsorption Capacity (mg/g)	Thermodynamic Parameters	Kinetic Model and Curve Fitting	Isotherm Model and Curve Fitting	Maximum Adsorption Conditions	References
SBA-15 bearing ethylenediamine propyle salicylaldimine	U(VI)			do	105.3	ΔH = 1.2 kJ/mol ΔS = 28.9 J/mol K ΔG = negative	Pseudo-second-order model, linear	Langmuir, linear	do	Dolatyari et al. (2016)
Polyethyleneimine-modified magnetic activated carbon	U(VI)	Surface area = 303 Pore size = 62		pH = 2–7 Dose = 50–500 mg Agitation speed = 120 rpm Concentration = 1–25 mg/l Contact time = 10–90 min Temperature = 20°C–50°C	115.31	ΔH = −71.26 kJ/mol ΔS = −21.25 J/mol K ΔG = negative	Pseudo-second-order model, linear	Langmuir model, linear	pH = 5 Temperature = 50°C	Saleh, Tuzen, et al. (2017)
Tributyl phosphate-coated hydroxyapatite	U(VI)	Surface area = 5.27	5.7 (pH_{IEP})	Dose = 5 g/l Agitation speed = 200 rpm Concentration = 10, 100, and 1000 ppb Contact time = 1 min, 1 h, and 1 day	38			Langmuir model, linear		Kim, Um, et al. (2017)
Fe_3O_4@C@layered double hydroxide	U(VI)	Surface area = 77 Pore volume = 0.243 Pore size = 10.9 (radii)		pH = 2–12 Dose = 0.1–3 g/l Concentration = 200 mg/l Contact time = up to 240 min Temperature = 25°C–55°C	174	ΔH = 22.4 kJ/mol ΔS = −111 J/mol K ΔG = negative	Pseudo-second-order model, linear	Langmuir model, linear	pH = 6 Dose = 3 g/l (most of the experiments at 1 g/l) Contact time = 180 min Temperature = 55°C	Zhang, Wang, et al. (2013)

(*continued*)

TABLE 3.10 (*Continued*)
Summary of Parameters and Optimized Conditions for Batch Adsorption of Uranium

Adsorbent	Adsorbate	Surface Area (m^2/g), Pore Volume (cm^3/g), Pore Size (nm)	pHzpc	Experimental Conditions	Adsorption Capacity (mg/g)	Thermodynamic Parameters	Kinetic Model and Curve Fitting	Isotherm Model and Curve Fitting	Maximum Adsorption Conditions	References
NH_2-functionalized ordered silica	U(VI)	Surface area = 500 Pore volume = 1 Pore size = 7.5		Dose = 0.1 g/l Agitation speed = 750 rpm Concentration = 100 mg/l Contact time = up to 150 min Temperature = 20°C	573					Huynh et al. (2017)
Ultrafine cellulose nanofibers	U(VI)				167					Ma et al. (2012)
Bovine serum albumin-coated graphene oxide	U(VI)			pH = 3–10 Dose = 0.5 g/l Concentration = 50–650 mg/l Contact time = 3–200 min	389				pH = 6 Concentration = 200 mg/l Contact time = 80 min	Yang, Liu, et al. (2017)

background $NaHCO_3$ concentration from 0.001 to 0.1 M decreased the adsorption of uranyl ions on hydroxyapatite from 100% to more than half. The decline was attributed to the increase in the formation of soluble uranium carbonate complex with an increase in ionic strength (Kim, Um, et al. 2017). The presence of bicarbonate in anaerobic conditions enhanced the reduction of U(VI) to U(IV) from 82% to 100% (Gallegos et al. 2013), which was evidenced by the XANES spectra. However, the percentage removal decreased slightly at pH 9. The presence of ions such as $NaHCO_3$, humic acid, and synthetic groundwater constituents did not significantly affect the percentage removal of uranyl ions under anoxic conditions on nZVI (Li, Wang, Yuan, et al. 2015).

Calcium ion hindered the adsorption of U(VI) on Mg/Al layered double hydroxide but only at a very high concentration (Ma, Huang, et al. 2015). A large percentage removal (95–99) of U(VI) was achieved at high Ca concentrations ($CaCl_2$:U ratio 1.5×10^3–2.1×10^4), and even at much larger concentrations ($CaCl_2$:U ratio 6×10^4), 75% removal was achieved. The high selectivity of U(VI) to Mg/Al LDH was attributed to stronger soft acid–soft base (UO_2^{2+}...S^{2-}) interaction as compared to the hard–soft in the case of calcium (Ca^{2+}...S^{2-}). The presence of calcium ions in the solution increased the adsorption of uranyl ions on magnetic calcium silicate hydrate with linearity (Zhang, Liu, Wang, et al. 2015). This was attributed to the ion exchange phenomenon of calcium and uranyl ions.

The adsorption on chitosan-coated attapulgite was enhanced by fulvic and humic acids below pH 6 and decreased at pH greater than 6 (Pan et al. 2017). This was attributed to the attachment of fulvic/humic acid (negative charge) below pH 6, which increased the amount of functional groups for the adsorption of uranyl ions. However, with a rise in the pH (> 6), the attachment was deterred due to electrostatic repulsion. Humic/fulvic acid then competed with the adsorbent for uranyl ions and formed soluble complexes with them.

The presence of Cu and Zn did not affect the adsorption of uranyl ions on magnetic calcium silicate hydrate, whereas aluminum affected the adsorption process (Zhang, Liu, Wang, et al. 2015). The effect of aluminumwas due to the modification of the structure of adsorbent via replacement of silicon with aluminum ions.

The XRD pattern of polysulfide Mg/Al LDH showed that the structure of the adsorbent is intact after adsorption (Ma, Huang, et al. 2015). The retention of d_{basal} spacing at 0.80 nm supported the statement. However, additional d_{basal} spacing appeared at 0.78/0.79 nm on increasing the calcium chloride concentration (Ca:U ratio > 5000). This was attributed to the intercalation of Cl^- ions into the LDH and formation of a separate Cl-LDH structure. The interaction of chloride ion was also observed in the case of NaCl (Cl 40,000 times) but not in the case of NO_3^- (20,000 times). The percentage removal of uranium remained above 97% in the presence of NaCl. The replacement of tetrasulfide ($(S_4)^{2-}$) of polysulfide Mg/Al LDH with Cl^- ion was held responsible for the high percentage removal of uranium in the presence of sodium chloride (Ma, Huang, et al. 2015).

3.9.4 Effect of Surface Modification

The functionalization or surface modification increase the adsorption capacity e.g. in case of hydroxyapatite surface modification with tributyl phosphate (Kim, Um, et al.

2017), attapulgite functionalization with chitosan (Pan et al. 2017), MIL-101 modification with ethylenediamine (Zhang, Zhang, et al. 2015), aminofunctionalization of silica (Huynh et al. 2017), and graphene surface modification with bovine serum albumin (Yang, Liu, et al. 2017). However in some cases, the adsorption capacity of functionalized groups was lower than that of unfunctionalized SBA-15 (Dolatyari et al. 2016). This was attributed to the hindrance by ligating groups for access to binding sites on the surface of SBA-15.

XRD was used to study the change in the structure on functionalization (Huynh et al. 2017; Kim, Um, et al. 2017). The absence of change in structure in the functionalized group can be estimated by the change in the XRD diffractogram. The nonsignificant change in diffraction angle (100) for SBA-15 functionalized and non-functionalized materials was attributed to non-alteration of cell parameters. However, the small change in cell parameters (11.1–11.2 and 11.3 nm) was attributed to the incorporation of the organic ligands (Huynh et al. 2017). The modification of attapulgite (XRD peak at 8.3°) with chitosan shifted the XRD peak to a higher 2θ (8.7°) (Pan et al. 2017).

The BET isotherm curve and type are also used to estimate the change in the structure post-functionalization (Huynh et al. 2017). The lowering of surface area and pore volume without change in the XRD and isotherm hysteresis was attributed to the modification of surface coverage rather than pore structure. The functional group caused steric hindrance. The introduction of amine group by the co-condensation method on SBA-15 led to the change of hysteresis (H1–H2) while maintaining the isotherm shape (type IV) along with a decrease in intensity (Huynh et al. 2017). However, all characteristic peaks were present. This was attributed to the less ordered solid with pore periodicity loss while maintaining the structure. The efficiency of the amine group on SBA-15 is estimated by the C, H, and N analysis. The grafting of amine caused the BET surface area of MIL-101(Cr) to decrease from 2852 to 512 m^2/g (Zhang, Zhang, et al. 2015). This was attributed to the substitution of water with a larger size group, i.e. ethylenediamine. In addition, pore size and pore volume also decreased.

The functionalization was also confirmed by FT-IR (Huynh et al. 2017). The FT-IR spectra depicted a decrease in intensity at 3734 cm^{-1} (isolated Si-OH group) and 3539 cm^{-1} (H-bonded) on SBA-15 upon functionalization. This was attributed to the use of surface hydroxide and silanol groups in functionalization.

The adsorption capacity shows the linear behavior with the -NH_2 group density on the surface of the adsorbent (Huynh et al. 2017). This suggested that functional group surface density is a key parameter controlling the adsorption process, and the grafting precursor (length of carbon) and proximity of the amine group do not have any role to play in the adsorption process. Similarly, the efficiency of ethylenediaminepropylesalicylaldimine-functionalized SBA-15 is more than that of N-propylsalicylaldimine-functionalized SBA-15 (Dolatyari et al. 2016). The presence of additional NH group is attributed to enhanced adsorption.

The functionalization is not linear with adsorption capacity in some cases; for example, the functionalization of phosphate coating on the surface of hydroxyapatite occurred more at pH 10 as compared to pH 7 (Kim, Um, et al. 2017). In spite of this, the adsorption of uranyl ions occurred more with the material prepared at pH 7 than at pH 10. This was attributed to the less conditioning of pH by the sample prepared at pH 7 than the sample prepared at acidic (4) and basic conditions (10).

The same functional group grafting by different routes also showed different adsorption capacity (Huynh et al. 2017). The adsorption capacity of the product formed by amino-functionalization on SBA-15 via co-condensation was lower as compared to the product prepared by the conventional synthesis method. This was attributed to the incorporation of functional groups into the walls of silica in the co-condensation process. This led to non-availability of groups for the adsorption of uranyl ions.

The reduction by oxygen-containing functional groups on carbon nanofibers was also postulated to have an effect on the adsorption of U(VI) (Sun et al. 2016). The peaks corresponding to the COOH group at $1710\,cm^{-1}$ reduced after the adsorption of uranyl ions. This was assigned to the interaction of the COOH group with uranyl ions (Sun et al. 2016). The reduction was also supported by XPS and XANES spectra. The XPS spectra also depicted the decrease in the intensity of O1s and its shift toward higher energy. This depicts the role of O in the adsorption of U(VI). The amount of CO(O) groups on loaded adsorbent was higher (6.89%) as compared to pristine material (5.69%). This was attributed to the oxidation of alcoholic groups to acid groups (Sun et al. 2016).

The surface modification of attapulgite with chitosan led to a change in the fitting of isotherm data from Langmuir to Freundlich (Pan et al. 2017). Bisphenol A of bisphenol A-embedded silica at pH 2–4 did not dissociate, whereas tannic acid of tannic acid-embedded silica dissociated into half (Yamazaki et al. 2015). This was attributed to the low adsorption capacity of bisphenol A-type adsorbent as compared to tannic acid-type adsorbent. The adsorption of uranyl ions on graphene oxide declined in the presence of common ions. However, the common ions did not affect the adsorption of uranyl ions after graphene oxide treatment with ammonia (Verma and Dutta 2015).

The coating of carbon on the surface of Fe_3O_4 reduced its surface area and pore volume from $74\,m^2/g$ to $13\,m^2/g$ and $0.097\,cm^3/g$ to $0.034\,cm^3/g$, respectively (Zhang, Wang, et al. 2013). In spite of this, the adsorption of uranium was higher after the coating of carbon. This was attributed to the increase in functional groups (FT-IR).

3.9.5 Stability

The stability of the adsorbent is shown by unchanged corresponding structural peaks of XRD and FT-IR (Ma, Huang, et al. 2015; Zhang, Zhang, et al. 2015). The visual change in morphology can be analyzed by SEM images. The unchanged morphology of Mg/Al LDH was evident by SEM images (Ma, Huang, et al. 2015). The presence of many small particles on the surface of adsorbent was attributed to UO_2S_4 and UO_2SO_4 (Ma, Huang, et al. 2015).

The Ca^{2+} and uranyl ions have similar charge and ionic radii (the uranyl radii are bigger with 0.004 nm) (Zhang, Liu, Wang, et al. 2015). These characteristics help to maintain the structure of the adsorbent after adsorption. The phenomenon of ion exchange between Ca^{2+} and uranyl ion aids in retaining the structure of the adsorbent.

3.9.6 Effect of Material and Temperature

The remediation of uranyl ions also varied with the nature of the adsorbent. The use of support with nZVI increased the adsorption efficiency. The nature of support also played a role in the remediation of uranyl. When Na-bentonite was used as the support for nZVI in near-zero dissolved oxygen conditions, 99.2% of removal was achieved as compared to Al-based bentonite (65.6%) (Sheng et al. 2014). This was attributed to the increased mass transfer in the former. In addition, the zeta potential of the material was also held responsible for it. The zeta potential of Na-bentonite was negative in the whole pH range, i.e., 2–10. However, the zeta potential of Al-bentonite was positive below pH 7 (below pH 7, uranium exists mostly as a positive cationic form). Hence, electrostatic interaction favored the adsorption on Na-bentonite. The nZVI alone remediated 48.3% of the uranyl ions.

Na-bentonite played multiple roles along with nZVI for remediation of uranyl ions (Sheng et al. 2014). It maintained the pH of the solution, prevented nZVI from agglomeration, and promoted transport of U(VI) to the surface of nZVI from solution. It was also postulated to carry away the reaction products. The SEM image of nZVI showed chain agglomerated products, whereas nZVI with support (Na- or Al-bentonite) showed well-dispersed nZVI on surface of support. The maintenance of pH by support was attributed to the release of proton from the aluminol or silanol group on consumption. The sodium content of nZVI-Na-bentonite decreased after adsorption. This was attributed to the exchange of ions between sodium and iron. The nZVI was also found in the chain like products in other studies (Ling and Zhang 2015).

The ZVI showed lower remediation capacity as compared to nZVI (Sheng et al. 2014). The adsorption on fine aquifer sediment for U(VI) was more than that of aquifer coarse sediment (Dong et al. 2012) with bulk sediment adsorption capability in between them. However, adsorption on per-unit surface area for coarse fraction was more than that of fine fraction. This was attributed to the amphoteric nature of the mineral surfaces. The coarse fraction is mainly composed of quartz (PZC=ca. 2–3), whereas the fine fraction is composed of kaolinite (PZC=5) and goethite (PZC=9.2). So, the fine fraction remains positively charged in adsorption pH conditions (3–5) and depresses the UO_2^{2+} adsorption.

The adsorption capacity of kaolinite is more than that of the aquifer fine fraction sediment at pH<4; however, it has a similar adsorption capability between pH 4 and 7 (Dong et al. 2012). Similarly, in the case of goethite adsorption, capability is similar to the fine fraction below pH 4 and more than the fine fraction at pH 4–7. However, the adsorption capability of the fine fraction did not lie between adsorption capability for kaolinite and goethite; which were it's chief constituents. This is attributed to the presence of common cations (Na, Al, K, Ca). These cations decreased the adsorption capability of fine fraction by aqueous complex formation and may compete for adsorption sites (Dong et al. 2012).

Temperature is suggested to affect adsorption by diffusion rate across the boundary layer and internal pores of the adsorbent (Dolatyari et al. 2016). The rise in temperature suggests the endothermic nature of adsorption (Zhang, Wang, et al. 2013; Dolatyari et al. 2016; Verma and Dutta 2015; Tan, Liu, et al. 2015; Tan, Wang, Liu,

Sun, et al. 2015). The positive enthalpy is attributed to dehydration of uranyl ions. The dehydration energy was more than that of adsorption energy (Dolatyari et al. 2016).

3.9.7 Mechanism

The adsorption of uranyl ions followed different mechanisms such as electrostatic attraction (Zhang, Wang, et al. 2013) and complexation (Huynh et al. 2017). The grafted SBA-15 for adsorption of uranyl ions had zero or slightly positive charge as compared to negative charge in the case of pristine SBA-15 (Huynh et al. 2017). This indicates that the highest adsorption at pH 6 on grafted samples was not due to the electrostatic interaction. The adsorption capacity varies linearly with the -NH_2 group density. The ratio of uranium ions adsorbed to amino groups on different grated samples of the adsorbent is varied from 2.1 to 2.4. This suggested the complexation between the solid and the aqueous species. In situ ATR-FT-IR and Raman study suggested inner-sphere interaction between U(VI) and adsorbent (NH_2-modified SBA-15) (Huynh et al. 2017). The adsorption of uranyl ions with Mg/Al layered double hydroxide occurred with redshift of peak at 963 cm^{-1} (Ma, Huang, et al. 2015), which depicts chemical bonding of UO_2^{2+} with $[S_4]^{2-}$. The change of redox state did not happen in this case.

The mechanism of adsorption is also varied with concentration (Chen, Zhuang, et al. 2014). The adsorption of uranium on $Mg(OH)_2$ occurs as monolayers on the smaller area of $Mg(OH)_2$. However, when the concentration is raised (100 mg/l), first adsorption occurs as monolayers and these distributed monolayers act as nucleation centers. These lead to precipitation of nanocrystal on the surface of the adsorbent. The adsorbent follows the Langmuir isotherm at low concentrations, and HRTEM analysis shows that previously unoccupied sites at lower concentrations still remain unoccupied at higher concentrations in spite of the rise in the adsorption capacity with a rise in concentration.

The adsorption of U(VI) with nZVI is proceeded in two steps: First, U(VI) is attracted by surface iron oxide/hydroxide, and afterward, U(VI) is reduced by nZVI. This is the reason for the temporal variation of uranium in the analysis of STEM-XDS image. The uranium adsorbed in the initial stage (1 h) is distributed well on the surface of the adsorbent. However, after 24 h it is concentrated near the center of nZVI (adsorbent) (Ling and Zhang 2015).

The XPS analysis for adsorption of uranyl ion on chitosan-modified attapulgite depicts a shift in binding energy of O^{2-} and -OH to a lower state after adsorption (Pan et al. 2017). This confirmed the participation of O^{2-} and -OH groups in adsorption.

The precipitation was observed in TEM images for uranium loaded on carbon nanofibers (Sun et al. 2016) and channels of modified SBA-15 (Huynh et al. 2017). The sample preparation for TEM analysis may be one of the reasons responsible for this (Huynh et al. 2017). The precipitation is postulated to occur in two steps, i.e. uranium adsorbed inside the pores followed by an increase in the concentration above the threshold. The irregular precipitate aggregation observed on carbon nanofibers showed the presence of schoepite and U_4O_9/UO_2. This was attributed to precipitation along with the reduction of U(VI). The presence of schoepite is evidenced by EXAFS and the corresponding Fourier transform.

3.9.8 Desorption

The desorption of uranyl ions can be done with hydrochloric acid (Saleh, Tuzen, et al. 2017; Yang, Liu, et al. 2017; Hamza, Gamal, et al. 2019), nitric acid (Dolatyari et al. 2016), sodium bicarbonate (Tan, Liu, et al. 2015; Kong et al. 2020), low pH (Huynh et al. 2017), and ammonium carbonate (Li, Wang, Yuan, et al. 2015). In some cases, the reuse is performed using multiple steps (Chen, Zhuang, et al. 2014; Verma and Dutta 2015). The recycling of used $Mg(OH)_2$ can be done in three steps, i.e., hydrothermal treatment with $NaHCO_3$, calcination, and dissolution in water (Chen, Zhuang, et al. 2014). In hydrothermal treatment, $Mg(OH)_2$ is transformed into $Na_2Mg(CO_3)_2$ and the calcination of $Na_2Mg(CO_3)_2$ transformed it into MgO and Na_2CO_3. When the calcined products are dissolved in water, they give nano-$Mg(OH)_2$. Ninety-six percent of the adsorbed uranium was removed, and enrichment multiple of 53 was achieved for uranium using hydrothermal treatment. The adsorbent can be reused up to three cycles with significant removal efficiency.

In desorption of ammonia modified graphene oxide for uranyl ions, the adsorbent first treated with 1 M nitric acid and afterward washed with ammonia for conditioning to reach pH 6. The adsorption efficiency after the first and second cycles was 83.55 and 82.16, respectively. The desorption efficiency of the subsequent cycles was 99.6, 98.7, and 98.4 (Verma and Dutta 2015).

The evidence of desorption can be studied with XPS studies (Yang, Liu, et al. 2017). The N and O1s XPS peaks after the desorption of the adsorbed sample with hydrochloric acid (recycled five times) shifted toward lower energy levels. This indicates regeneration of the COOH groups and nitrogen-containing groups (Yang, Liu, et al. 2017). The adsorption capacity decreased after desorption, e.g. in Fe_3O_4@C@Ni-Al layered double hydroxide (Zhang, Wang, et al. 2013), nZVI, nZVI-Na-bentonite, and nZVI-Al-bentonite (Sheng et al. 2014).

The adsorption of U(VI) on attapulgite is reversible; however, on surface modification with chitosan, the process becomes irreversible (Pan et al. 2017). The irreversible desorption/adsorption became reversible/partially reversible by the presence of humic/fulvic acid in the adsorbate solution.

4 Remediation of Essential Elements Exerting Toxicity on Excessive Exposure (Mn, Co, Cu, Zn, Se) Via Batch Adsorption in Response to Variable Factors and Elucidation of the Mechanism for the Batch Adsorption Process

Deepak Gusain
Durban University of Technology

Shikha Dubey and Yogesh Chandra Sharma
IIT(BHU), Varanasi

Faizal Bux
Durban University of Technology

CONTENTS

This chapter includes discussion on factors affecting adsorption of essential elements. The requirement of these metals in human diet is required for some functions as follows: cobalt is a part of vitamin B12, copper is a part of several metalloenzymes such as cytochrome c oxidase, manganese deficiency is associated with skeletal abnormalities, selenium is a part of several proteins and its deficiency leads to Keshan disease, and zinc is also a part of several metalloenzymes. The aforementioned elements are essential, but excessive intake leads to negative effects on health. Excessive cobalt intake (> 10 mg/l) is associated with heart failure, excessive copper intake is related to gastrointestinal distress, nausea and vomiting, manganese affects the brain (manganism) and also causes dilation of blood vessels and induces hypotension, excessive intake of selenium leads to neurological and dermal effects, and excessive intake of zinc is associated with affecting the pancreas and brain. So, it is necessary to keep the exposure of elements under recommended guideline values.

4.1 MANGANESE

Manganese is an essential element, but chronic exposure to manganese is suspected to cause neurotoxicity (Tokar et al. 2015; Ijomone et al. 2019; Adhikari et al. 2019). WHO has not derived a formal guideline value in its recent guidelines for drinking water quality because the health-based value of 0.4 mg/l is well above the concentration normally present in drinking water (WHO 2017). Manganese finds its application in the metallurgical, pharmaceutical, and food industries. Manganese predominates in water in the +2 oxidation state in a pH range of 4–7 (Patil et al. 2016). An overview of the experimental parameters and optimized conditions from batch adsorption experiments for manganese is presented in Table 4.1.

4.1.1 Effect of pH

The percentage removal of manganese was low in acidic pH and increased with increase of pH (Idris 2015; Ganesan, Kamaraj, Sozhan, et al. 2013; Etale et al. 2016; Taffarel and Rubio 2010; Lee et al. 2009; Al-Wakeel et al. 2015). The low percentage removal at acidic pH was due to competition from protons and repulsion from protonated sites. In most of the cases, the increase of adsorption occurred up to a near neutral pH, i.e. 6 (Al-Wakeel et al. 2015), 7 (Idris 2015), and 8.5 (Lee et al. 2009), and afterwards the decline in percentage removal happened. The decline in percentage removal after the optimum pH was due to the formation of insoluble $MnOH^+$ or $Mn(OH)_2$, which have low affinity to bind to the adsorption sites (Idris 2015). In some cases, the adsorption increased with pH in the full pH range studied such as 2–12 (Ganesan, Kamaraj, Sozhan, et al. 2013), 3–9 (Etale et al. 2016), and 4–8 (Taffarel and Rubio 2010). The pH of the solution after adsorption of manganese with manganese oxide-coated zeolite decreased (Taffarel and Rubio 2010). This is attributed to the ion exchange phenomenon. The decline in pH after adsorption was higher, at operating pH higher than 6. This was due to the release of one proton with each manganese ion adsorbed; however, at pH higher than 6, $Mn(OH)^+$ is proposed to occur, which leads to the release of two H^+ ions for each manganese ion adsorbed. The manganese solution (30 mg/l) was reported to form a yellow color at initial pH 10.02, and the color changed with a decline in pH to 8.34. This phenomenon is attributed to the formation of hydrated MnO_2 (Khobragade and Pal 2016). The variation in optimum pH values for the batch adsorption process of manganese along with other elements is presented in Figure 4.1.

4.1.2 Effect of Coexisting Ions

The adsorption of manganese with manganese oxide-coated anthracite (Cerrato et al. 2011) and with manganese-coated sand (Lee et al. 2009) increased in the presence of sodium hypochlorite or free chlorine. This was attributed to continuous regeneration of the manganese-coated surface caused by oxidation. The result of manganese adsorption with manganese oxide-coated anthracite investigated by XPS (X-ray photoelectron spectroscopy) (Cerrato et al. 2011) showed that after adsorption, Mn is oxidized to Mn(IV); the oxidized form is insoluble and probably in the form of MnO_2. Once it is adsorbed and oxidized, the manganese cannot dissolve back into water due to its insolubility. However, the absence of free chlorine also leads to the absence of oxidation of manganese into insoluble Mn(IV). Hence, a high amount of Mn is recovered during the desorption process. In addition to this, the higher dissolved oxygen also promotes the uptake of the manganese (Cerrato et al. 2011).

TABLE 4.1
Summary of Parameters and Optimized Conditions for Batch Adsorption of Manganese

Adsorbent	Adsorbate	Surface Area (m^2/g), Pore Volume (cm^3/g), Pore Size (nm)	pHzpc	Experimental Conditions	Adsorption Capacity (mg/g)	Thermodynamic Parameters	Kinetic Model and Curve Fitting	Isotherm Model and Curve Fitting	Maximum Adsorption Conditions	References
Diethylenetriamine-functionalized MCM-41	Mn(II)	Surface area = 318 Pore volume = 0.63 Pore size = 6.18		pH = 1–9 Dose = 0.5 g/l Agitation speed = 250 rpm Concentration = 10–200 μg/cm^3 (isotherm study) Contact time = 1–40 min Temperature = 298–328 K	88.9	ΔH° = 37.37 kJ/mol ΔS° = 202.8 J/mol K ΔG° = negative	Pseudo-second-order model Linear	Langmuir model	pH = 7 Temperature = 328 K	Idris (2015)
Sodium hydroxide-modified natural zeolite	Mn(II)	Surface area = 49.8–66.7 Pore volume = 0.17–0.43 Pore volume = 15.3–34.6		pH = 6 Agitation speed = 200 rpm Concentration = 25–250 mg/l Contact time = 8 h Temperature = 25°C	28.9–66.1			Langmuir model		Ates and Akgül (2016)
Oxidized multiwalled carbon nanotube	Mn(II)			pH = 2–12 Dose = 0.67 g/l Agitation speed = 180 rpm Concentration = 0.25–2 mg/l Contact time = 14 h Temperature = 303–343 K	3.50	ΔH° = 3.143 kJ/mol ΔS° = 0.7174 J/mol K ΔG° = negative	Pseudo-second-order model Linear	Langmuir adsorption isotherm Linear	Contact time = 354 min (~6 h) Temperature = 343 K	Ganesan, Kamaraj, Sozhan, et al. (2013)
Iron-impregnated pumice	Mn(II)		6.9	pH = 7 Dose = 5–20 g/l Agitation speed = 150 rpm Concentration = 20–100 mg/l Contact time = 24 h Temperature = 25°C	7.14			Langmuir and Temkin Linear	dose = 15 g/l	Çifçi and Meriç (2017)

(Continued)

TABLE 4.1 (*Continued*)
Summary of Parameters and Optimized Conditions for Batch Adsorption of Manganese

Adsorbent	Adsorbate	Surface Area (m^2/g), Pore Volume (cm^3/g), Pore Size (nm)	pHzpc	Experimental Conditions	Adsorption Capacity (mg/g)	Thermodynamic Parameters	Kinetic Model and Curve Fitting	Isotherm Model and Curve Fitting	Maximum Adsorption Conditions	References
Sodium dodecyl sulfate (adsolubilization)	Mn(II)			pH = 4–8 Dose = 3–30 g/l Agitation speed = 150 rpm Concentration = 20–100 mg/l Contact time = up to 70 min Temperature = 30°C			Pseudo-second-order model	Freundlich model Linear	pH = 6–7 Dose = 20 g/l Contact time = 30 min	Khobragade and Pal (2016)
Maghemite nanoparticles	Mn(II)	Surface area = 40.8 Pore volume = 11.3		pH = 3–9 Dose = 1–10 mg/l Concentration = 0.0456–0.5406 mg/l Contact time = up to 60 min Temperature = 25°C	4.03				pH = 9 Dose = 3 mg/l Contact time = 30 min	Etale et al. (2016)
Manganese-coated sand	Mn(II)		6	pH = 5–10.5 Dose = 5 g/l Concentration = 2 mg/l Contact time = 24 h Temperature = 25°C	10.93–20.21 (in presence of $NaClO_4$) Column results				pH = 8.5	Lee et al. (2009)
Manganese oxide-coated zeolite	Mn(II)			pH = 4–8 Dose = 2.5 g/l Agitation speed = 50 rpm Concentration = 2 mg/l Contact time = up to 150 min Temperature = 25°C	0.259 meq Mn^{2+}/g		Pseudo-second-order model Linear	Langmuir and Freundlich model Linear	pH = 8	Taffarel and Rubio (2010)
Glycine-modified chitosan	Mn(II)			pH = 1–7 Dose = 1 g/l Agitation speed = 200 rpm Concentration = 0.005 M Contact time = up to 150 min Temperature = 25°C	71.4	$\Delta H°$ = 5.74 kJ/mol $\Delta S°$ = 78.49 J/mol K ΔG = negative	Pseudo-second-order Linear	Langmuir model Linear	pH = 6	Al-Wakeel et al. (2015)

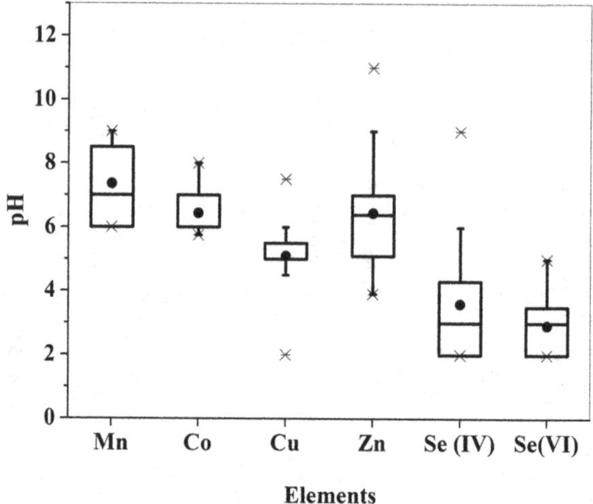

FIGURE 4.1 Box plot of optimum pH for adsorption vs essential metal graph of literature surveyed (literature surveyed is in Tables 4.1–4.5. In places where the pH range is specified for optimum pH conditions, both upper and lower limits of pH are taken into consideration).

4.1.3 Effect of Surface Modification and Temperature

The sodium hydroxide treatment of zeolite (Ates and Akgül 2016), iron impregnation of pumice (Çifçi and Meriç 2017), and sodium dodecyl modification of alumina lead to increased removal capacity (Khobragade and Pal 2016). The treatment of the zeolite with sodium hydroxide leads to increase of the uptake capacity only up to treatment with 1.5 M sodium hydroxide, and at 2 M, the sodium hydroxide uptake capacity declined.

The treatment of the adsorbent with a chemical reagent leads to a change in the surface and structural properties. The surface modification of silica leads to a decline in surface area after modification with diethylenetriamine. This was due to occupation of the surface with functional groups (Idris 2015). The acid treatment of multiwalled carbon nanotubes leads to the introduction of the functional groups (Ganesan, Kamaraj, Sozhan, et al. 2013). The treatment of the zeolite with sodium hydroxide leads to decationization, dealumination, and desilication (Ates and Akgül 2016). The increase of temperature also leads to the increase in the porosity and activation of the multiwalled carbon nanotube surface along with accessibility to more surface area due to the presence of carbon nanotubes as concentric cylinders (Ganesan, Kamaraj, Sozhan, et al. 2013).

4.1.4 Desorption

The desorption of Mn was achieved with hydrochloric acid (Al-Wakeel et al. 2015) and disodium ethylenediaminetetraacetate (Khobragade and Pal 2016). The desorption of Mn(II) from sodium dodecyl sulfate-modified alumina with 0.2 M disodium ethylenediaminetetraacetate showed better desorption efficiency (92%) than with hydrochloric acid (Khobragade and Pal 2016).

4.2 COBALT

Cobalt is essential in small amounts, but at higher concentrations, it causes low blood pressure, paralysis, diarrhea, and bone defects (Tokar et al. 2015; Fang et al. 2014). Cobalt finds its application in grinding wheels, painting on glass and porcelain, manufacturing of Vitamin B12, enamels and semiconductors (Fang et al. 2014), battery recharging, and galvanization (Anirudhan et al. 2016). An overview of the experimental parameters and optimized conditions from batch adsorption experiments for cobalt is presented in Table 4.2.

4.2.1 Effect of pH

The cobalt removal increased with increase in pH from 3 to 8 with a nanocellulose/nanobentonite composite (Anirudhan et al. 2016), 2 to 7 with polyaniline/polypyrrole copolymer nanofibers (Javadian 2014), and 2 to 6 with activated carbon (Kyzas et al. 2016).

The pHzpc was 4.2 for the nanocellulose/nanobentonite composite. At pH lower than pHzpc, the surface of the adsorbent was positively charged, and the extent of adsorption was lower due to less electrostatic attraction. In addition to this, the hydronium ions also compete with the cobalt (II) ions in the pH range of 3–4.5. Above pHzpc, the surface becomes negatively charged, which leads to increased electrostatic force of attraction between cobalt (II) ions and the adsorbent surface. The lower percentage removal at low pH with activated carbon in addition to electrostatic repulsion is attributed to the suppression of the complexation between the cobalt ions and the functional groups on the surface of the activated carbon (Kyzas et al. 2016).

The pH range studied was different in different studies. The authors have set different pH values to set for precipitation of cobalt such as 5.75 (Ramos et al. 2016), 7 (Lingamdinne et al. 2016), and 8 (Anirudhan et al. 2016). Fang et al. (2014) used the stability constants to determine the relative proportion of cobalt (II) ions in the solution. It depicts that the major forms of cobalt(II) species at pH less than 8.2 are Co^{2+} and $Co(OH)^{+}$ (Fang et al. 2014).

4.2.2 Effect of Coexisting Ions

The chloride and nitrate declined the adsorption of the cobalt on the multiwalled carbon nanotube/iron oxide composite (Wang et al. 2011). The formation of soluble complexes of cobalt with chloride and nitrate is held responsible for this. The radical radius order is chloride < nitrate < perchlorate. The smaller radical radius species took up more exchange sites and caused greater decline in the removal of the cobalt ions. Similarly, in a study of the effect of cations, i.e. Li, Na, and K on cobalt adsorption, the adsorption of cobalt was more affected by the element having a smaller hydration radius.

The adsorption of cobalt is also influenced by humic/fulvic acid (Wang et al. 2011). The adsorption of cobalt on multiwalled carbon nanotubes enhanced at pH less than 5.4, reduced between pH 5.4 and 7, and further increased with pH more than 7 in the presence of humic/fulvic acid.

TABLE 4.2
Summary of Parameters and Optimized Conditions for Batch Adsorption of Cobalt

Adsorbent	Adsorbate	Surface Area (m^2/g), Pore Volume (cm^3/g), Pore Size (nm)	pHzpc	Experimental Conditions	Adsorption Capacity (mg/g)	Thermodynamic Parameters	Kinetic Model and Curve Fitting	Isotherm Model and Curve Fitting	Maximum Adsorption Conditions	References
Lemon peel as biosorbent (batch and column study)	Co			pH = 6 Dose = 10 g/l Agitation speed = 200 rpm Concentration = 0–1000 mg/l Contact time = 10 h	22	ΔH° = −21.2 kJ/mol ΔS° = 54.61 J/mol K ΔG° = negative (wrongly calculated)	Pseudo-second-order model (linear)	Langmuir model Linear	Dose = 10 g/l Agitation speed = 200 rpm Contact time = 10 h	Bhatnagar et al. (2010)
Ligand-anchored functional mesoporous silica	Co sorption	S_{BET} = 449 m^2/g Pore volume = 0.53 Pore size = 5.6		pH = 2–12 Dose = 5 g/l Concentration = 2–75 mg/l (isotherm study) Contact time = 30 min	157				pH = 9.5 Adsorption + precipitation	Shahat et al. (2015)
Modified activated carbon cloth	Co	Surface area = 1870 Pore volume = 0.827		pH = 3–7 Agitation speed = 150 rpm Concentration = c.a. 10^{-5} M Contact time = 48 h	0.05258 mmol/g				Contact time = 24 h	Duman and Ayranci (2010)

(continued)

TABLE 4.2 (*Continued*)
Summary of Parameters and Optimized Conditions for Batch Adsorption of Cobalt

Adsorbent	Adsorbate	Surface Area (m^2/g), Pore Volume (cm^3/g), Pore Size (nm)	pHzpc	Experimental Conditions	Adsorption Capacity (mg/g)	Thermodynamic Parameters	Kinetic Model and Curve Fitting	Isotherm Model and Curve Fitting	Maximum Adsorption Conditions	References
GO-NH_2	Co	Surface area = 320		pH = 3–9 Dose = 0.3 g/l Concentration = 30 mg/l (pH study) Contact time = 180 min Temperature = 298–328 K	116.35	ΔH° = −10.77 kJ/mol ΔS° = 35.49 J/mol ΔG° = negative	Pseudo-second-order model Linear	Langmuir model	pH = 7 Contact time = 2 h (>9 0% removal in 5 min) Temperature = 298 K	Fang et al. (2014)
Activated carbon	Co	Surface area (PoP400) = 904.5 Surface area (PoP600) = 1041.43		pH = 2–6 Dose = 1 g/l (pH study) Concentration = 200 mg/l (pH study) Contact time = c.a. up to 24 h Temperature = 25°C –45°C	373 (for PoP400) and 405 (PoP400)	ΔH° = 26–4 kJ/mol (from initial concentration 10–700 mg/l for PoP400) and 52.94–15.92 kJ/mol another variant (PoP600) ΔS° = 15–128 J/mol K for PoP400 and 56–226 J/mol K ΔG° = negative	Pseudo-first-order model	Langmuir model	pH = 6 Contact time = 3 h Temperature = 65°C	Kyzas et al. (2016)
Tartaric acid-functionalized nanosilica	Co			pH = 1–7 Dose = 0.1–2 g/l Concentration = 10^{-3}–0.05 mol/l Contact time = up to 60 min	111.1	ΔH° = positive ΔS° = positive ΔG° = negative	Pseudo-second-order model Linear	Langmuir model Linear	pH = 6 Dose = 0.3 g/l Contact time = 20 min	Mahmoud, Yakout, et al. (2016)

(*continued*)

TABLE 4.2 (*Continued*)
Summary of Parameters and Optimized Conditions for Batch Adsorption of Cobalt

Adsorbent	Adsorbate	Surface Area (m^2/g), Pore Volume (cm^3/g), Pore Size (nm)	pHzpc	Experimental Conditions	Adsorption Capacity (mg/g)	Thermodynamic Parameters	Kinetic Model and Curve Fitting	Isotherm Model and Curve Fitting	Maximum Adsorption Conditions	References
Oxalic acid-functionalized nanosilica	Co			Do	58.82	Do	Do	Do	Do	Mahmoud, Yakout, et al. (2016)
Carboxylate-functionalized sugarcane bagasse	Co		2.92	pH = 2–5.75 Dose = 0.2 g/l Agitation speed = 130 rpm Concentration = 0.05–1.12 mmol/l (isotherm study) Contact time = 10 h Temperature = 25°C	0.561 mmol/g		Pseudo-second-order model nonLinear	Langmuir model nonLinear	pH = 5.75 Contact time = 3 h	Ramos et al. (2016)
Graphene oxide	Co	Surface area = 2.8 Pore volume = 0.01 Pore size = 14.34	3.8–3.9 (referenced)	pH = 2–8 Dose = up to 1 g/l Concentration = 15 mg/l (isotherm study) Contact time = 80 min Temperature = 293–323 K	21.28	ΔH° = 0.588 kJ/mol ΔS° = 5 J/mol K ΔG° = negative	Pseudo- second-order model Linear	Freundlich model Linear	pH = 5–6 (5.5) Dose = 1 g/l Contact time = 60 min	Lingamdinne et al. (2016)

(continued)

TABLE 4.2 (*Continued*)
Summary of Parameters and Optimized Conditions for Batch Adsorption of Cobalt

Adsorbent	Adsorbate	Surface Area (m^2/g), Pore Volume (cm^3/g), Pore Size (nm)	pHzpc	Experimental Conditions	Adsorption Capacity (mg/g)	Thermodynamic Parameters	Kinetic Model and Curve Fitting	Isotherm Model and Curve Fitting	Maximum Adsorption Conditions	References
Nanocellulose/ nanobentonite composite	Co		4.2	pH = 3–8 Dose = 0.05–5 g/l Agitation speed = 100–200 rpm Concentration = 50–200 mg/l (isotherm study) Contact time = 2 h Temperature = 20°C–50°C	350.8	$\Delta H°$ = 4.368 kJ/mol $\Delta S°$ = 525.34 J/K mol $\Delta G°$ = negative	Pseudo- second-order Nonlinear	Sips model Nonlinear	pH = 6 Dose = 2 g/l Agitation speed = 200 rpm Temperature = 50°C	Anirudhan et al. (2016)
Glycine-modified chitosan biosorption	Co			pH = 7 and 9 Dose = 5 g/l Agitation speed = 150 rpm Concentration = 250 ppm Contact time = 20–100 min Temperature = 20°C–45°C	59.1 (pH = 7)	$\Delta H°$ = 14.08 and 24.83 kJ/mol (pH 7 and 9, respectively) $\Delta S°$ = −29.6 and 60.6 J/ Kmol (respectively) $\Delta G°$ = negative	Pseudo-second-order model Linear	Langmuir model Linear		Negm et al. (2015)
Chloroacetic acid-modified chitosan biosorption	Co			Do	82.9 (pH = 7)	24.26–25.03	Pseudo-second-order model Linear	Langmuir model Linear		Negm et al. (2015)

(*continued*)

TABLE 4.2 (*Continued*)
Summary of Parameters and Optimized Conditions for Batch Adsorption of Cobalt

Adsorbent	Adsorbate	Surface Area (m^2/g), Pore Volume (cm^3/g), Pore Size (nm)	pHzpc	Experimental Conditions	Adsorption Capacity (mg/g)	Thermodynamic Parameters	Kinetic Model and Curve Fitting	Isotherm Model and Curve Fitting	Maximum Adsorption Conditions	References
Polyaniline/ polypyrrole copolymer nanofiber	Co	Surface area = 72		pH = 2–7 Dose = 0.2–3 g/l Agitation speed = 300 rpm Concentration = 100–500 mg/l (isotherm study) Contact time = 3–15 min Temperature = 15°C–35°C	185.18	ΔH° = 36.081 kJ/K mol ΔS° = 0.142 kJ K/ mol ΔG° = negative	Pseudo-second-order model linear		pH = 7 Dose = 2.2 g/l Contact time = 11 min Temperature = 35°C	Javadian (2014)
Fe-Mn binary oxide adsorbent	Co	Surface area = 316.76 Pore size = 5.34		pH = 4, 6 and 8 Dose = 1 g/l Agitation speed = 160 rpm Concentration = 5.89 mg/l in most studies and 0.589–58.9 mg/l in isotherm study Contact time = up to 24 h Temperature = 30°C	20.25			Freundlich model Linear	pH = 6 Contact time = 2 h	Jiang et al. (2015)

(*continued*)

TABLE 4.2 (*Continued*)
Summary of Parameters and Optimized Conditions for Batch Adsorption of Cobalt

Adsorbent	Adsorbate	Surface Area (m^2/g), Pore Volume (cm^3/g), Pore Size (nm)	pHzpc	Experimental Conditions	Adsorption Capacity (mg/g)	Thermodynamic Parameters	Kinetic Model and Curve Fitting	Isotherm Model and Curve Fitting	Maximum Adsorption Conditions	References
Palygorskite	Co	Surface area = 143		pH = 2–12 Dose = 5 g/l Concentration = 30–100 mg/l (isotherm study) Contact time = 24 h Temperature = 35°C	8.88		Pseudo-second-order model Linear	Langmuir isotherm Linear	pH = c.a. 4.5–8	He et al. (2011)
Manganese antimonate	Co	Surface area = 170.7 Pore volume = 0.14		pH = up to 7 Dose = 5 g/l Agitation speed = 300 rpm Concentration = 2–10 mg/l (isotherm study) Contact time = 40 h Temperature = 283–323 K	30.2	ΔH° = 39.1 kJ/K mol ΔS° = 216.8 J K/mol ΔG° = negative			pH = 3–7 Temperature = 323 K	Zhang, Wei, et al. (2016)

In addition to pH, the humic/fulvic acid effect on cobalt adsorption is also affected by the concentration of the humic/fulvic acid. The adsorption of cobalt increased up to 16.5 mg/l concentration of the humic acid/fulvic acid, but after 16.5 mg/l, the adsorption of cobalt behaved invariably.

4.2.3 Mechanism and Desorption

The mechanism of adsorption of cobalt by an adsorbent was suggested as ion exchange followed by complexation (Anirudhan et al. 2016; Ramos et al. 2016). The adsorbent surface contains a carboxyl group (Anirudhan et al. 2016). Anirudhan et al. (2016) used FT-IR as evidence for the complexation mechanism. The FT-IR depicted a peak at 940 cm^{-1}, which indicates the formation of Co(II)-O bond. However, Ramos et al. (2016) suggest splitting of band at 1685 cm^{-1}, which corresponds to the C=O group in the carboxyl group. The authors suggest that splitting of the band at 1685 cm^{-1} is attributed to the deprotonation of carboxylate groups during the adsorption process. This evidence leads to the conclusion that the hydronium ions are first exchanged by metal ions, and it is followed by complex formation with metal ions. The mechanism of adsorption was investigated with XPS analysis (Jiang et al. 2015) such as the adsorption of cobalt with Fe-Mn binary oxide. The acidic solutions were used as desorbing agents (Anirudhan et al. 2016; Ramos et al. 2016; Fang et al. 2014; Duman and Ayranci 2010; Javadian 2014).

4.3 COPPER

Copper occurs in +2 and +1 oxidation states (RSC 2020d). Copper in excess concentration causes gastrointestinal disturbances, damage to liver and kidneys, and anemia (Gloria et al. 2011). People affected with Wilson disease are more sensitive to copper contamination in water and require medical assistance (Gloria et al. 2011; Członkowska et al. 2018). Copper is released into the environment by industrial and nonindustrial sources (Awual et al. 2013). Copper finds its way into the environment from oil and coal combustion, brick and pipe manufacturing, urban wastewater, fertilizers, pasture manure, motor vehicles, and copper water pipes (Fuge 2013). An overview of the experimental parameters and optimized conditions from batch adsorption experiments for copper is presented in Table 4.3.

4.3.1 Effect of pH and Morphology

The effect of pH on a metal ion is elaborated as a competition of the metal ion with the hydronium ion and consumption of the metal ion as a metal hydroxide after a certain pH. The percentage removal of copper increases with increasing pH (Hao et al. 2010; Sheng et al. 2010; Wu et al. 2012; Hamdaoui 2017; Kannamba et al. 2010; Li et al. 2010). The optimum amount of percentage removal is found in the conditions of pH 5–6 (Kannamba et al. 2010; Li et al. 2010). The low adsorption capacity at lower pH is attributed to brittleness due to swelling of chitosan tripolyphosphate beads, in addition to repulsion between surface groups and surface charge of the adsorbent (Ngah and Fatinathan 2010). In some cases, the adsorption decreased (SenthilKumar et al. 2011; Sani et al. 2017), remained constant (Sheng et al. 2010), and increased (Wu et al. 2012) further on raising the pH after optimum pH. The decrease and constancy in percentage removal were attributed to precipitation by authors (SenthilKumar et al. 2011; Wu et al. 2012; Sheng et al. 2010). The pH range studied was different in different studies. The authors have recommended different pH values for precipitation of copper, i.e. 4–4.5 (Liu, Zhu, et al. 2013), 5.8 (Li et al. 2010), 6 (Hao et al. 2010; Ngah and Fatinathan 2010; Wu et al. 2012; Pizarro et al. 2015; Bakhtiari and Azizian 2015; Kannamba et al. 2010; Rajput et al. 2017), and 7.5 (Sheng et al. 2010; Hamdaoui 2017).

The pH after adsorption of copper ions on multiwalled carbon nanotubes was lower than the initial pH, which declines further on increase of percentage removal (Wu et al. 2012). The decline in pH is attributed to the release of hydronium ion from the surface of the multiwalled carbon nanotubes into the solution. On increase of pH, the degree of deprotonation of the multiwalled carbon nanotubes increased and more hydronium ions were released into the solution and more copper ions were adsorbed (Sheng et al. 2010). The platelet-shaped thiol-functionalized SBA-15 has faster initial adsorption than a fibrous and rod-like morphology (Lee et al. 2016). This is attributed to the short channeling pores. The low particle size particles have larger adsorption capacity due to high specific surface area (Hossain et al. 2012).

4.3.2 Mechanism

Adsorption of copper on amino-functionalized Fe_3O_4 is explained on the basis of protonation and deprotonation of amino-functionalized Fe_3O_4 particles (Hao et al. 2010; Lee et al. 2016). The conditions favored the formation of NH_2 to form NH_3^+; hence, a very low amount of NH_2 groups are available for adsorption of Cu^{2+}. In addition to this, repulsion occurred between Cu^{2+} and adsorbate NH_3^+ (Hao et al. 2010). Kannamba et al. (2010) used FT-IR and XPS data to explain the chemisorption and complexation mechanism of adsorption between copper and chemically modified chitosan. The peak present at 601 cm^{-1} in the FT-IR spectrum after adsorption was due to the formation of a N-Cu bond (N was an element of the adsorbent). Similarly, the binding energy peak of Cu2p was present in the XPS data of the adsorbent. The binding energy peak of C and O did not show any characteristic change after copper adsorption; however, N binding energy peaks for amine and amide shift by 0.52 and 0.43 eV, respectively. Similarly, peaks corresponding to the binding energy of sulfur shift by 0.5 eV. The change in shift of binding energy on nitrogen and sulfur was attributed to bond formation of copper with nitrogen and sulfur (Kannamba et al. 2010). The adsorption of copper on γ-aminopropyltriethoxysilane-modified chrysotile was attributed to ion exchange, complexation, and protonation processes (Liu, Zhu, et al. 2013). The ion exchange mechanism is also postulated on the basis of decrease of pH of solution after adsorption attributed to the increase in hydronium ions (Ngah and Fatinathan 2010). However, in addition to hydronium ions, calcium, magnesium, sodium, and potassium ions were also measured after adsorption (Ngah and Fatinathan 2010).

TABLE 4.3
Summary of Parameters and Optimized Conditions for Batch Adsorption of Copper

Adsorbent	Adsorbate	Surface area (m^2/g), Pore Volume (cm^3/g), Pore Size (nm)	pHzpc	Experimental Conditions	Adsorption Capacity (mg/g)	Thermodynamic Parameters	Kinetic Model and Curve Fitting	Isotherm Model and Curve Fitting	Maximum Adsorption Conditions	References
Magnetic calcium alginate hydrogel beads	Cu			pH = 2–6 Dose = 2–6 g/l Agitation speed = 150 rpm Concentration = 250–750 ppm Contact time = 6 h Temperature = 30°C	159.24		Pseudo-second-order model Linear		pH = 2 Dose = 2 g/l Concentration = 250 mg/l	Zhu et al. (2014)
Superparamagnetic maghemite nanoparticle	Cu	Surface area = 79.35 Pore volume = 0.18 Pore size = 9.11	6.3	pH = 2–6 Dose = 0.05–0.15 g/l Concentration = 5–20 ppm Contact time = 3 h Temperature = 25°C–45°C	34.0		Pseudo-second-order model Linear	Langmuir and Freundlich Nonlinear	pH = 5 Dose = 0.1 g/l Temperature = 25°C	Rajput et al. (2017)
Cedar bark	Cu			pH = 2–6 Dose = 1.5–6.5 g/l Agitation speed = 300 rpm Concentration = 50–300 mg/l Contact time = 480 min Temperature = 25°C	16.39			Langmuir and Harkins-Jura isotherm Linear	pH = 5 Dose = 5.5 g/l Concentration = 200 mg/l Contact time = 180 min	Hamdaoui (2017)

(continued)

TABLE 4.3 (*Continued*)
Summary of Parameters and Optimized Conditions for Batch Adsorption of Copper

Adsorbent	Adsorbate	Surface area (m^2/g), Pore Volume (cm^3/g), Pore Size (nm)	pHzpc	Experimental Conditions	Adsorption Capacity (mg/g)	Thermodynamic Parameters	Kinetic Model and Curve Fitting	Isotherm Model and Curve Fitting	Maximum Adsorption Conditions	References
ZnO/montmorillonite composite	Cu	Surface area = 267.1 Pore volume = 0.485 Pore size = 15.3 Å		pH = 1–6 Dose = 0.8–4 g/l Concentration = 50–150 mg/l Contact time = 15–105 min	54.06		Pseudo-second-order model Linear	Langmuir model Linear	pH = 4 Dose = 1.2 g/l Concentration = 50 mg/l Contact time = 90 min	Sani et al. (2017)
Hydrogel-supported nanosized hydrous manganese dioxide	Cu			pH = 1–5 Dose = 2 g/l Agitation speed = 160 rpm Concentration = 0.25–10 mmol/l (isotherm study) Contact time = 24 h Temperature = 25°C	0.857 mmol/g		Pseudo-second-order model Linear	Langmuir model Linear	pH = 5	Zhu and Li (2015)
Carbon nanotube/calcium alginate composite	Cu	Surface area = 76 Pore volume = 0.37		pH = 2–7 Dose = 0.5 g/l Concentration = 5–40 mg/l (isotherm study) Contact time = 2 h Temperature = 25°C	67.9			Both Langmuir and Freundlich model Linear	pH = 5	Li et al. (2010)

(continued)

TABLE 4.3 (*Continued*)
Summary of Parameters and Optimized Conditions for Batch Adsorption of Copper

Adsorbent	Adsorbate	Surface area (m^2/g), Pore Volume (cm^3/g), Pore Size (nm)	pHzpc	Experimental Conditions	Adsorption Capacity (mg/g)	Thermodynamic Parameters	Kinetic Model and Curve Fitting	Isotherm Model and Curve Fitting	Maximum Adsorption Conditions	References
Garden grass	Cu	Surface area =21.28 Pore volume =0.03 Pore size =37.23 Å		pH=1–8 Dose=0.5–20 g/l Agitation speed =120 rpm Concentration =10–100 mg/l Contact time=6 h Temperature =20°C–70°C	58.34	ΔH°=−0.0021 kJ/mol ΔS° =0.0034 J/mol K ΔG° =positive	Pseudo-first-order model Nonlinear	Langmuir model Nonlinear	pH=6 Dose=0.5 g/l Concentration =10 mg/l Temperature =20°C	Hossain et al. (2012)
TiO_2 (anatase)	Cu			pH=3–9 Dose=0.5 g/l Agitation speed =150 rpm Concentration =0.1–20 mg/l (isotherm study) Contact time =24 h Temperature =25°C		ΔH° =19.36 kJ/mol ΔS° =117 J/mol K ΔG° =negative	Pseudo-second-order and Elovich model	Freundlich model Nonlinear	pH=6	Özlem Kocabaş-Ataklı and Yürüm (2013)
Cashew nut shell	Cu	Surface area =395 Pore volume =0.4732 Pore size =5.89		pH=2–8 Dose=1–6 g/l Agitation speed =120 rpm Concentration =10–50 mg/l Contact time =5–60 min Temperature =30°C–60°C	20	ΔH° =−11.41 to −9.35 ΔS° =−20.3 to 22.5 J/mol K ΔG° =negative	Pseudo-second-order model	Langmuir and Freundlich model Linear	pH=5 Dose=3 g/l Concentration =10 mg/l Contact time =30 min Temperature=30°C	SenthilKumar et al. (2011)

(*continued*)

TABLE 4.3 (*Continued*)
Summary of Parameters and Optimized Conditions for Batch Adsorption of Copper

Adsorbent	Adsorbate	Surface area (m^2/g), Pore Volume (cm^3/g), Pore Size (nm)	pHzpc	Experimental Conditions	Adsorption Capacity (mg/g)	Thermodynamic Parameters	Kinetic Model and Curve Fitting	Isotherm Model and Curve Fitting	Maximum Adsorption Conditions	References
γ-alumina nanoparticle	Cu	Surface area =160 Pore size =20		pH = 3–8 Dose = up to 8 g/l Agitation speed = 150 rpm Concentration = 25–200 mg/l (isotherm study) Contact time = 4 h Temperature = 283–313 K	28.6	5.944 kJ/mol ΔG° = negative ΔS° = positive	Pseudo-second-order model Linear	Freundlich model Linear	pH = 5 Dose = 3 g/l Contact time = 4 h Temperature = 283 K	Fouladgar et al. (2015)
MOF-5	Cu	Surface area = 885.5 Pore volume = 0.434 Pore size = 1.04	7.2	pH = 3–5.2 Dose = 0.4 g/l Agitation speed = 150 rpm Concentration = up to 450 mg/l (isotherm study) Contact time = 3 h Temperature = 25°C	290		Pseudo-first-order Nonlinear	Langmuir-Freundlich model (L-F) Nonlinear	pH = 5.2	Bakhtiari and Azizian (2015)
Amino-Fe(III) functionalized mesoporous silica	Cu and tetracycline	Surface area = 296 Pore volume = 0.49		pH = 3–9 Dose = 1 g/l Agitation speed = 150 rpm Concentration = 0–0.25 mM (isotherm study) Contact time = 24 h Temperature = 25°C	112 mmol/kg			Freundlich model Nonlinear	pH = 5.5	Zhang, Liu, Wu, et al. (2015)

(continued)

TABLE 4.3 (*Continued*)
Summary of Parameters and Optimized Conditions for Batch Adsorption of Copper

Adsorbent	Adsorbate	Surface area (m^2/g), Pore Volume (cm^3/g), Pore Size (nm)	pHzpc	Experimental Conditions	Adsorption Capacity (mg/g)	Thermodynamic Parameters	Kinetic Model and Curve Fitting	Isotherm Model and Curve Fitting	Maximum Adsorption Conditions	References
3-aminopropyl-triethoxysilane-functionalized fly ash	Cu	Surface area =415.25 Pore volume =0.602 Pore size =43.57 Å		pH=2–8 Dose=10–60 g/l Agitation speed =150 rpm Concentration =10–45 mg/l (isotherm study) Contact time =24 h (isotherm study)	4.1			Freundlich model Linear	pH=5–5.8	Pizarro et al. (2015)
γ-aminopropyltriethoxysilane-modified chrysotile	Cu			pH=2–4 Dose=1 g/l Agitation speed =100 rpm Concentration =10–100 mmol/l Contact time =10 h Temperature =293–333 K	1.574 mmol/g	ΔH° =24.096 kJ/mol ΔS° =123 J/ mol K ΔG° =negative	Pseudo-second-order model Linear	Langmuir model	pH=4 Concentration =10 mmol/l (in terms of percentage removal) 100 mmol/l (in terms of adsorption capacity) Temperature =333 K	Liu, Zhu, et al. (2013)
Tetrakis(3-carboxysalicylidene) naphthalene–1,2,5,5–tetramine (TSNT) ligand-modified mesoporous silica	Cu	Surface area =510 Pore volume =0.63 Pore size =7.3		pH=2–9.5 Dose=0.5 g/l Agitation speed =110 rpm Concentration =0.5–60 mg/l (isotherm study) Contact time =up to 80 min Temperature =25°C	199.2			Only Langmuir model used to linear fit data	pH=5.2 Contact time =60 min	Awual et al. (2014)

(continued)

TABLE 4.3 (*Continued*)
Summary of Parameters and Optimized Conditions for Batch Adsorption of Copper

Adsorbent	Adsorbate	Surface area (m^2/g), Pore Volume (cm^3/g), Pore Size (nm)	pHzpc	Experimental Conditions	Adsorption Capacity (mg/g)	Thermodynamic Parameters	Kinetic Model and Curve Fitting	Isotherm Model and Curve Fitting	Maximum Adsorption Conditions	References
NH_2-functionalized SBA-15	Cu	Surface area =504 Pore volume =0.67 Pore size=6	7.3 isoelectric point	pH=2–6 Dose=1 g/l Agitation speed =150 rpm Concentration =20–60 mg/l (isotherm study) Contact time =up to 3 h Temperature =30°C	0.27 mmol/g			Langmuir model Linear	pH=5	Lee et al. (2016)
SH-functionalized SBA-15	Cu	Surface area =504 Pore volume =0.67 Pore size =6.4	2.8 isoelectric point	Do	0.127 mmol/g			Langmuir model Linear	pH=6	Lee et al. (2016)
Chitosan tripolyphosphate beads	Cu	Surface area =0.2667 Pore size =12.67	4	pH=1–6 Dose=0.5–20 g/l Agitation speed =400 rpm Concentration =20–300 mg/l (equilibrium study) Contact time =10–180 min Temperature =300–343 K	26.06	ΔH° =32.64 kJ/mol ΔS° =118.44 J/K mol ΔG° =negative	Pseudo-second-order and intraparticle diffusion model	Langmuir, Freundlich and Dubinin-Radushkevich model Nonlinear	pH=4.5 Dose=4 g/l Contact time =100 min Temperature =343 K	Ngah and Fatinathan (2010)

(*continued*)

TABLE 4.3 (*Continued*)
Summary of Parameters and Optimized Conditions for Batch Adsorption of Copper

Adsorbent	Adsorbate	Surface area (m^2/g), Pore Volume (cm^3/g), Pore Size (nm)	pHzpc	Experimental Conditions	Adsorption Capacity (mg/g)	Thermodynamic Parameters	Kinetic Model and Curve Fitting	Isotherm Model and Curve Fitting	Maximum Adsorption Conditions	References
Graphene oxide	Cu			pH = 1–7 Dose = 1–3 g/l concentration = 25–250 mg/l (equilibrium study) Contact time = up to 150 min	117.5			Freundlich model Linear	pH = 5.3 Dose = 1 g/l Contact time = 120 min	Wu et al. (2012)
Multiwalled carbon nanotube	Cu	Surface area = 197 Pore size = 3.6	5.3	pH = 3–10 Dose = 1 g/l Concentration = 1–20 mg/l (isotherm study) Contact time = 2 days Temperature = 293–333 K	7.776 (Langmuir capacity at 20 mg/l concentration)	ΔH° = 6.42–61.38 kJ/mol ΔS° = 70–281 J/mol K ΔG° = negative		Freundlich model Nonlinear	pH = 7.5	Sheng et al. (2010)
Amino-functionalized Fe_3O_4	Cu		5.8	pH = 2–6 Dose = 0.02-1 g/l Concentration = 0.1–5 mg/l (isotherm study) Contact time = up to 80 min Temperature = 300–333 K	25.77–26.58	27.18 kJ/mol ΔS° = 126.1 J/K mol ΔG° = negative	Pseudo-second-order model Linear	Langmuir model Linear	pH = 6 Dose = 0.1 g/l Contact time = 5 min Temperature = 333 K	Hao et al. (2010)

4.3.3 Desorption

The acidic solutions (Awual et al. 2013; Hao et al. 2010; Ngah and Fatinathan 2010; Hossain et al. 2012) and EDTA (Kannamba et al. 2010; Ngah and Fatinathan 2010; Cao et al. 2019) were used for desorption of copper. The acidic solutions and EDTA differ in desorption mechanism (Kannamba et al. 2010). EDTA serves as a better desorbing agent as compared to hydrochloric acid for desorption of copper from magnetic calcium alginate/maghemite hydrogel beads. The reason is attributed to EDTA acting as a complex-forming agent, whereas hydrochloric acid acting as a cation exchange agent (Zhu et al. 2014). The molar strength of regenerating solutions also affects the desorption capacity (Kannamba et al. 2010). The high desorption efficiency at lower pH was due to strong competitive adsorption, which is attributed to high hydronium concentration (Awual et al. 2013). The adsorption capacity was not significantly affected in some cases (Awual et al. 2013; Hao et al. 2010; Zhu et al. 2014; Sani et al. 2017), and in some cases, it dropped after regeneration (Ngah and Fatinathan 2010; Hossain et al. 2012).

4.4 ZINC

Zinc is a micronutrient and found in different effluents (Wang, Yuan, et al. 2013). The element causes muscular stiffness, irritation, and nausea above acceptable concentrations (Meitei and Prasad 2014). Zinc finds its way into the environment from pharmaceutical wastewater, paints, insecticides, and cosmetics (Afroze et al. 2016). An overview of the experimental parameters and optimized conditions from batch adsorption experiments for zinc is presented in Table 4.4.

4.4.1 Effect of pH

The adsorption of zinc increases with pH (Zhang, Li, et al. 2010; Lu and Chiu 2006; Afroze et al. 2016; Feng et al. 2010; Bogusz et al. 2015). The dependence of metal adsorption on pH is based on two things, i.e. chemical behavior of the metal and chemical behavior of the functional groups at the surface of the adsorbent (Afroze et al. 2016). Zinc undergoes hydrolysis at higher pH, i.e. greater than 6 (Zhang, Li, et al. 2010; Adeli et al. 2017) or predominant form of zinc as $Zn(OH)^+$ above pH 5.1 (Afroze et al. 2016). So, Zn(II) undergoes less adsorption at higher pH. At higher pH, conversion of Zn(II) to $Zn(OH)^+$ leads to the decrease in charge of the metallic ion. The decrease in charge led to less adsorption of Zn(II). However, the total removal of zinc increased on raising the pH up to 8 (Zhang, Li, et al. 2010). The another reason for increase of percentage removal is also attributed to the adsorbent (Lu and Chiu 2006). The adsorbent surface becomes more negative with increase of pH, leading to increase of electrostatic attraction.

The precipitation pH reported for zinc was pH 8 (Feng et al. 2010). The predominant species of zinc between 8 and 12 were $Zn(OH)^+$, $Zn(OH)^{\circ}_2$, and $Zn(OH)^-_3$. The zinc removal between pH 8 and 12 was attributed to the precipitation of $Zn(OH)_2$ and adsorption of $Zn(OH)^+$ and $Zn(OH)^-_3$ (Lu and Chiu 2006; Sen and Gomez 2011). However, at pH 12, predominant species were OH^-, $Zn(OH)^{3-}$, and $Zn(OH)_4^{2-}$. The competition between them and decrease of electrostatic attraction were responsible for decline in removal at pH 12 (Lu and Chiu 2006).

TABLE 4.4
Summary of Parameters and Optimized Conditions for Batch Adsorption of Zinc

Adsorbent	Adsorbate	Surface Area (m^2/g), Pore Volume (cm^3/g), Pore Size (nm)	pHzpc	Experimental Conditions	Adsorption Capacity (mg/g)	Thermodynamic Parameters	Kinetic Model and Curve Fitting	Isotherm Model and Curve Fitting	Maximum Adsorption Conditions	References
Eucalyptus sheathiana bark	Zn	Surface area =6.55 Pore volume =0.003	5	pH=2–5.5 Dose =0.2–0.6 g/l Agitation speed =120 rpm Concentration =20–70 mg/l (isotherm) Contact time =100 min Temperature =30°C, 45°C and 60°C	128.21	ΔH° =−32.43 kJ/mol ΔS° =−90 J/mol K ΔG° =negative	Pseudo-second-order model linear	Freundlich model Linear	pH=5.1 Dose =0.2–0.6 g/l Contact time =100 min Temperature =30°C	Afroze et al. (2016)
Eucalyptus sheathiana bark modified with NaOH	Zn	Surface area =20.33 Pore volume =0.011		Do	250	ΔH° =−6.19 kJ/mol ΔS° =−20 J/mol K ΔG° =negative	Pseudo-second-order model Linear	Freundlich model Linear	Do	Afroze et al. (2016)
Oxidized CNT sheets	Zn		3.9	pH=7 Dose=2 g/l Concentration =100–1200 mg/l Contact time =up to 72 h Temperature =25°C	58		Pseudo-second-order model at low concentration and pseudo first order model at high initial concentration linear	both Langmuir and Freundlich model Linear	Contact time =48 h	Tofighy and Mohammadi (2011)

(continued)

TABLE 4.4 (*Continued*)
Summary of Parameters and Optimized Conditions for Batch Adsorption of Zinc

Adsorbent	Adsorbate	Surface Area (m^2/g), Pore Volume (cm^3/g), Pore Size (nm)	pHzpc	Experimental Conditions	Adsorption Capacity (mg/g)	Thermodynamic Parameters	Kinetic Model and Curve Fitting	Isotherm Model and Curve Fitting	Maximum Adsorption Conditions	References
Bentonite	Zn			pH = 3.81–7.69 Dose = 0.25–0.75 g/l Agitation speed = 80 rpm Concentration = 10–90 mg/l (isotherm study) Contact time = up to 120 min Temperature = 30°C–65°C	68.49	$K_L = q_e/Ce$ ΔH° = −12.31 kJ/mol ΔH° = negative ΔS° = −0.03285 J/mol K	Overall pseudo-second-order model with linear curve fitting and two-step process	Langmuir model Linear	pH = 6.76 Dose = 0.25 g/l (in terms of adsorption capacity) Agitation speed = 80 rpm Contact time = 80 min Temperature = 30°C	Sen and Gomez (2011)
Industrial biochar from wheat straw	Zn	Surface area = 26.3		pH = 1–7.5 Dose = 4 g/l Agitation speed = 120 rpm Concentration = 5–500 mg/l (isotherm study) Contact time = up to 26 h Temperature = 22°C	40.18		Pseudo-second-order model Linear	Langmuir model	pH = 7 Contact time = 24 h	Bogusz et al. (2015)
Lab-prepared biochar from Sida hermaphrodita	Zn	Surface area = 27.1		Do	45.12		Pseudo-second-order model Linear	Langmuir model Linear	Do	Bogusz et al. (2015)

(*continued*)

TABLE 4.4 (*Continued*)
Summary of Parameters and Optimized Conditions for Batch Adsorption of Zinc

Adsorbent	Adsorbate	Surface Area (m^2/g), Pore Volume (cm^3/g), Pore Size (nm)	pHzpc	Experimental Conditions	Adsorption Capacity (mg/g)	Thermodynamic Parameters	Kinetic Model and Curve Fitting	Isotherm Model and Curve Fitting	Maximum Adsorption Conditions	References
Graphene oxide	Zn		Zeta potential measured	pH = 2–10 Dose = 0.066–0.33 g/l Agitation speed = 170 rpm Concentration = 10–100 mg/l (isotherm) Contact time = up to 480 min Temperature = 20°C–45°C	245.7	ΔH° = −2.171 kJ/mol ΔS° = 134–137 J/K mol ΔG° = negative	Pseudo-second-order model Linear	Langmuir model Linear	pH = 7 Dose = 0.066 g/l Contact time = 120 min Temperature = 20°C	Wang, Yuan, et al. (2013)
Nanozinc oxide	Zn			pH = 4–8 Dose = 0.025 g/l Agitation speed = 200 rpm Concentration = 100–600 mg/l (isotherm) Contact time = up to 2 h Temperature = 30°C–70°C	357	ΔH° = −22.23 kJ/mol ΔS° = −70.64 J/K mol ΔG° = negative	Pseudo-second-order model Linear	Langmuir model Linear	pH = 5.5 Temperature = 30°C	Sheela et al. (2012)

(continued)

TABLE 4.4 (*Continued*)
Summary of Parameters and Optimized Conditions for Batch Adsorption of Zinc

Adsorbent	Adsorbate	Surface Area (m^2/g), Pore Volume (cm^3/g), Pore Size (nm)	pHzpc	Experimental Conditions	Adsorption Capacity (mg/g)	Thermodynamic Parameters	Kinetic Model and Curve Fitting	Isotherm Model and Curve Fitting	Maximum Adsorption Conditions	References
PVA/EDTA chelating agent	Zn			pH = 1–8 Dose = 0.05 – 0.8 g/l Agitation speed = 100 rpm Concentration = 10–100 mg/l (isotherm) Contact time = 5–90 min Temperature = 25°C–35°C	39.92 (calculated from maximum conditions)	ΔH° = 35.89 kJ/mol ΔS° = 188.35 J/K mol ΔG° = negative	Pseudo-second-order model Linear	No clear distinction suggested by the author Based on R^2 it was Dubinin-Radushkevich	pH = 6 Dose = 0.6 g/l Contact time = 25 min	Zhang, Li, et al. (2010)
Activated carbon from watermelon shell	Zn	Surface area = 710 Pore volume = 0.263	3.05	pH = 4.5 Dose = 10 g/l Concentration = 10–500 mg/l (isotherm) Contact time = 2 weeks Temperature = room temperature	11.31			Langmuir model	Concentration = 50 mg/l	Moreno-Barbosa et al. (2013)
Activated carbon from walnut shell (GACN)	Zn	Surface area = 789 Pore volume = 0.304	4.5	Do	6.0			Langmuir model	Concentration = 10 mg/l	Moreno-Barbosa et al. (2013)

(continued)

TABLE 4.4 (*Continued*)
Summary of Parameters and Optimized Conditions for Batch Adsorption of Zinc

Adsorbent	Adsorbate	Surface Area (m^2/g), Pore Volume (cm^3/g), Pore Size (nm)	pHzpc	Experimental Conditions	Adsorption Capacity (mg/g)	Thermodynamic Parameters	Kinetic Model and Curve Fitting	Isotherm Model and Curve Fitting	Maximum Adsorption Conditions	References
Sulfurized activated carbon	Zn	Surface area =500 (calculation method is different)	4.3	pH=2–10 Dose=up to 8 g/l Agitation speed =200 rpm Concentration =50–250 mg/l Contact time =up to 300 min Temperature= 30°C	147		Pseudo-second-order model Linear	Langmuir model Linear	pH=6.5 Dose =4 and 6 g/l for real waste water Concentration =250 mg/l (on the basis of adsorption capacity) Contact time =240 min	Anoop Krishnan et al. (2016)
Biochar from pine (softwood)	Zn	Surface area =219.35 Pore volume =0.125		pH=2–6 Dose=25 g/l Agitation speed =120 rpm Concentration =0–3.5 mM (isotherm study) Contact time=24 h Temperature=25°C Salinity=up to 0.3 M	1			Langmuir model Nonlinear	pH=4 (on the basis of adsorption capacity) Salinity =0.01 M (on the basis of adsorption capacity)	Jiang, Huang, et al. (2016)
Biochar from jarrah (hardwood)	Zn	Surface area =309.29 Pore volume =0.146		Do	2.31			Freundlich model Nonlinear	Do	Jiang, Huang, et al. (2016)

(*continued*)

TABLE 4.4 (*Continued*)
Summary of Parameters and Optimized Conditions for Batch Adsorption of Zinc

Adsorbent	Adsorbate	Surface Area (m^2/g), Pore Volume (cm^3/g), Pore Size (nm)	pHzpc	Experimental Conditions	Adsorption Capacity (mg/g)	Thermodynamic Parameters	Kinetic Model and Curve Fitting	Isotherm Model and Curve Fitting	Maximum Adsorption Conditions	References
Fungal dead mass composite with bentonite	Zn			pH = 2–8 Dose = 0.05–0.3 g/l Agitation speed = 120 rpm Concentration = 50–225 mg/l Contact time = up to 120 min Temperature = 30°C–58°C Salt and surfactant do not affect the adsorption process	78.5	ΔH° = 44.097 kJ/mol ΔG° = negative ΔS° = 173.84 J/K mol	Pseudo-second-order model Linear	Langmuir model Linear	pH = 5 Dose = 0.05 g/l Concentration = 200 mg/l Contact time = 30 min Temperature = 51°C	Rashid et al. (2016)
Metakaolin-based geopolymer (MKG) (batch and column study)	Zn	Surface area = 39.24		pH = 2–8 Dose = 0.4–2 g/l Agitation speed = 200 rpm Concentration = 25–600 mg/l Contact time = 5–90 min Temperature = 10°C–50°C	74.53		Pseudo-second-order	Langmuir model	pH = 6.39 Dose = 2 g/l Temperature = 25°C Contact time = 40 min	Kara et al. (2017)

(*continued*)

TABLE 4.4 (*Continued*)
Summary of Parameters and Optimized Conditions for Batch Adsorption of Zinc

Adsorbent	Adsorbate	Surface Area (m^2/g), Pore Volume (cm^3/g), Pore Size (nm)	pHzpc	Experimental Conditions	Adsorption Capacity (mg/g)	Thermodynamic Parameters	Kinetic Model and Curve Fitting	Isotherm Model and Curve Fitting	Maximum Adsorption Conditions	References
Biogenic elemental selenium nanoparticles	Zn			pH = 3–7 Dose = 5 ml of 0.22 g/l solution Concentration = 5.8–215 mg/l Contact time = 1–960 min	62.1			Langmuir model	pH = 3.9 Contact time = 240 min (equilibrium time, maximum adsorption at 120 min) concentration = 5.8 mg/l	Jain et al. (2015)
Magnetic hydroxyapatite	Zn	Surface area = 142.5		pH = 4–10 Dose = 0.05–5 g/l Concentration = 10^{-4}–10^{-2} mol/l (isotherm study) Contact time = 15 min–2 days Temperature = 25°C	2.15 mmol/g (140.63 mg/g)		Pseudo-second-order model Linear	Langmuir model Linear	pH = 8 Dose = 0.1 g/l Contact time = 24 h	Feng et al. (2010)
Graphene oxide	Zn			pH = 4–8 Dose = 0.1 g/l Concentration = 1 mg/l (pH study) Contact time = up to 120 min Temperature = 25°C	345		Pseudo-second-model Nonlinear	Langmuir model Nonlinear	pH = 5 Contact time = 30 min	Sitko et al. (2013)

(*continued*)

TABLE 4.4 (*Continued*)
Summary of Parameters and Optimized Conditions for Batch Adsorption of Zinc

Adsorbent	Adsorbate	Surface Area (m^2/g), Pore Volume (cm^3/g), Pore Size (nm)	pHzpc	Experimental Conditions	Adsorption Capacity (mg/g)	Thermodynamic Parameters	Kinetic Model and Curve Fitting	Isotherm Model and Curve Fitting	Maximum Adsorption Conditions	References
Sodium dodecyl-coated magnetite nanoparticles	Zn			pH = 2–6 Dose = up to 1 g/l Concentration = 5–50 mg/l (isotherm study) Contact time = 5 min in most studies equilibrium achieved in 1 min Temperature = 25°C SDS amount = up to 60 mg/20 ml of SDS Salt = up to 10%	59.2			Langmuir model Linear	pH = 6 Dose = 1 g/l SDS amount = 37.5 mg/20 ml Salt = 0%	Adeli et al. (2017)
Single-walled carbon nanotube (CNTs)	Zn	Surface area = 423 Pore volume = 0.43		pH = 1–12 Dose = 0.5 g/l Agitation speed = 180 rpm Concentration = 10–80 mg/l (isotherm study) Contact time = 0–360 min Temperature = 25°C	43.66			Langmuir model	pH 8 at 10 mg/l initial concentration pH 7 at 80 mg/l initial concentration Contact time = 60 min	Lu and Chiu (2006)

(continued)

TABLE 4.4 (*Continued*)
Summary of Parameters and Optimized Conditions for Batch Adsorption of Zinc

Adsorbent	Adsorbate	Surface Area (m^2/g), Pore Volume (cm^3/g), Pore Size (nm)	pHzpc	Experimental Conditions	Adsorption Capacity (mg/g)	Thermodynamic Parameters	Kinetic Model and Curve Fitting	Isotherm Model and Curve Fitting	Maximum Adsorption Conditions	References
Multiwalled carbon nanotube	Zn	Surface area =297 Pore volume =0.38		Do	32.8			Langmuir model	pH 9–11 at 10 mg/l initial concentration pH 7 at 80 mg/l initial concentration Contact time=60 min	Lu and Chiu (2006)
Cross-linked magnetic chitosan-phenylthiourea	Zn	Surface area =64.5		pH=1–5 Dose=0.1 g/l Agitation speed =150 rpm Concentration =10–400 mg/l (isotherm study) Contact time =up to 120 min Temperature =293–313 K	52		Pseudo-second-order model Linear	Langmuir model Linear	pH=5 Contact time =100 min Temperature =293 K	Monier and Abdel-Latif (2012)
Zinc oxide	Zn			pH=4–8 Dose=0.5 mg/l Agitation speed =200 rpm Concentration =100–600 mg/l (isotherm study) Contact time =up to 2 h Temperature =30°C–70°C	357	ΔH°=−22.23 kJ/mol ΔS°= negative ΔG°=negative	Pseudo-second-order model Linear	Langmuir model Linear	pH=8 Temperature =30°C	Sheela et al. (2012)

The deprotonation of the functional groups on the surface of the adsorbent also helped in increased adsorption of zinc with an increase of pH (Bogusz et al. 2015; Monier and Abdel-Latif 2012). In case of biochar, the increase in pH also increased the removal of zinc from aqueous solution (Bogusz et al. 2015). The rise in pH causes the carboxyl group on the surface of the adsorbent to be deprotonated along with binding of metal ions. In addition to this, there was increase in pH of the solution from 7 to 7.5 on application of the biochar adsorbent. This was attributed to ash content (present within biochar) release into the solution.

In the case of adsorption of zinc with magadiite, the difference in adsorption at pH 6.5 and 6.9 was also attributed to competition from hydronium ions (Ide et al. 2011). The titration curve suggests that the hydronium ion replaced the sodium ion of the adsorbent below pH 6.5. Hence, at pH 6.2, the zinc faced the competition from the hydronium ion, which led to its decreased adsorption than at pH 6.9. In addition, the basal spacing also played a major role. The larger amount of adsorption of zinc with magadiite at pH 6.9 for mixed electrolyte solution than at pH 6.2 for aqueous sodium chloride solution was attributed to larger basal spacing (1.64 nm) in the mixed electrolyte solution than in the aqueous sodium chloride (1.59 nm) solution. This led to better intercalation of the zinc ion in the mixed electrolyte.

4.4.2 Effect of Coexisting Ions

The effect of salt addition on sodium dodecyl sulfate-coated iron oxide caused the reduction in removal of zinc (Adeli et al. 2017). The competitive adsorption between sodium and metal ion via the ion exchange mechanism is responsible for decrease in adsorption (Adeli et al. 2017). In addition, salt reduced the adsorption capacity in other adsorbents (Afroze et al. 2016). The charge, weight, atomic radius, and molecular weight of the salt also affect the adsorption capacity of zinc on the adsorbent. The divalent and trivalent species of the salt decreased the removal efficiency more as compared to monovalent species. The effect of ionic strength is attributed to the difference in ionic osmotic pressure, which decreased on increase of ionic strength of the solution (Afroze et al. 2016).

The decrease in adsorption of zinc by biochar occurred in the presence of chloride; however, the effect of nitrate is less influential than that of chloride. The chloride was predicted to form complexes with aliphatic carbonaceous groups present on the surface of the adsorbent. The complexes formed compete with the adsorption sites on the surface of adsorbent, whereas chloride and nitrate ions did not compete. The two different biochar sources showed different interfering capacities. The high content of oxygen groups on the surface of biochar was predicted to be responsible for the higher adsorption capacity, which led to the closely firmed resistance against the interfering ions. This makes the adsorbent less prone to attack by interfering ions as compared to biochar with low oxygen content (Bogusz et al. 2015).

4.4.3 Mechanism

The XPS analysis helped in understanding the mechanism of adsorption of zinc on graphene oxide (Sitko et al. 2013). The analysis of C1s and O1s spectra showed differences before and after adsorption. The difference between the oxygen peaks suggests involvement of the oxygen-containing functional group taking part in adsorption.

The 29Si magic angle spinning NMR spectra of the adsorbent were used to estimate the reason for better adsorption of zinc as compared to cadmium and potassium (Ide et al. 2011). The Q3 and Q4 peaks for cadmium- and potassium-adsorbed samples were lower and comparable for zinc in comparison to the original adsorbent. Hence, the results suggest that intercalation of the cations other than the zinc ion led to significant alteration in the layered structure of magadiite along with suppression of metal adsorption.

4.4.4 Desorption

The hydrochloric acid can be applied in case of desorption of zinc from PVA/EDTA resin (Zhang, Li, et al. 2010), sulfurized activated carbon (Anoop Krishnan et al. 2016), and polymeric resin (Zheng, Li, et al. 2017). In spite of high desorption of zinc from $MnFe_2O_4$ and $CoFe_2O_4$ nanoparticles by HCl, HNO_3 was selected as HCl leads to higher degradation of the adsorbent (Asadi et al. 2020). In the case of zinc desorption from cross-linked magnetic chitosan-phenylthiourea, EDTA is more effective than hydrochloric acid (Monier and Abdel-Latif 2012). The reason is the desorption mechanism; the hydrochloric acid desorbs the material by ion exchange, whereas the EDTA forms a complex with the metal.

4.5 SELENIUM

Selenium is an essential element. Selenium predominates as selenate and selenite (Fordyce 2007). Its deficiency and excess lead to serious health issues like Keshan disease and selenosis, respectively (Tokar et al. 2015). The concentration of less than 40 μg/d leads to its deficiency but the concentration of more than 400 μg/d leads to toxicity (Fordyce 2007). Selenium finds its source into the environment via geological sources, mining, and municipal waste (Tokar et al. 2015; Jyothi et al. 2020). In addition, selenium finds its application in photocopiers, antifungals, antioxidants, and pigments and as a catalyst (Fordyce 2007). An overview of the experimental parameters and optimized conditions from batch adsorption experiments for selenium is presented in Table 4.5.

4.5.1 Effect of pH

The adsorption of selenite decreases with increase of pH (Larimi et al. 2013; Lu, Fu, et al. 2017; Lu, Yu, et al. 2017; Svecova et al. 2011; Li, Wang, O'Connell, et al. 2015; Gezer et al. 2011). The adsorption of selenate also decreases with increase of pH (Lu, Yu, et al. 2017; Svecova et al. 2011; Li, Wang, O'Connell, et al. 2015; Fu et al. 2014; Qin et al. 2017). The reduction in the removal capacity of selenium with increase of pH is assigned to deprotonation of the surface and decrease of electrostatic attraction (Svecova et al. 2011; Li, Wang, O'Connell, et al. 2015; Sheng, Linghu, et al. 2016; Sun et al. 2015). The increase in competition from OH^- is also held responsible for low percentage removal at higher pH (Tian et al. 2017; Lu, Fu, et al. 2017).

The maximum amount of percentage removal for selenite and selenate with polyamine-modified magnetic graphene oxide was found in the acidic region (Lu, Yu, et al. 2017). The adsorption of selenite with $MgAl_2O_4$ increased up to optimum pH and declined on further increasing the pH (Tian et al. 2017). The decrease in percentage removal for selenite was nonlinear or nongradual. The decline in percentage removal for selenite with titanate nanotubes on increase in pH from 3 to 7 was 5%; however, after increase of pH from 7 to 9, the adsorption declined from c.a. 90% to 35% (Sheng, Linghu, et al. 2016). However, selenate showed a gradual decrease in percentage removal on increase of pH from 4 to 9.

In some cases, the removal efficiency remains unaltered in a wide range of pH; three case studies in respect to the statement are discussed here. The removal efficiency of selenite with nZVI on increasing pH (2–12) did not vary (with elimination of the effect of dissolved oxygen by purging the solution with N_2 for half an hour before batch experiments) (Xia et al. 2017), but the percentage removal declined gradually in the case of hematite under similar experimental conditions. The percentage removal of selenite on "magnetic nanoparticle–graphene oxide composite" remains unaffected with change of pH from 2 to c.a. 10 (Fu et al. 2014), but on increasing the pH further up to 11, it decreased sharply. Similarly, the percentage removal of selenite on iron oxide and silicon oxide was not affected up to pH 8, but on further increasing the pH, it declined (Sheha and El-Shazly 2010).

The pH_{zpc} in the case of removal of selenium acts as an important factor in electrostatic interaction (Lu, Fu, et al. 2017). At pH below the pHzpc, the surface of the adsorbent is positive and hence attracts the negatively charged oxyanions (Chen and An 2012), and at pH above pHzpc, it contains negative charge and hence repels the selenite ions.

The pK_a value of the adsorbate and adsorbent plays a major role in the adsorption capacity of selenium species (Larimi et al. 2013; Lu, Yu, et al. 2017). The pKa value for hydrolysis of H_2SeO_3 was 2.6. The decrease in percentage removal below pH 2.4 was attributed to the formation of uncharged H_2SeO_3 as compared to SeO_3^{2-} (Larimi et al. 2013).

The pKa of polyamine-modified magnetic graphene oxide is 9.2, which leads to the positive charge on the surface of the adsorbent below pH 9.2 and negative charge above pH 9.2. At pH above 9.2, there was hardly any electrostatic attraction between the adsorbate and the adsorbent, which leads to a decrease in adsorption capacity (Lu, Yu, et al. 2017).

The adsorption of selenate with nZVI (Das et al. 2017) and carbon nanosphere (Li, Wang, O'Connell, et al. 2015) and selenite with Fe_3O_4-precipitated mesoporous magnetic carbon microspheres (Lu, Fu, et al. 2017) occurred with increase in pH. The increase of pH of the solution after adsorption of selenate with nZVI was attributed to the formation of the corrosion products, i.e. magnetite, lepidocrocite, goethite, and wustite. The magnetite and lepidocrocite were chief products (Das et al. 2017). However, a major change in pH is not universal with adsorption of selenite and selenate, e.g. change of pH in the case of adsorption of selenite and selenate on titanium dioxide was negligible (Svecova et al. 2011).

The adsorption of selenite on titanium dioxide at pH 2 was four times that at pH 5; however, the XPS analysis suggested it to be in the order of two (Svecova et al. 2011).

The reason is attributed to a number of factors such as high surface sensitivity of the XPS, reorganization of the surface complex at the surface of the adsorbent occurred by drying under vacuum, and leaching of the material during sample preparation for XPS.

The increase of experimental pH in the removal of selenate with Al-substituted ferrihydrite affects the formation of an inner/outer sphere complex (Johnston and Chrysochoou 2016). The inner sphere complex formation decreases for selenium adsorption on ferrihydrite and 24 mol% Al ferrihydrite, with increase of pH (Johnston and Chrysochoou 2016). The increased proportion of Al in Al-ferrihydrite suppresses the formation of the inner sphere complex. This is attributed to the weakening of interaction between ferrihydrite and oxyanion (Johnston and Chrysochoou 2016).

4.5.2 Zeta Potential

The zeta potential acts as a determining factor in the adsorption of Se(VI) (Lu, Yu, et al. 2017; Dong, Chen, et al. 2016). The higher adsorption capacity of aluminum-based bentonite than sodium-based bentonite was more attributed to the difference in zeta potential (Dong, Chen, et al. 2016). The zeta potential of sodium-based bentonite is negative, and in the case of aluminum-based bentonite, it changes from negative to positive with decrease in pH (Dong, Chen, et al. 2016). The increase of pH causes the decline of zeta potential (Lu, Yu, et al. 2017), which is attributed to the decrease in percentage removal on increase of pH.

4.5.3 Effect of Coexisting Ions

The effect of the presence of sulfate is more pronounced in the case of selenite than selenate adsorption (Staicu et al. 2017). The effect of sulfate as compared to nitrate, bicarbonate, and chloride on hindering remediation of selenate with ZVI/Al bentonite is more pronounced (Dong, Chen, et al. 2016). The bentonite (Al) in the composite of bentonite and ZVI acts as a buffering agent against competition from counter ions, i.e. nitrate, bicarbonate, and chloride but not against sulfate (Dong, Chen, et al. 2016).

There were also other studies depicting the reduction of the percentage removal of selenate due to sulfate (Li, Wang, O'Connell, et al. 2015) and nitrite (Das et al. 2017). The reason in the case of adsorption with nZVI is attributed to passivation of the surface of nZVI, which leads to decrease in electron transfer on the surface and hence decrease in removal efficiency (Das et al. 2017).

The removal of selenite with Fe_3O_4-precipitated mesoporous magnetic carbon microspheres was not affected by the presence of sulfate, nitrate, chloride, and carbonate, but phosphate decreased the adsorption capacity (Lu, Fu, et al. 2017). This is attributed to the formation of the inner sphere complex by phosphate on Fe_3O_4 attached on the surface of carbon microspheres (Lu, Fu, et al. 2017). Similarly, the sulfate did not affect the percentage removal of selenite on $MgAl_2O_4$. However, nitrate, carbonate, and phosphate adversely affect the adsorption of selenite on $MgAl_2O_4$ (Tian et al. 2017).

The nitrate, sulfate, and chloride application in acidic form increases the adsorption of selenite on amino-functionalized magnetic particles. The anion with the least charge density is hypothesized to have the most pronounced effect on adsorption. The effect of nitrate is more pronounced than those of chloride and sulfate (Larimi et al. 2013).

The adsorption of selenate decreases in natural water (Li, Wang, O'Connell, et al. 2015; Li, Farmen, et al. 2017). The adsorption of selenate decreased on changing distilled water (97%) with well or canal water (2%–3%) (Li, Wang, O'Connell, et al. 2015). This is attributed to the presence of the competing ions. However, such drastic effect did not occur with arsenate adsorption under similar experimental conditions.

4.5.4 Effect of Materials and Adsorbates

The removal efficiency of ZVI for selenite (240 mg/g) was much larger as compared to hematite (Xia et al. 2017), whereas the efficiency of FeOOH for Se(VI) adsorption was higher than that of Fe_3O_4 (Qin et al. 2017). The carbon nanosphere showed higher efficiency for removal of selenate as compared to graphene oxide (Li, Wang, O'Connell, et al. 2015). The water molecules and functional group form a hydrogen bond between the basal plane of graphene oxide (tertiary alcohols and epoxide), which maintains the stacked nature of graphene oxide. The selenate ions compete with water molecules via intercalation in between the layers. This hindrance is reasoned for its less adsorption capacity for selenate (Li, Wang, O'Connell, et al. 2015). In addition to the earlier fact, low specific surface area and agglomeration above 50 mg/l also account for low adsorption capacity.

The iron hydroxide complex made up of an Fe:OH ratio of 1:1 and 1:2 has higher efficiency as compared to the iron hydroxide complex made up of an Fe:OH ratio of 2:1, 1:3, and 1:4 (Zhang, Fu, et al. 2017). The reason is attributed to competition from the hydroxide ion and agglomeration in the case of the iron hydroxide complex (1:4). In the case of Fe:OH ratio 2:1, structural instability is taken into account for lower percentage removal.

The geometry (Staicu et al. 2017; Li, Farmen, et al. 2017), pore size (Howarth et al. 2015), doping (Kameda et al. 2014), and size of materials (Fu et al. 2014) affect the adsorption of selenium. The layered form of Mg-Al-CO_3 removes selenate within 2 min as compared to 120 min by periclase form (Li, Farmen, et al. 2017). The insufficient surface sites along with the slower nature of adsorption in the interlayer structure are responsible for lower adsorption capacity. However, on increasing the contact time up to 22 h, the adsorption capacity of periclase was higher than that of the layered structure. The reason is attributed to the phenomenon of interlayer anion exchange and accommodation into the layered structure.

The atomic geometry of the selenate and selenite plays a role in the adsorption capacity on FerrIX resin (Staicu et al. 2017). The selenate has a tetrahedral structure with selenium inside the structure and oxygen at the apices of the structure. The selenites have a pyramidal structure with Se occupying one apice and the rest three apices by the oxygen. The structure led to different charge distributions on the species with selenite over two planes and over one plane in the case of selenite. This is attributed to superior electrostatic interaction for selenate as compared to selenite.

The faster uptake of selenate and selenite on NU-1000 in comparison to UiO-66 and its derivatives is attributed to the bigger pore size (12 and 30 Å tetrahedral and

TABLE 4.5
Summary of Parameters and Optimized Conditions for Batch Adsorption of Selenium

Adsorbent	Adsorbate	Surface Area (m^2/g), Pore Volume (cm^3/g), Pore Size (nm)	pHzpc	Experimental Conditions	Adsorption Capacity (mg/g)	Thermodynamic Parameters	Kinetic Model and Curve Fitting	Isotherm Model and Curve Fitting	Maximum Adsorption Conditions	References
ZVI (CGPM)	Se(VI)	Surface area = 2.5		pH = 6 Dose = up to 5 g/l Agitation speed = 300 rpm Concentration = 1 mg/l Contact time = up to 200 h			Pseudo-first order Nonlinear			Das et al. (2017)
ZVI (PM)	Se(VI)	Surface area = 2.3		Do			Pseudo-first order Nonlinear			Das et al. (2017)
ZVI (QMP)	Se(VI)	Surface area = 2.3		Do			Pseudo-first order Nonlinear			Das et al. (2017)
ZVI	Se(VI)	Surface area = 0.01		pH = 6 Dose = 0–1 mM (Fe concentration) Agitation speed = 300 rpm Concentration = 1 mg/l Contact time = up to 24 h	49.2–49.9					Qin et al. (2017)
Mg-Al-CO_3 LDH (batch and column study)	Se(VI)	Surface area = 91		Dose = 1 g/l Agitation speed = 300 rpm Concentration = 100 mg/l Contact time = up to 120 min	89.5					Li, Farmen, et al. (2017)
NU-1000	Se(VI)	Surface area = 1240		pH = 6 Dose = 2 g/l Concentration = 100 mg/l Contact time = up to 72 h	85					Howarth et al. (2015)
NU-1000	Se(IV)	Do		Do	95					Howarth et al. (2015)

(continued)

TABLE 4.5 (*Continued*)
Summary of Parameters and Optimized Conditions for Batch Adsorption of Selenium

Adsorbent	Adsorbate	Surface Area (m^2/g), Pore Volume (cm^3/g), Pore Size (nm)	pHzpc	Experimental Conditions	Adsorption Capacity (mg/g)	Thermodynamic Parameters	Kinetic Model and Curve Fitting	Isotherm Model and Curve Fitting	Maximum Adsorption Conditions	References
NZVI/CNT composite	Se(IV)	Surface area = 61.1	7.5	pH = 4.3–7.6 Agitation speed = 150 rpm Contact time = up to 120 min Temperature = 25°C					pH = 4.3–6	Sheng, Alsaedi, et al. (2016)
NZVI	Se(IV)	Surface area = 47.4	8	Do					pH = 4.3	Sheng, Alsaedi, et al. (2016)
Polyamine-modified magnetic graphene oxide	Se(IV)			pH = 3.1–10.2 Dose = 15–25 μg/l Agitation speed = 300 rpm Concentration = 4 mg/l (kinetic and pH study) Contact time = up to 60 min Temperature = room temperature	120.1		Difficult to apply either first and second order model	Freundlich Nonlinear	pH = 3.1 Contact time = 30 min	Lu, Yu, et al. (2017)
Polyamine modified magnetic graphene oxide	Se(VI)			Do	83.7		Difficult to apply either first- and second-order model	Freundlich model Nonlinear	Do	Lu, Yu, et al. (2017)
Magnetic carbon microspheres	Se(IV)		5.7	pH = 3–11 Dose = 1 g/l Agitation speed = 120 rpm Concentration = 10 mg/l Contact time = 90 min	9.98				pH = 3	Lu, Fu, et al. (2017)
Carbon nanosphere	Se(IV)	Surface area = 864 Pore volume = 0.22	ISP= 6.16	pH = 5.5 Dose = 0.44 g/l Concentration = 1 mg/l Contact time = up to 24 h					Contact time = 22 h	Li, Wang, O'Connell, et al. (2015)

(continued)

TABLE 4.5 (*Continued*)
Summary of Parameters and Optimized Conditions for Batch Adsorption of Selenium

Adsorbent	Adsorbate	Surface Area (m^2/g), Pore Volume (cm^3/g), Pore Size (nm)	pHzpc	Experimental Conditions	Adsorption Capacity (mg/g)	Thermodynamic Parameters	Kinetic Model and Curve Fitting	Isotherm Model and Curve Fitting	Maximum Adsorption Conditions	References
Ferrous hydroxide complex	Se(IV)			Dose = 35–140 mg/l Concentration = 3–150 mg/l Contact time = up to 60 min	256.41		Pseudo-second-order model	BET isotherm model, nonlinear, not any preferred mechanism	Dose = 140 mg/l Contact time = 30 min	Zhang, Fu, et al. (2017)
nZVI	Se(IV)	Surface area = 25–35		pH = 2–13 Dose = 0.2–1.5 g/l Agitation speed = 180 rpm Concentration = 0.1–580 mg/l Contact time = up to 2880 min Temperature = 22°C	240				pH independent Dose = 0.5 g/l Contact time = 60 min for 0.5–0.7 g/l	Xia et al. (2017)
Hematite	Se(IV)			pH = 2–13 Dose = 0.5 g/l Agitation speed = 180 rpm Concentration = 120 mg/l Contact time = 24 h Temperature = 22°C	20.8				pH = 2 (but nonsignificant difference at removal between various pH values)	Xia et al. (2017)
Nanocrystalline aluminum oxide chitosan bead	Se(IV)		7.25–8	pH = c.a. 4.3–10 Dose = 1.75 g/l Agitation speed = 150 rpm Concentration = 1 mg/l Contact time > 150 h	11.08		Pseudo-second-order model Nonlinear (no comparison)	Langmuir model Nonlinear (no comparison)	pH = c.a. 4.3–5 Time = 36 h	Yamani et al. (2014)

(continued)

TABLE 4.5 (*Continued*)
Summary of Parameters and Optimized Conditions for Batch Adsorption of Selenium

Adsorbent	Adsorbate	Surface Area (m^2/g), Pore Volume (cm^3/g), Pore Size (nm)	pHzpc	Experimental Conditions	Adsorption Capacity (mg/g)	Thermodynamic Parameters	Kinetic Model and Curve Fitting	Isotherm Model and Curve Fitting	Maximum Adsorption Conditions	References
Nanocrystalline aluminum oxide chitosan bead	Se(VI)		7.25–8	Do	20.11		Pseudo-second-order model Nonlinear (no comparison)	Langmuir model Nonlinear (no comparison)	pH = c.a. 4.3–5 Contact time = 36 h	Yamani et al. (2014)
Nanocrystalline aluminum oxide	Se(IV)	Surface area = 40	8.5	pH = c.a. 4.3–10 Dose = 0.145 g/l Agitation speed = 150 rpm Concentration = 1 mg/l Contact time > 150 h	10.88		Pseudo-second-order model Nonlinear (no comparison)	Langmuir model Nonlinear (no comparison)		Yamani et al. (2014)
Nanocrystalline aluminum oxide	Se(VI)	Do	8.5	Do	9.35		Pseudo-second-order model Nonlinear (no comparison)	Langmuir model Nonlinear (no comparison)		Yamani et al. (2014)
TiO_2 (rutile) (batch and column study)	Se(IV)	Surface area = 5	4.5–5	pH = 2–12 Dose = 20 g/l Concentration = 6–600 ppm Contact time = 100 h	0.0133 mmol/g			Freundlich model Nonlinear (no comparison)	pH = 2	Svecova et al. (2011)
TiO_2 (rutile) (batch and column study)	Se(VI)	Do	4.5–5	pH = 2–12 Dose = 20 g/l Concentration = 7–700 ppm Contact time = 100 h	0.0034 mmol/g			Freundlich model Nonlinear	pH = 2	Svecova et al. (2011)

(*continued*)

TABLE 4.5 (*Continued*)
Summary of Parameters and Optimized Conditions for Batch Adsorption of Selenium

Adsorbent	Adsorbate	Surface Area (m^2/g), Pore Volume (cm^3/g), Pore Size (nm)	pHzpc	Experimental Conditions	Adsorption Capacity (mg/g)	Thermodynamic Parameters	Kinetic Model and Curve Fitting	Isotherm Model and Curve Fitting	Maximum Adsorption Conditions	References
Gracilaria-modified biochar	Se(VI)			Dose = 10 g/l Agitation speed = 100 rpm Concentration = 100 µg/l Contact time = 1 h Temperature = 15°C	0.03 mM/g					Johansson et al. (2015)
Mg-Al layered double hydroxide	Se(VI)			Dose = 4 g/l Concentration = 1 mM Contact time = 120 min Temperature = 30°C			Pseudo-second-order model Linear	Langmuir model Linear	Contact time = 60 min	Kameda et al. (2014)
Fe-doped Mg-Al layered double hydroxide	Se(VI)			Dose = 4 g/l Concentration = 0.5–25 mM (isotherm study) Contact time = 120 min Temperature = 10°C–60°C	1.4 mmol/g		Pseudo-second-order model Linear	Langmuir model Linear	Temperature = 60°C	Kameda et al. (2014)
Thiourea formaldehyde resin	Se(IV)			pH = up to 6 Dose = 0.4–2 g/l Concentration = 50–1000 mg/l (isotherm study) Contact time = up to 70 h	833.3			Langmuir model Linear	Acidity = 3 M HCl	Gezer et al. (2011)
Thiourea formaldehyde resin	Se(VI)			pH = up to 6 Dose = 0.4-2 g/l Concentration = 100–500 mg/l (isotherm study) Contact time = up to 70 h	526.3			Langmuir model Linear	Acidity = 5 M HCl	Gezer et al. (2011)

(continued)

TABLE 4.5 (*Continued*)
Summary of Parameters and Optimized Conditions for Batch Adsorption of Selenium

Adsorbent	Adsorbate	Surface Area (m^2/g), Pore Volume (cm^3/g), Pore Size (nm)	pHzpc	Experimental Conditions	Adsorption Capacity (mg/g)	Thermodynamic Parameters	Kinetic Model and Curve Fitting	Isotherm Model and Curve Fitting	Maximum Adsorption Conditions	References
Fe_3O_4 chitosan nanocomposite hollow fibers	Se(IV)		6.85	pH = 3–11 Dose = 0.05–0.6 g/l Agitation speed = 150 rpm Concentration = 50–450 µg/l Contact time = 30–150 min	1.34 µg/mg		Pseudo-second order	No preferred model Linear	pH = 3.67 Dose = 92.97 mg/l Concentration = 291.36 µg/l Contact time = 109.22 min	Seyed Dorraji et al. (2017)
Mg Al layered double hydroxide	Se(IV)	Surface area = 301.3 Pore size = 169.1	12.2	Dose = 20 g/l Agitation speed = 30 rpm Concentration = 0.1–10 meq/l Contact time = 24 h Temperature = 20°C	494–563 meq/kg			Dual mode Langmuir-Freundlich model Nonlinear		Constantino et al. (2017)
Mg Al layered double hydroxide	Se(VI)	Do	12.2	Do	609–659 meq/kg			Dual mode Langmuir-Freundlich model Nonlinear		Constantino et al. (2017)
$MgAl_2O_4$	Se(IV)			pH = 3–11 Dose = 0.2 g/l Concentration = 0.01–100 mg/l Contact time = up to 180 min Temperature = 298 K	179.59		Pseudo-second order linear	Freundlich model Linear	pH = 5 Contact time = 120 min	Tian et al. (2017)

(continued)

TABLE 4.5 (*Continued*)
Summary of Parameters and Optimized Conditions for Batch Adsorption of Selenium

Adsorbent	Adsorbate	Surface Area (m^2/g), Pore Volume (cm^3/g), Pore Size (nm)	pHzpc	Experimental Conditions	Adsorption Capacity (mg/g)	Thermodynamic Parameters	Kinetic Model and Curve Fitting	Isotherm Model and Curve Fitting	Maximum Adsorption Conditions	References
Biosynthesized nZVI	Se			pH = 3–11 Dose = 10–100 mg/l Agitation speed = 100–200 rpm Concentration = 1–2 mg/l Contact time = 15–120 min					pH = 3 Concentration = 2 mg/l Agitation speed = 200 rpm Dose = 100 mg/l	Adio et al. (2017)
Dendrimer functionalized graphene oxide	Se(IV)		ISP = 7.9	pH = 3–8 Dose = 0.2 g/l Agitation speed = 300 rpm Concentration = up to 75 mg/l Contact time = 24 h	60.89				pH = 3	Xiao et al. (2016)
Dendrimer functionalized graphene oxide	Se(VI)		ISP = 7.9	pH = 3–8 Dose = 0.2 g/l Agitation speed = 300 rpm Concentration = up to 82 mg/l Contact time = 24 h	77.85			Langmuir model Nonlinear	pH = 3	Xiao et al. (2016)
Titanate nanotube	Se(IV)	surface area = 388.2	5	pH = 3.5–9.5 Contact time = up to 180 min Temperature = 25°C	c.a. 67		Pseudo-second order Linear No comparison		pH = 3.5–6 Contact time = 120 min	Sheng, Linghu, et al. (2016)
Titanate nanotube	Se(VI)	Surface area = 388.2	5	pH = 3–8 Contact time = up to 180 min Temperature = 25°C	c.a. 45		Pseudo-second order Linear No comparison		pH = 3.5–4 Contact time = 120 min	Sheng, Linghu, et al. (2016)

(continued)

TABLE 4.5 (*Continued*)
Summary of Parameters and Optimized Conditions for Batch Adsorption of Selenium

Adsorbent	Adsorbate	Surface Area (m^2/g), Pore Volume (cm^3/g), Pore Size (nm)	pHzpc	Experimental Conditions	Adsorption Capacity (mg/g)	Thermodynamic Parameters	Kinetic Model and Curve Fitting	Isotherm Model and Curve Fitting	Maximum Adsorption Conditions	References
$CuFe_2O_4$	Se(IV)	Surface area =148.9	8.1	pH=2–11 Dose=0.4 g/l Agitation speed=200 rpm Concentration=1–25 mg/l (isotherm study) and 10 mg/l in kinetic study Contact time=up to 12 h Temperature=25°C	14.1		Pseudo-second-order model		pH=2 Contact time =300 min	Sun et al. (2015)
$CuFe_2O_4$	Se(VI)	Surface area =148.9	8.1	Do	5.97		Do		Do	Sun et al. (2015)
$CoFe_2O_4$	Se(IV)	Surface area =201.8	8	Do	11.6		Do		Do	Sun et al. (2015)
$CoFe_2O_4$	Se(VI)	Surface area =201.8	8	Do	5.55		Do		Do	Sun et al. (2015)
$MnFe_2O_4$	Se(IV)	Surface area =48.1	6	Do	3.9		Do		Do	Sun et al. (2015)
$MnFe_2O_4$	Se(VI)	Surface area =48.1	6	Do	5.27		Do		Do	Sun et al. (2015)
Zero valent iron with weak magnetic field	Se(VI)	Surface area =0.3015		pH=6 Dose=1 g/l Agitation speed=310 rpm Concentration=10–100 mg/l (kinetic study) Contact time=up to 12 h Temperature=25°C	36.9				Contact time =1 h for concentration 10 mg/l Contact time =5 h for concentration 20–100 mg/l	Liang, Guan, et al. (2015)

(continued)

TABLE 4.5 (*Continued*)
Summary of Parameters and Optimized Conditions for Batch Adsorption of Selenium

Adsorbent	Adsorbate	Surface Area (m^2/g), Pore Volume (cm^3/g), Pore Size (nm)	pHzpc	Experimental Conditions	Adsorption Capacity (mg/g)	Thermodynamic Parameters	Kinetic Model and Curve Fitting	Isotherm Model and Curve Fitting	Maximum Adsorption Conditions	References
Iron oxide–graphene oxide composite	Se(IV)			pH = 2–11 Dose = 1 g/l Agitation speed = 300 rpm Concentration = 300 ppb Contact time = up to 12 h Temperature = room temperature	23.80			Langmuir model	pH = 2–9 Contact time = 30 s	Fu et al. (2014)
Iron oxide–graphene oxide composite	Se(VI)			Do	15.11			Langmuir model	pH = 2–3 Contact time = 30 s	Fu et al. (2014)
Mesoporous conjugate	Se(IV)	Surface area = 421 Pore volume = 0.51 Pore size = 6	6.52	pH = 1–9 Concentration = 1–80 mg/l (isotherm study) Contact time = up to 90 min	103.73			Langmuir model No comparison	pH = 2.5 Contact time = 70 min	Rabiul Awual et al. (2014)
Magnetite supported on zeolite (magnetite: zeolite=2:1)	Se(IV)	Surface area = 74–126		pH = 0.5–13.5 Dose = 0.5 g/l Agitation speed = 160 rpm Concentration = 5 mg/l Contact time = up to 24 h Temperature = 25°C	23		Pseudo-second order	Langmuir	pH = 3–5	Verbinnen et al. (2013)
Magnetite supported on zeolite (magnetite: zeolite = 2:1)	Se(VI)	Surface area = 74–126		Do			Pseudo-second order	Freundlich	pH = c.a. 2	Verbinnen et al. (2013)

octahedral pores, respectively, and aperture of the same size) of the former as compared to that of the latter (8 and 11 Å of tetrahedral and octahedral pores, respectively, and an aperture of 7 Å) (Howarth et al. 2015). The bigger pore size facilitates faster diffusion of selenite (4.8 Å) and selenate (5.2 Å) in NU-1000 in comparison to UiO-66 and its derivatives.

The doping of magnesium aluminum layered double hydroxide enhanced the removal capacity of Se(VI) (Kameda et al. 2014). The size of Fe_3O_4 in the Fe_3O_4-graphene oxide composite affects the adsorption of selenite and selenate (Fu et al. 2014). The independent iron oxide size also affects the adsorption of selenite and selenate. The effect is more pronounced in the case of selenite (Fu et al. 2014). The larger size of Fe_3O_4 (400 nm) did not remove any selenate, whereas a small size (10–20 nm) removes mostly less than 20%. However, in the case of selenite, the small size of Fe_3O_4 (10–20 nm) removes more than 90% in the entire pH range studied (2–10) (Fu et al. 2014).

The bentonite(Al) acts as a buffering system in the case of the composite of bentonite (Al) with zero valent iron (Dong, Chen, et al. 2016). The corrosion of Fe leads to decrease in concentration of the H^+ ion, which is compensated by H^+ ion release from Al-OH or Si-OH, whereas at higher pH, Fe^{2+} reacts with the OH^- ion with the intermediate product as Fe(OH), and its final transformation to Fe_2O_3.

The adsorption of selenite was more than that of selenate on polyamine-modified magnetic graphene oxide (Lu, Yu, et al. 2017), thiourea formaldehyde resin (Gezer et al. 2011), rutile TiO_2 (Svecova et al. 2011), and UiO-66 (Howarth et al. 2015). The higher adsorption capacity can be due to the higher charge density of selenite as compared to selenate (Kameda et al. 2014).

4.5.5 Effect of Temperature

The adsorption of selenite on magnetite increases with increase in temperature (Larimi et al. 2013). The hydration of selenite is attributed to this phenomenon. The selenite ions are hypothesized to be dissociated from the hydration complex before adsorption, and the energy is provided by temperature. Similarly, the adsorption of selenate increases with increase in temperature on iron-doped magnesium aluminum layered double hydroxide (Kameda et al. 2014). The removal of selenite also decreased with increase in temperature (Sheha and El-Shazly 2010).

4.5.6 Pretreatment

The pretreatment of the selenium solution with $BaCl_2$ increases the percentage removal. This is attributed to the removal of competing ions like sulfate if present in the aqueous solution (Li, Farmen, et al. 2017; Staicu et al. 2017). Similarly, the pretreatment with barium salt is also suggested in remediation with nano zerovalent iron (Das et al. 2017). The material recovered after sonication of the spent ZVI used for the adsorption of selenate showed higher efficiency than the pristine material (Tang et al. 2014). However, functionalization with the amino group on magnetite decreased its adsorption efficiency as compared to its pristine form (Larimi et al. 2013).

4.5.7 Role of Iron and Oxygen

The application of ZVI under anaerobic conditions removes selenate up to 9.8%. But on addition of Fe^{2+} ions, the removal reached 100% (Tang et al. 2014). The reason is attributed to the conversion of the oxidative layer into an electron transfer medium. The adsorption of selenite was followed by its reduction to selenite and then to Se(0) under anaerobic conditions. This is evidenced by increase of selenate and selenite reaching up to 21% and 41%, respectively, at 10 h on the adsorbent. On further increasing time, their concentration (selenate and selenite) decreased. However, the concentration of Se(0) increases up to the whole time period, i.e. 24 h.

The selenite reduction immobilization is also suggested to be the main mechanism for the removal of selenite with ferrous hydroxide complexes (Zhang, Fu, et al. 2017), and passivation of the nZVI surface causes decrease in percentage removal of selenate (Das et al. 2017).

Addition of Fe^{2+} ions to the ZVI-selenate anaerobic system maintains the pH of the system. Without the addition of Fe(II), the pH of the system increased from 4 to 10. The addition of Fe(II) equilibrates the pH at 6, until exhaustion of Fe(II). After this, the pH increased to 9.5. The effect of Fe(II) on sequestration of Se(VI) was higher at high oxygen concentration (Qin et al. 2017).

Qin et al. (2017) suggested that oxygen plays an important role in sequestration of Se(VI) with nZVI. The removal of Se(VI) was negligible under anaerobic conditions. The oxygen is suggested to generate the iron oxide/hydroxide at the surface of ZVI. The EXAFS spectra suggested that the adsorption of selenate on ZVI leads to ZVI transformation into magnetite, maghemite, and lepidocrocite. The absence of iron oxide/hydroxide limits the transfer of electrons from nZVI to selenium for reduction and hence its sequestration.

However, Fe(II) adsorbed on magnetite and FeOOH used for the sequestration of Se(VI) under anoxic conditions suggested no change in concentration of Fe(II). This suggested that the removal happened due to Fe_3O_4 and FeOOH (Qin et al. 2017).

However, in the case of removal of Se(VI) with FeS colloids under anoxic conditions, no change in oxidation state was observed, whereas in the case of selenite, there were changes in the oxidation state of Se and Fe in FeS (Han, Batchelor, et al. 2013).

4.5.8 Mechanism

The selenite reduction immobilization is suggested to be the main mechanism for the removal of selenite on ZVI (Zhang, Fu, et al. 2017). The passivation of the nZVI surface causes decrease in percentage removal (Das et al. 2017). The nZVI has a core–shell structure with an interfacial layer of iron oxide (Xia et al. 2017). The core–shell nature of nZVI was estimated by SEM-HADF (high angular dark field imaging in scanning electron microscopy). The brightness of the image was used to estimate the oxygen at the periphery. The center is bright in nature, whereas the peripheral dim nature was attributed to low atomic oxygen.

The outer shell iron oxide attracts Se(IV), while the inner ZVI core supplies electrons for reduction of Se(IV) to Se(0) and Se(-II). The reduction of Se(IV) is confirmed by EELS spectra, and the oxidation of iron is confirmed by the presence of magnetite and maghemite by XRD (X-ray diffraction). The lack of bonding between Fe and Se is confirmed by STEM-EDS(Scanning transmission electron microscopy-energy dispersive X-ray spectroscopy) mapping (Xia et al. 2017). The Fe and O are present only in the background of the Se sphere, and this suggests the lack of bonding between Se and Fe.

After adsorption of selenite with nZVI, the nZVI retains its shape but selenite is found in spherical and needle shapes (Xia et al. 2017). The spherical- and needle-shaped materials are amorphous and gray trigonal allotropes of selenium. First the spherical selenium material formed, which was converted to a larger structure by Oswald ripening, and afterward it transformed to a more thermodynamically stable needle-shaped material.

The anion exchange and hydrogen interaction by amine groups are suggested for adsorption of selenite and selenate on UiO-66 (metal organic framework) (Howarth et al. 2015). Based on diffuse reflectance infrared Fourier transform spectroscopy, it can be suggested that SeO_4^{2-} and SeO_3^{2-} anions also undergo ion exchange with NU-1000 (metal organic framework) and replace two hydroxyl groups (Howarth et al. 2015). The adsorption of selenate on iron-doped magnesium aluminum layered double hydroxide occurred with reduction of Se(VI) to Se(IV). Afterward, it is adsorbed by the anion exchange mechanism (Kameda et al. 2014). The reduction is helpful due to the higher charge density of selenite (SeO_3^{2-}) as compared to selenate (SeO_4^{2-}) (Kameda et al. 2014). The anion exchange between chloride and selenate is confirmed by the increase of basal spacing in XRD (Kameda et al. 2014).

Se ions would be adsorbed as Se and polymeric Se_8 on the gold surface (Jia et al. 2013). The presence of Se can also be detected by cyclic voltammogram analysis. It even distinguishes between polymeric and atomic Se adsorbed on the gold surface (Jia et al. 2013). The adsorption of Se on the Au surface is suggested to be the same as that of S on Au (Jia et al. 2013). Low energy electron diffraction is not able to decipher any feature for adsorption (Jia et al. 2013).

Sun et al. (2015) suggested the follow-up of a pseudo-second-order model as a proxy for chemical sorption. Jia et al. (2013) authenticated the chemisorbed species by the absence of a new XPS peak. The authors suggested if gold selenide formed, then there will be a presence of a new peak. The strong interaction for Se chemisorption on Au is also analyzed by valence band study. The Se adsorption on the Au surface led to the disappearance of peaks at 0.45 and 8 eV. The disappearance of peaks suggested toward chemisorption of Au with Se.

The selenite forms an inner sphere complex with Fe_3O_4 of the Fe_3O_4 carbon microsphere composite (Lu, Fu, et al. 2017). Svecova et al. (2011) suggested the formation of an inner sphere complex on the basis of Se-Ti distance by EXAFS data. The presence and absence of the second shell of Ti scattering in Se(IV or VI) adsorption of titanium nanotubes indicate formation of the inner sphere complex or outer sphere complex, respectively (Sheng, Linghu, et al. 2016).

In the raw ATR FT-IR (Attenuated total reflection Fourier-transform infrared) spectrum of ferrihydrite and 24% ferrihydrite after selenium adsorption, there were no signs of frequency vibration on the adsorbed selenium (Johnston and Chrysochoou 2016). However, the second derivative plot suggests one additional Se-O vibration peak at position 910 cm^{-1}. The author suggested it to be indicative of the inner sphere complex and suppressed in raw ATR FT-IR spectra by aluminum.

The change of oxidation state can be predicted by XANES spectra. The adsorption of Se(IV) and Se(VI) on titanium nanotubes suggested nonchanging of the valence state during adsorption by XANES spectra (Sheng, Linghu, et al. 2016).

4.5.9 Desorption

The regeneration of the adsorbent is achieved by calcination (Tian et al. 2017) and use of alkaline treatment (Lu, Yu, et al. 2017). The alkaline pH (10.1) causes the desorption of Se(IV) and Se(VI) from polyamine-modified magnetic graphene oxide. More than 80% of adsorption was achieved after three cycles of desorption (Lu, Yu, et al. 2017). The regeneration of the adsorbent was performed by calcination of the material at 500°C, and it was used four times without any significant decline in adsorption capacity (Tian et al. 2017). In some cases, the material is reused without desorption (Lu, Fu, et al. 2017). The desorption of polymeric Se_8 from the gold surface requires higher potential as compared to Se (Jia et al. 2013).

5 Impact of Factors on Remediation of Miscellaneous (Fe, Cs) and Nontoxic Elements (Sc, Ti, Ga, Ge) Via Batch Adsorption Process

Deepak Gusain
Durban University of Technology

Shikha Dubey and Yogesh Chandra Sharma
IIT(BHU), Varanasi

Faizal Bux
Durban University of Technology

CONTENTS

This chapter includes factors affecting the adsorption of essential elements such as iron, minor toxic metals such as cesium and elements such as scandium, titanium, gallium and germanium whose toxicity is still unknown. The increase of iron content in the body leads to cardiovascular disease and exposure to nonradioactive cesium causes irritation to the gastrointestinal tract. The toxicity of scandium, titanium, gallium and germanium is still not fully understood. The effect of various parameters such as pH and other factors affecting the adsorption of the respective elements are discussed in this chapter.

5.1 IRON

Iron exists in ferric and ferrous forms (Tokar et al. 2015). It is available at concentrations ranging from 0.5 to 50 mg/l in natural fresh water (WHO 2017). It is an essential element, and a value of 2 mg/l does not pose any hazard to the health. WHO has not set up a guideline value for iron in drinking water.

5.1.1 Effect of pH

The adsorption of ferric ions (Bhattacharyya and Gupta 2009) and ferrous ions (Jiang, Zhang, et al. 2016) increased with increase of the pH. The high protonation of the adsorbent at low pH and competition from the proton were held responsible for this trend. In the case of the sulfonated chitin, the low percentage removal of ferric ions at lower pH was attributed to the reduced availability of SO_3^- ions at lower pH (Dong, Zhang, et al. 2016).

5.1.2 Effect of Coexisting Ions

Arsenate suppresses the adsorption of the ferrous ions on the goethite and hematite (Catalano et al. 2011). This was due to competitive adsorption. The adsorption of ferrous ion on aluminum hydroxide occurred by the formation of Fe(II)-Al(III) layered double hydroxide (Zhu and Elzinga 2015). The arsenate hindered the formation of Fe(II)-Al(III)-LDH. Chloride, nitrate, sulfate and bicarbonate ions did not significantly affect the adsorption of the ferric ions by protonated chitin, carboxylated chitin and aminated chitin (Kousalya et al. 2010).

The Fe^{2+}-ionic imprinted polyamine-functionalized silica gel has high selectivity for ferrous ions in the presence of cerium or praseodymium (Wang, Li, et al. 2013). This was due to the unmatched size, shape, and spatial arrangement of the active sites in comparison to cerium and praseodymium ions. The ionic radii of the cerium and praseodymium were larger (103 and 101 pm) than that of the Fe^{2+} ion (74 pm), which does not allow the entry of the former ions into the pores.

5.1.3 Effect of Surface Modification

The acid activation of the montmorillonite and kaolinite increased the adsorption efficiency of the ferric ion from 18.2 to 19.8 mg/g and 7.5 to 8.7 mg/g, respectively (Bhattacharyya and Gupta 2006). However, the calcined product of tetrabutylammonium bromide-modified montmorillonite and the calcined product of tetrabutylammonium bromide-modified kaolinite have lower efficiency for ferric ions in comparison to montmorillonite and kaolinite (Bhattacharyya and Gupta 2009). This was attributed to the blockage of pores by the residual of tetrabutylammonium bromide, inactiveness of adsorption sites for the metal ion due to negative charge elimination or establishment of new linkages having no affinity for the metal ions.

5.1.4 Effect of the Material

The adsorption of the ferrous and ferric ions depends on the material applied for the adsorption. The adsorption of Fe(II) on low structural Fe-content montmorillonite was lower than that on high structural Fe content montmorillonite (Soltermann et al. 2014). However, the adsorption under reducing conditions was the same. Similarly, the removal efficiency of the ferric ions with natural bentonite was lower than that of natural quartz (Al-Anber 2010).

5.1.5 Effect of Temperature

The adsorption capacity of the ferric ions increased with natural bentonite and quartz (Al-Anber 2010) and sulfonated cellulose (Dong, Zhang, et al. 2016) and decreased with egg shells (Yeddou and Bensmaili 2007) on increasing the temperature. The egg shells are made up of calcium carbonate along with a small amount of organic matter. The declined adsorption capacity with increase of temperature is attributed to the increased solubility of ferric carbonate at higher temperature. The change of temperature also changes the rate of sorption of ferric ions with eggshell (Yeddou and Bensmaili 2007). The increase of temperature from 20°C to 50°C leads to a decline in the adsorption rate from 0.474 to 0.285 g/mg min.

5.1.6 Miscellaneous

The presence of iron ions in the solution leads to a change in the structure of the smectites and ferrihydrite (Jones et al. 2017; Boland et al. 2014). The adsorption of the ferrous ions on smectites leads to a change in its lattice arrangement (Jones et al. 2017). The presence of ferrous ions enhanced the transformation of ferrihydrite to goethite, via lepidocrocite (Boland et al. 2014). The transformation was even faster at high pH conditions as compared to low pH conditions. The presence of the ferrous ions induced the release of zinc and nickel ions from zinc- and nickel-substituted goethite and hematite by electron transfer and atomic exchange mechanisms (Frierdich and Catalano 2012).

5.2 CESIUM

Cesium is naturally found in rocks and soil at low concentrations as isotope ^{133}Cs (ATSDR 2004). Cesium finds its application in scintillation counters, atomic clocks and vacuum tubes (RSC 2020c; ATSDR 2004). However, two radioactive isotopes ^{134}Cs and ^{137}Cs are discharged into the environment via nuclear power plants (ATSDR 2004). The ^{134}Cs and ^{137}Cs have a half-life of 2 and 30 years, respectively, and the decay of these radioactive elements occurred with emission of beta particles and gamma radiation. A number of people in Brazil who handled radioactive cesium became sick from exposure to radiation, but people exposed to radioactive cesium after the nuclear bombing were not expected to have the same effects due to a small amount of exposure. An overview of the experimental parameters and optimized conditions from batch adsorption experiments for cesium is presented in Table 5.1.

5.2.1 Effect of pH

The adsorption of cesium in respect to the pH depends on the adsorption mechanism and adsorbent applied for removal. The adsorption capacity with the prussian blue-based adsorbent was maximum near the neutral pH, i.e. pH 7 (Jang and Lee 2016; Yang, Sun, et al. 2014; Lujanienė et al. 2017). The decrease of adsorption in acidic and basic media was attributed to the instability of the adsorbent or dissolution of the prussian blue.

The adsorption of cesium with trititanate nanofibers and trititanate nanotubes (Yang et al. 2011), ethylamine-modified montmorillonite (Long et al. 2013), copper ferrocyanide (Han, Zhang, et al. 2013), potassium nickel hexacyanoferrate (Qing et al. 2015), poly(b-cyclodextrin)/bentonite composite (Liu, Xie, et al. 2017), Fe_3O_4@WO_3 (Mu et al. 2017), nanostructured potassium copper hexacyanoferrate-cellulose hydrogel (Kim, Kim, et al. 2017) and graphene oxide (Tan et al. 2016) was low at lower pH (i.e. acidic) and increased with increase of pH. The low removal capacity at lower pH was attributed to competition from hydronium ions (Long et al. 2013; Han, Zhang, et al. 2013), protonation of adsorption sites (Long et al. 2013), decrease of negative charge (Liu, Xie, et al. 2017), and exchange of sodium ions with hydronium ions in the case of trititanate nanofibers and trititanate nanotubes (Yang et al. 2011). The protonation and deprotonation of the surface group constants in graphene oxide were estimated with potentiometric titration (Tan et al. 2016). The decline of percentage removal in highly alkaline pH, i.e. 12 (Han, Zhang, et al. 2013) > 10 (Yang, Yu, et al. 2017) is attributed to the decomposition of the adsorbent. In the case of adsorption of cesium with magnetic potassium titanium hexacyanoferrate, the removal declined at pH above 11, attributed to the formation of uncharged Cs(OH)(aq) species (Zhang, Zhao, et al. 2015).

The adsorption efficiency measured in terms of distribution coefficient on the supramolecular material increased with increase of nitric acid concentration from 0.3 to 3 M (Zhang, Wang, et al. 2017). The behavior was due to the absence of

TABLE 5.1
Summary of Parameters and Optimized Conditions for Batch Adsorption of Cesium

Adsorbent	Adsorbate	Surface Area (m^2/g), Pore Volume (cm^3/g), Pore Size (nm)	pHzpc	Experimental Conditions	Adsorption Capacity (mg/g)	Thermodynamic Parameters	Kinetic Model and Curve Fitting	Isotherm Model and Curve Fitting	Maximum Adsorption Conditions	References
Chabazite-type zeolite	Cs	Surface area = 16.8 (N_2 low) Surface area = 336 (CO_2 flow)		pH = dose = 20 g/l Agitation speed = rpm concentration = 10–100 mmol/l Contact time = up to 180 min Temperature = 293, 323 and 353 K	2.8 mmol/g	ΔH° = 2.52 kJ/mol ΔG° = negative ΔS° = 11.5 kJ/mol		Langmuir	Temperature = 323 K Contact time = 60 min	Du et al. (2017)
Magnetic prussian blue	Cs	Surface area = 322.19		pH = 2–9 Dose = 2 g/l Agitation speed = 120 and 150 rpm Concentration = 1–50 mM (isotherm study) Contact time = up to 24 h Temperature = 10°C–30°C	280.82	ΔH° = −7.89 kJ/mol ΔS° = −1.44 J/mol K Delta G = negative	Pseudo-second-order model Linear	Temkin model Nonlinear	pH = 7 Contact time = 4 h Temperature = 10°C	Jang and Lee (2016)
Nickel oxide-grafted andic soil	Cs	Surface area = 64.3 Pore volume = 0.1298 Pore size = 6	5		16.1		Difficult to differentiate between first and second order	Langmuir model		Ding et al. (2013)

(continued)

TABLE 5.1 (*Continued*)
Summary of Parameters and Optimized Conditions for Batch Adsorption of Cesium

Adsorbent	Adsorbate	Surface Area (m^2/g), Pore Volume (cm^3/g), Pore Size (nm)	pHzpc	Experimental Conditions	Adsorption Capacity (mg/g)	Thermodynamic Parameters	Kinetic Model and Curve Fitting	Isotherm Model and Curve Fitting	Maximum Adsorption Conditions	References
Solid-state supramolecular adsorbent BnPC6DTXAD-7 (BnPDTX7)	Cs	Surface area = 187.6 Pore volume = 0.6435 Pore size = 30.66		Acid concentration = 0.4–6 M HNO_3 Dose = 50 g/l Agitation speed = 150 rpm Concentration = 0.5 mM Contact time = up to 120 min Temperature = 298–318 K	0.1396 mM/g	ΔH° = −22.19 kJ/mol ΔS° = −36.45 J/mol K ΔG° negative		Langmuir model	Acid concentration = 3 M HNO_3 Contact time = 60 min Temperature = 298 K	Zhang, Wang, et al. (2017)
Prussian blue/Fe_3O_4/ graphene oxide composite	Cs			pH = 4–10 Dose = 1.66 g/l Agitation speed = 200 rpm Concentration = 50–300 mg/l Contact time = up to 48 h Temperature = 298 K	55.56			Langmuir model Linear	pH = 7 Contact time = 12 h	Yang, Sun, et al. (2014)
Fe_3O_4@SiO_2@ potassium titanium ferrocyanide	Cs			pH = dose = 2.5 g/l Agitation speed = 180 rpm Concentration = 0.15–1 g/l Contact time = 5 min–4 h Temperature = 25°C	43.09		Pseudo-second order Linear	Langmuir model Nonlinear	contact time = 30 min	Yi et al. (2014)
Copper hexacyanoferrate–polymer composite	Cs	Surface area = 479.1 Pore volume = 1.767		pH = 1.8–12.2 Dose = 100 g/l Agitation speed = concentration = 5–35 mg/l contact time = up to 240 min Temperature = 293 K	1.56		Pseudo-second order Linear	Langmuir model Linear	pH = 9 Contact time = 240 min	Dwivedi et al. (2013)

TABLE 5.1 (*Continued*)
Summary of Parameters and Optimized Conditions for Batch Adsorption of Cesium

Adsorbent	Adsorbate	Surface Area (m^2/g), Pore Volume (cm^3/g), Pore Size (nm)	pHzpc	Experimental Conditions	Adsorption Capacity (mg/g)	Thermodynamic Parameters	Kinetic Model and Curve Fitting	Isotherm Model and Curve Fitting	Maximum Adsorption Conditions	References
Graphene oxide	Cs	Surface area = 139		pH = 2–9 Dose = 0.5 g/l Concentration = 10 mg/l Contact time = up to 48 h Temperature = 298–338 K	40	ΔH° = 10.35 kJ/mol ΔS° = 110.99 J/mol K ΔG° = negative	Pseudo-second-order model Linear	Langmuir model Linear	pH > 6 Temperature = 338 K	Tan et al. (2016)
MWCNT-reinforced zeolite-A beads	Cs			Dose = 1.375 g/l Agitation speed = 300 rpm Concentration = 200 mg/l Contact time = 12 h Temperature = room temperature	113		Pseudo-second-order model	Langmuir model Linear		Vipin et al. (2016)
Magnetic potassium titanium hexacyanoferrate	Cs			pH = 1–10 Dose = 0.075–1.525 g/l Agitation speed = 300 rpm Concentration = 4–16 mg/l Contact time = 60 min Temperature = 278–318 K		ΔH° = 35.64 kJ/mol ΔS° = 257.24 J/mol K ΔG° = negative	Pseudo-second-order model Linear	Langmuir model Linear	pH = c.a. 5–8 Temperature = 318 K	Zhang, Zhao, et al. (2015)
Ethylamine-modified montmorillonite	Cs			pH = 1–10 Dose = 2 g/l Concentration = 20–340 mg/l Contact time = 240 min Temperature = 30°C–60°C	80.27	ΔH° = −2.67 kJ/mol ΔS° = 55.35 J/K mol ΔG° = negative	Pseudo-second-order model Linear	Langmuir model Linear	pH = 4–10 Contact time = 45 min	Long et al. (2013)

(continued)

TABLE 5.1 (*Continued*)
Summary of Parameters and Optimized Conditions for Batch Adsorption of Cesium

Adsorbent	Adsorbate	Surface Area (m^2/g), Pore Volume (cm^3/g), Pore Size (nm)	pHzpc	Experimental Conditions	Adsorption Capacity (mg/g)	Thermodynamic Parameters	Kinetic Model and Curve Fitting	Isotherm Model and Curve Fitting	Maximum Adsorption Conditions	References
Phosphate-modified montmorillonite	Cs		4	pH = 2–12 Dose = 80 mg/l Agitation speed = 150 rpm Concentration = 20–300 mg/l (isotherm study) Contact time = 5–120 min Temperature = 288–308 K		ΔH° = −14.90 kJ/mol ΔS° = −4.75 kJ mol negative		Freundlich model Nonlinear	pH = 11 Temperature = 288 K	Ma et al. (2011)
Copper ferrocyanide	Cs			pH = c.a. 2–12 Dose = 80 mg/l Agitation speed = 150 rpm Contact time = 5–120 min Temperature = 288–308 K		ΔH° =9.426 kJ/mol ΔG° = negative ΔS° =0.1091 kJ/mol	Pseudo-second-order model Linear	Langmuir, Freundlich and D-R isotherm	pH = 9 Contact time = 90 min	Han, Zhang, et al. (2013)
Potassium nickel hexacyanoferrate	Cs			pH = 1.7–6.5 Dose = 1.44 mg/l Concentration = 0.075 mol/l Contact time = 8 h Temperature = 25°C			Pseudo-second-order Nonlinear	Freundlich model Nonlinear	pH = 6.5 (max at pH 1.7)	Qing et al. (2015)
Carboxymethyl cellulose sodium/prussian blue composite loaded with lanthanum(III)	Cs			pH = 3–11 Dose = 1 g/l Agitation speed = 150 rpm Concentration = 20 mg/l Contact time = c.a. 1400 min Temperature = 288–328 K	35.22	ΔH° = 16.64 kJ/mol ΔS° = 67.56 J/K mol ΔG° = negative	Pseudo-second-order Linear and nonlinear	Freundlich model Linear and nonlinear	Dose = 1 g/l on economic basis Contact time = 1000 min	Zong et al. (2017)

(*continued*)

TABLE 5.1 (*Continued*)
Summary of Parameters and Optimized Conditions for Batch Adsorption of Cesium

Adsorbent	Adsorbate	Surface Area (m^2/g), Pore Volume (cm^3/g), Pore Size (nm)	pHzpc	Experimental Conditions	Adsorption Capacity (mg/g)	Thermodynamic Parameters	Kinetic Model and Curve Fitting	Isotherm Model and Curve Fitting	Maximum Adsorption Conditions	References
Poly(b-cyclodextrin)/bentonite composite	Cs			pH = 3–11 Dose = 5–40 g/l Agitation speed = 100 rpm Concentration = 100 mg/l Contact time = 120 min Temperature = 293–313 K		ΔH° = 70.10 kJ/mol ΔS° = 298.26 J/K mol ΔG° = negative	Pseudo-second order	Freundlich model Linear	pH = 9–11 Contact time = 60 min Temperature = 313 K	Liu, Xie, et al. (2017)
Nanomanganese oxide	Cs	Surface area = 165	2.7	pH = 3–8 Dose = c.a. 7 g/l Agitation speed = 200 rpm Concentration = 190 μg/l Contact time = up to 4 h	72.5				pH = 3–9 Contact time = 2 h	Al Abdullah et al. (2016)
Mesoporous magnetic AMP polyhedric composite	Cs	Surface area = 165.2 Pore size = 7		pH = 2–12 Dose = 1.66 g/l Agitation speed = 200 rpm Concentration = 25–200 mg/l Contact time = c.a. 1440 min Temperature = room temperature	83.33		Pseudo-second-order model Linear	Langmuir model Linear	pH = 2–10 (most experiments at pH 7) Contact time = 5 min	Yang, Yu, et al. (2017)
Fe_3O_4 @WO_3	Cs	Surface area = 63		pH = 1–11 Dose = 1 g/l Agitation speed = 120 rpm Concentration = 25–300 mg/l Contact time = 1–2880 min Temperature = 293–313 K	53.175		Pseudo-second order Linear	Langmuir model Linear	pH = optimum pH 4.5 but the removal was same in pH 4–8 Contact time = 180 min Temperature = 293 K	Mu et al. (2017)

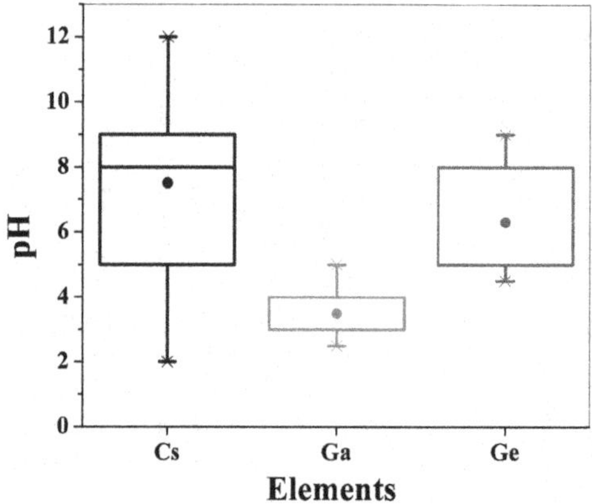

FIGURE 5.1 Box plot of optimum pH for adsorption vs miscellaneous element graph of literature surveyed (literature surveyed is in Tables 5.1 and 5.2. In places where the pH range is specified for optimum pH conditions, both upper and lower limits of pH are taken into consideration).

competition from protons till 3 M nitric acid concentration and the presence of competition after 3 M concentration.

In some cases, pH has an insignificant effect on the removal of the cesium, e.g. in the usage of a mesoporous magnetic ammonium 12-molybdophosphate polyhedric composite (pH range 2–10), carboxymethyl cellulose sodium/prussian blue composite (pH range 3–11) (Zong et al. 2017), and magnetic potassium titanium hexacyanoferrate (pH range 1–9) (Zhang, Zhao, et al. 2015) as an adsorbent. The variation in optimum pH values for the batch adsorption process of cesium along with gallium and germanium is presented in Figure 5.1.

5.2.2 Effect of Coexisting Ions

The effect of the presence of the ions is studied on the basis of distribution coefficient (Jang and Lee 2016; Ding et al. 2013; Yi et al. 2014). The higher value of distribution coefficient suggests better adsorption performance and values above 5000 and 50,000 suggest excellent and good adsorption capacity, respectively (Jang and Lee 2016). The value of the distribution coefficient for magnetic prussian blue in the presence of sodium and magnesium chloride was $3–6 \times 10^5$, indicating nonhindrance by sodium and magnesium ions.

The distribution coefficient of nickel, iron, strontium, molybdenum, zirconium, barium, neodymium and sodium ions was less than 10 ml/g for Fe_3O_4@SiO_2@potassium titanium ferrocyanide (Yi et al. 2014) as compared to 7849 ml/g for cesium ion. The low value of distribution coefficient for interfering ions was attributed to charge and size mismatch.

The decline in removal efficiency was pronounced by the presence of potassium ions as compared to other ions, e.g. sodium ions (Ding et al. 2013; Han, Zhang, et al. 2013), sodium and calcium ions (Long et al. 2013), sodium, lithium, calcium and magnesium ions (Zong et al. 2017) and sodium, lithium, magnesium and calcium ions (Mu et al. 2017). This was due to the closer ionic radii and hydration energy (Ding et al. 2013; Han, Zhang, et al. 2013; Long et al. 2013; Mu et al. 2017). The effect of ammonium ions was more than that of potassium ions in the case of cesium adsorption with potassium nickel hexacyanoferrate (Qing et al. 2015). Similarly, on the basis of similar ionic radii, the presence of rubidium and barium only affected the adsorption of cesium, out of 28 metals studied for interference study (Zhang, Wang, et al. 2017).

The adsorption of cesium with prussian blue declined in the presence of organic acids (Mihara et al. 2016). However, the organic acids did not significantly affect the adsorption of cesium with prussian blue-impregnated alginate gel. This was due to the protection of prussian blue from organic acids. However, adsorption declined in the presence of sodium and potassium ions.

5.2.3 Effect of Surface Modification

The surface area and pore volume decreased after modification of akadama clay with nickel oxide (Ding et al. 2013). The pores were modified from 20 and 80 to 60 nm in diameter. In spite of this, the adsorption capacity increased after modification. The modification also led to increase of the negative charge on the surface of the adsorbent, which was the reason for increased adsorption capacity. Along with this, the cation exchange capacity declined after modification, which is contrary to the increase of negative charge (decrease of zeta potential). The decline of cation exchange capacity is attributed to the decrease of surface area and pore volume.

5.2.4 Effect of Temperature

The percentage removal for cesium with chabazite-type zeolite increased with increase of temperature from 293 to 323 K, but on further increase of the temperature to 353 K, the percentage removal declined (Du et al. 2017). This is attributed to motion of cesium ions and expansion of the framework with change of temperature. The increase of temperature led to increase in the motion of the cesium ions, and this lead to the difficulty of cesium ions to be adsorbed. In addition, the increase of the temperature also causes the expansion of the framework of the chabazite-type zeolite. The expansion led to the increase of the exposure of the adsorption sites and led to the increase of the adsorption. The aforementioned factors in the adsorption of cesium on chabazite-type zeolite led to counter each other, and this is the reason for the increase of removal up to 323 K and the decline afterward.

The increase of temperature from 293 to 323 K and 353 K leads to the decline of equilibrium time from 80 to 20 min (Du et al. 2017). The reason is attributed to the hydration sheath around the cesium ions. The energy was not sufficient at 293 K for removal of the hydration sheath. The increase of temperature provided the required energy, and this led to decrease of the equilibrium time. The increase of temperature also declined the adsorption capacity of cesium in some cases, e.g. adsorption by magnetic prussian blue (Jang and Lee 2016), supramolecular materials (Zhang, Wang, et al. 2017) and ethylamine-modified montmorillonite (Long et al. 2013).

5.2.5 Effect of Radiation

The stability of the structure for $[(CH_3)_2NH_2][UO_2(L_1)]\cdot DMF\cdot 6.5H_2O$ and $[(CH_3)_2NH_2][UO_2(L_2)]\cdot 0.5DMF\cdot 15H_2O$ (here H3L1 and H3L2 stands for 1,3,5-tri(4′-carboxylphenyl)benzoic acid and 3,5-di (4′-carboxylphenyl) benzoic acid, respectively) was maintained after β and γ irradiation. This indicates the higher radiation resistance of the adsorbent materials. The adsorption capacity of cesium with carboxymethyl cellulose sodium/prussian blue composite declined from 85.88% to 82.02% on increase of the irradiation dose from 0 to 400 kGy. In addition, there was no significant change in the FT-IR spectrum of the sample after irradiation.

5.2.6 Mechanism

Ion exchange mechanism was used for adsorption of cesium onto ethylamine-modified montmorillonite (Long et al. 2013). This mechanism was applied by estimating calcium ions in the solution after adsorption. However, the amount of calcium ions dissolved into the solution was less than the amount of cesium ions in the solution. This indicates the presence of other mechanisms also, in addition to ion exchange.

The grafting of andic soil with nickel oxide led to decrease in the cation exchange capacity, along with increase of adsorption capacity (Ding et al. 2013). This suggested that the ion exchange phenomenon was not the dominant process.

The mechanism of adsorption of cesium with graphene oxide was simulated using FITEQL v 4.0 program. The surface complexation modeling using the diffuse layer model suggests the follow-up of the outer sphere complex model at pH smaller than 5 and the inner sphere complex model at pH higher than 6 (Tan et al. 2016).

The rate-limiting step for adsorption of cesium ions onto potassium nickel hexacyanoferrate was estimated by the Weber and Morris model and Boyd model (Qing et al. 2015). The Weber and Morris model suggests the involvement of the film diffusion in controlling the adsorption kinetics as the plot of q_t vs $t^{1/2}$ was not linear. To further confirm the rate-limiting step, Boyd's plot was used for fitting of kinetic data. The Boyd's plot was neither linear nor passing through the origin. This led to the conclusion that adsorption of cesium followed intraparticle diffusion and film diffusion together.

5.2.7 Desorption

The desorption of cesium was achieved with hydrochloric acid (Ding et al. 2013; Al Abdullah et al. 2016; Chen, Kang, et al. 2014; Awual et al. 2020), ammonium chloride (Yang, Yu, et al. 2017), potassium chloride, sodium hydroxide (Ding et al. 2013), and ammonium nitrate (Mu et al. 2017). The desorption of cesium from nickel oxide-grafted andic soil was c.a. 70%, 37%, and 77% with hydrochloric acid, sodium hydroxide, and potassium chloride, respectively (Ding et al. 2013). Desorption with hydrochloric acid was attributed to the electrostatic nature of adsorption and with sodium hydroxide to a phenomenon apart from the electrostatic nature. The relatively similar amount of desorption by potassium chloride and hydrochloric acid suggests the ion exchange phenomenon. So, the desorption study of nickel oxide-grafted andic soil suggested the electrostatic and ion exchange nature of adsorption.

5.2.8 Miscellaneous

The cucurbit[6]uril-based supramolecular assembly exhibits high selectivity of the cesium ions even in the presence of sodium, rubidium, and potassium ions (Chen, Kang, et al. 2014). The mixing of the 4,4′, 4″-benzene-1,3,5-triyl tribenzoic acid (H3BTB) and supramolecular assembly (Q[6] leads to the formation of the $\{(NH_4)_3Q[6]_{1.5}BTB\}$ under alkaline conditions. The addition of sodium, rubidium, and potassium ions leads to the formation of a structure similar to $\{(NH_4)_3Q[6]_{1.5}BTB\}$. However, with cesium ions, the structure was the same, but ammonium ions were replaced with cesium ions. The selective replacement was attributed to the suitable size of the cesium(I) ion (1.67 Å). The replacement also occurred in the presence of 4,4′, 4″-benzene-1,3,5-triyl tribenzoic acid only. The 4,4′, 4″-benzene-1,3,5-triyl tribenzoic acid acts as a structure-directing agent and leads to the formation of a supramolecular assembly and helps in selective binding of the cesium ion. The selective binding is also evidenced by EDS (energy-dispersive X-ray spectroscopy) and ICP (inductively coupled plasma) analysis.

The activity of cesium reduced from 2993 to 28 Bq/l after treatment with the Fe_3O_4@SiO_2@potassium titanium ferrocyanide adsorbent (Yi et al. 2014). The adsorption of cesium with nanomanganese oxide was studied in terms of Bq/g (Al Abdullah et al. 2016).

5.3 SCANDIUM

Scandium finds its application in mercury vapor lamps, Russian MIG fighter planes and tracers in the oil industry (RSC 2020h). Due to the limited use, there have been hardly any reports of scandium pollution till date. However, the adsorption studies were available mostly for its extraction from rare-earth element solutions.

The percentage removal of scandium increased with increase of pH (Ramasamy et al. 2017; Ma et al. 2014; Zhao et al. 2016). The high competition of the hydronium ion at low pH was held responsible for the low removal at low pH (Ma et al. 2014).

In the case of scandium adsorption on "silica doped with bifunctional ionic liquid", the adsorption of the scandium decreased with increasing concentration of the nitric acid (Turanov et al. 2016), and scandium is separated from lanthanides(III) by simply adjusting the pH of the aqueous phase. The adsorption of scandium from the rare-earth element solution with lysine-modified SBA-15 increased with increase of pH, but below pH 5, there was no adsorption of any rare-earth element (Ma et al. 2014).

In some cases, the maximum adsorption capacity was achieved at lower pH, e.g. the percentage removal of scandium from the rare-earth element solution with "PAN immobilized chemically onto APTES-functionalized silica gel" (PAN = 1-(2-pyridylazo)-2-napththol and APTES = 3-Aminopropyl triethoxy silane) was c.a. 100% at pH 4 and did not change on increase of pH up to 7. However, the immobilized material changes under different conditions, e.g. percentage removal with "acetylacetone immobilized chemically onto APTES-functionalized silica gel" increased on increase of pH from 4 to 7 (Ramasamy et al. 2017). In addition, the percentage removal declined in a natural water solution of rare-earth element with "PAN immobilized chemically onto APTES-functionalized silica gel" at pH 4, and the percentage removal increased on increase of pH under real water conditions (Ramasamy et al. 2017).

The percentage removal of scandium was 93% from the rare-earth element solution with "PAN immobilized chemically onto APTES-functionalized silica gel" in a time period of 15 min (Ramasamy et al. 2017). The faster removal was attributed to the smaller ionic radius of scandium. The rapid adsorption of scandium with lysine-modified SBA-15 is attributed to the –CO group acting as an effective adsorption site (Ma et al. 2014).

The lysine-modified SBA-15 has selective adsorption at pH 5 toward scandium at a low concentration of the rare-earth element solution (Ma et al. 2014). This was due to the presence of interaction strength between the functional groups (-NH_2 and -CO) and metal ions and smaller ionic radius of scandium. This led to higher polarizability of the scandium ion. The adsorbent dose also plays a role in selectivity like in the case of adsorption of scandium removal from red mud solution with "tri-butyl phosphate-modified activated carbon"; at an adsorbent dosage of 6.25 g/l, the percentage removal of scandium was higher than the rest of the elements. This helps in choosing the optimum dose in respect of selectivity for scandium in the presence of other ions. The increase of the adsorbent dose leads to escalation in removal efficiency for other elements (Hualei et al. 2008). The desorption of the scandium can be achieved with acidic media, e.g. nitric acid and (Turanov et al. 2016) sulfuric acid (Zhao et al. 2016).

5.4 TITANIUM

Titanium finds its application in a wide number of products like aircraft, golf clubs, laptops, power plant condensers, hulls of ships, submarines, paint, enamels and sunscreens. There is no known biological role of titanium (RSC 2020j). In addition, it is nontoxic in aqueous media. So there are only a few reports for titanium adsorption from water like its nanomaterial removal (Kiser et al. 2009). In addition, there were some theoretical studies for adsorption of titanium on other solid surfaces (Ciszewski et al. 1998; Gale et al. 1999; Kucharczyk et al. 2010). The adsorption of titanium on tungsten leads to decrease in work function (Szczudło et al. 2001).

5.5 GALLIUM

Gallium has no biological role and is considered as a nontoxic element (RSC 2020e). However, there are studies on recovery of gallium along with arsenic (gallium arsenide) from the semiconductor industry (Sturgill et al. 2000) and from bayer liquor (Zhao et al. 2012). An overview of the experimental parameters and optimized conditions from batch adsorption experiments for gallium is presented in Table 5.2.

TABLE 5.2

Summary of Parameters and Optimized Conditions for Batch Adsorption of Gallium and Germanium

Adsorbent	Adsorbate	Surface Area (m^2/g), Pore Volume (cm^3/g), Pore Size (nm)	pHzpc	Experimental Conditions	Adsorption Capacity (mg/g)	Thermodynamic Parameters	Kinetic Model and Curve Fitting	Isotherm Model and Curve Fitting	Maximum Adsorption Conditions	References
Bentonite	Ga			pH = 1–3 Dose = 5–35 g/l Agitation speed = 300 rpm Concentration = 5.74×10^{-3} M Contact time = 5–300 min Temperature = 20°C–70°C					pH = 2.5 Dose = 35 g/l Temperature = 20°C	Chegrouche and Bensmaili (2002)
Quaternary amphoteric starch	Ga			pH = 1–3 Dose = 10 g/l Concentration = 57–570 mg/l Contact time = 2 h Temperature = 30°C–60°C Sodium chloride concentration = up to 1 M	0.48 meq/g	ΔH° = 7.65 Kcal/mol ΔS° = 20.18–20.48 cal/mol ΔG° = positive			pH = 0 M HCl Concentration = 57 mg/l Temperature = 60°C Sodium chloride concentration = 0 M	Chan (1993)
Tertiary amphoteric starch				Do	0.54 meq/g	ΔH° = 7.84 Kcal/mol ΔS° = 21.12–21.19 cal/mol ΔG° = positive			Do	Chan (1993)

(continued)

TABLE 5.2 (*Continued*)
Summary of Parameters and Optimized Conditions for Batch Adsorption of Gallium and Germanium

Adsorbent	Adsorbate	Surface Area (m^2/g), Pore Volume (cm^3/g), Pore Size (nm)	pHzpc	Experimental Conditions	Adsorption Capacity (mg/g)	Thermodynamic Parameters	Kinetic Model and Curve Fitting	Isotherm Model and Curve Fitting	Maximum Adsorption Conditions	References
Nano-TiO_2	Ga			pH = 1–5 Dose = 10 g/l Concentration = up to 25 mg/l (isotherm study) Temperature = 2°C–40°C	8.92	ΔH° = 18.96–24.03 kJ/mol ΔG° negative Δ°S = 110.1–140.2 J/K mol	Pseudo- second order linear	Linear Dubinin-Radushkevich isotherm (only DR isotherm is analyzed)	pH = 3–5 (pH 3 chosen in the study) Temperature = 40°C	Zhang, Zhu, et al. (2010)
Nano-SiO_2 (batch and column study)	Ga			pH = 1–14 Dose = 1 g/l Concentration = 10–50 mg/l (isotherm) Contact time = up to 4 min Temperature = 275–323 K	5.77 and 4.6 in column study	Δ°H = 6.18–8.72 kJ/mol Δ°S = 97–102 J/mol K Δ°G = negative	Pseudo-second order linear (no comparison with first order)	Langmuir Linear (no comparison with other isotherms)	pH = 3–4 and 8–12 Contact time = 1 min Temperature = 323 K	Zhang et al. (2011)
Alum water treatment sludge	Ga	Surface area = 47.17	5.6	pH = 3–10 Dose = 10 g/l Concentration = 1 mg/l Contact time = 2 h Temperature = 25°C	28.74		Pseudo-second-order model	Freundlich model	pH = 3–4	Hua et al. (2015)
Bauxite	Ga	Surface area = 28.97	7.9	Do	19.72		Pseudo-second-order model	Freundlich model	Do	Hua et al. (2015)

(*continued*)

TABLE 5.2 (*Continued*)
Summary of Parameters and Optimized Conditions for Batch Adsorption of Gallium and Germanium

Adsorbent	Adsorbate	Surface Area (m^2/g), Pore Volume (cm^3/g), Pore Size (nm)	pHzpc	Experimental Conditions	Adsorption Capacity (mg/g)	Thermodynamic Parameters	Kinetic Model and Curve Fitting	Isotherm Model and Curve Fitting	Maximum Adsorption Conditions	References
Blast furnace slag	Ga	Surface area =3.37	5.8	Do	3.21		Pseudo-second-order model	Freundlich model	Do	Hua et al. (2015)
Bauxite processing residue sand	Ga	Surface area =17.13	6.6	Do	8.70		Pseudo-second-order model	Freundlich model	Do	Hua et al. (2015)
Activated carbon (catechol-functionalized Ge)	Ge	Surface area =1580 Pore volume =0.64	3.8	pH=5, 8 and 10 Dose=4 g/l Concentration =50 mg/l Contact time =up to 420 min Temperature =23°C	4.6				pH=5 Contact time=60 min	Marco-Lozar et al. (2007)
Activated carbon T (catechol functionalized Ge)		Surface area=1540 Pore volume=0.62	7.1	Do	8.7				pH=5 Contact time=420 min	Marco-Lozar et al. (2007)
Catechol-functionalized nanosilica	Ge			pH=2–12 Dose=2 gm/l Agitation speed =300 rpm Concentration =5–50 mg/l (isotherm) contact time=2 h Temperature =25°C	6.07		Pseudo-second-order model linear	Langmuir model linear	pH=4.5 Contact time=30 min	Cui et al. (2016)

5.5.1 Effect of pH and Coexisting Ions

The adsorption of gallium increased on raising the pH up to 3 or 4 (Hua et al. 2015; Zhang et al. 2011; Zhang, Zhu, et al. 2010); after that, it starts to precipitate as $Ga(OH)_3$ (Hua et al. 2015). After pH 7, it forms soluble $Ga(OH)^{-}{}_4$. In some cases, the adsorption of Gallium declined after pH 7 (Hua et al. 2015; Zhang et al. 2011). This was attributed to the isoelectric point of the adsorbent (Zhang et al. 2011). The surface of the adsorbent was positive below the isoelectric point and negative above it. The positively charged gallium species occurred below pH 4. Hence, adsorbents having isoelectric point in the acidic region prefer to adsorb more in the acidic region than in the basic region. The absence of gallium ions in the solution reduced the adsorption of aluminum or vice versa with graphene oxide (Jankovský et al. 2015).

5.5.2 Effect of Temperature and Desorption

The increase in temperature causes percentage removal to be decreased (Chegrouche and Bensmaili 2002) or increased (Chou et al. 2010) varying from adsorbent to adsorbent. The reason for increase in adsorption with rise in temperature was credited to the increase in rate of diffusion of particles across the external surface and into the interior pores of the adsorbent molecules (Chou et al. 2010). Sodium hydroxide was applied for the desorption of gallium from nano-TiO_2 (Zhang, Zhu, et al. 2010).

Sodium hydroxide and HCl were applied for the desorption of gallium from nano-TiO_2 (Zhang, Zhu, et al. 2010) and polyacrylic acid-functionalized graphene oxide, respectively (Zhang, Liu et al. 2019).

5.6 GERMANIUM

Germanium has no known toxicity toward human beings, but being effective against bacteria, it has been utilized in many fields such as electronics and infrared spectroscopy (RSC 2020f; Cui et al. 2016). An overview of the experimental parameters and optimized conditions from batch adsorption experiments for germanium is presented in Table 5.2.

5.6.1 Effect of pH and Desorption

The Ge–catechol complex is not stable at pH = 1. The adsorption for Ge was performed by previously converting germanium to a complex (Marco-Lozar et al. 2007). The optimum pH was found to be pH 5 (Marco-Lozar et al. 2007) and 4.5 (Cui et al. 2016). The higher adsorption at acidic pH is credited to the positive surface charge on the adsorbent below pHzpc, and after pHzpc, it was negative, and there is repulsion between the adsorbent and the adsorbate (Marco-Lozar et al. 2007). Apart from electrostatic interaction, adsorption of germanium on catechol-modified silica is also attributed to complex formation (Cui et al. 2016). Hydrochloric acid is used for germanium–catechol complex desorption from activated carbon (Marco-Lozar et al. 2007).

6 Impact of Factors on Remediation of Anions (Fluoride, Nitrate, Perchlorate, and Sulfate) Via Batch Adsorption Processes

Deepak Gusain
Durban University of Technology

Shikha Dubey and Yogesh Chandra Sharma
IIT(BHU), Varanasi

Faizal Bux
Durban University of Technology

CONTENTS

Fluoride in small concentrations prevents tooth decay, but its exposure in higher concentration causes fragile bones and decrease in fertility. Nitrate is among the chief factors in eutrophication, and perchlorate causes hindrance of iodine uptake by the thyroid gland. This chapter includes factors affecting the adsorption of anions such as fluoride, nitrate, perchlorate, and sulfate. The effect of various parameters such as pH and coexisting ions on the adsorption of respective elements and the mechanism of adsorption are discussed in this chapter.

6.1 FLUORIDE

Fluoride is a common contaminant existing in a number of minerals like fluorspar, cryolite, and fluorapatite (WHO 2017). In addition to natural sources, various industries such as glass, semiconductor, electroplating, and aluminum also contribute to the release of fluoride into the environment (Shen et al. 2003; Bhatnagar et al. 2011). Fluoride is used to combat dental caries (WHO 2017). The effect of fluoride is both beneficial and harmful depending on the concentration and duration of exposure. The recommended value for artificial fluoridation of water supplies was 0.5–1 mg/l (WHO 2017). The excess of fluoride leads to osteoporosis and negative effects on the brain, thyroid, pineal gland, and kidney (Bhatnagar et al. 2011; Agalakova and Nadei 2020). The WHO has set up a guideline value 1.5 mg/l for fluoride (WHO 2017). An overview of the experimental parameters and optimized conditions from batch adsorption experiments for fluoride is presented in Table 6.1.

6.1.1 Effect of pH

The maximum amount of fluoride adsorption was achieved at different pH values with different adsorbents. The optimum pH depends on the mechanism of adsorption and speciation of the fluoride species. The adsorption was maximum at acidic pH in the case of biochar (Mohan et al. 2012), mesoporous alumina (Kundu et al. 2017), Fe-doped graphene oxide (Sharma et al. 2017), iron-aluminum oxide nanoparticles anchored on graphene oxide (Liu, Cui, et al. 2016), pectin/Al_2O_3-ZrO_2 core/shell beads (Zhu et al. 2016), La-Zr composite (Chen, Zhang, Li, et al. 2016), zirconium phosphate (Zhang, Li, et al. 2017), Ce-chitosan composite (Zhu et al. 2017), hydroxyl aluminum oxalate (Wu et al. 2016), Al_2O_3/expanded graphite (Jin et al. 2015), Zr(IV)-immobilized chitosan (Liu, Zhang, et al. 2015), and cetyltrimethyl ammonium bromide-coated hydroxyapatite (Prabhu and Meenakshi 2014).The maximum amount of adsorption in the acidic region was attributed to the increased protonation (Mohan et al. 2012; Parashar et al. 2016) or positive charge (Kundu et al. 2017; Zhang, Li, et al. 2012) or zeta potential (Sharma et al. 2017). The decreased adsorption in alkaline pH is also attributed to the increase in competition for active sites between fluoride and hydroxyl ions (Ma et al. 2017; Zhu et al. 2016; Chen et al. 2013; Chen, Shu, et al. 2017; Zhang, Li, et al. 2012).

However, the decrease in adsorption on decreasing the pH to less than 3 is attributed to the formation of the neutral hydrofluoric acid (Liu, Cui, et al. 2016; Chen, Zhang, Li, et al. 2016; Li, Zhu, et al. 2017; Chen, Shu, et al. 2017; Parashar et al. 2016; Jin et al. 2015) or in some cases decomposition of the adsorbent (Ma et al. 2017). In some cases, adsorption of fluoride was pH independent in the long pH range of 3–10 with polypyrrole/TiO_2 composite (Chen, Shu, et al. 2017) and akaganeite-anchored graphene oxide, 2.75–10.3 in the case of goethite-anchored graphene oxide (Kuang et al. 2017), 3–9 with Fe-Al-La trimetal hydroxide (Li, Zhu, et al. 2017), 3.5–8.5 in the case of polypyrrole/hydrous tin oxide composite (Parashar et al. 2016), 3–12 in the case of dicalcium phosphate in laid-in chitosan bead (Shen et al. 2016) and 3–10 in the case of sulfate-doped hydroxyapatite (Chen, Zhang, He, et al. 2016). Adsorption in the case of pH independent adsorption process is governed by ion exchange process (Chen, Shu, et al. 2017; Parashar et al. 2016; Chen, Zhang, He, et al. 2016). In addition, the pH change occurred due to the ion exchange process governing the adsorption process (Mohan et al. 2012; Liu, Cui, et al. 2016). However, all ion exchange processes are not pH independent (Dhillon et al. 2015; Wang, Yu, et al. 2017; Tang and Zhang 2016), and pH changes with Mg/Fe/La mixed oxide adsorbent after adsorption, which is attributed to the conversion of Mg/Fe/La mixed oxides to Mg/Fe/La mixed hydroxides (Wu et al. 2017).

The maximum adsorption of fluoride with lithium/aluminum layered double hydroxide was near the neutral pH, and the adsorption declined on increasing or decreasing the pH away from the optimum pH (Zhou et al. 2011). The isoelectric point of the adsorbent was 7.2, the surface was positively charged at pH lower than 7.2, and acidic pH conditions affected the stability of the adsorbent and formation of HF. The pH conditions above 7.2 promoted negative charge on the surface of the adsorbent. Hence, the adsorption was maximum near neutral pH and governed by the electrostatic nature of adsorption. Similarly, the adsorption of fluoride with Fe-Ca-Zr hybrid metal oxide depicts maximum adsorption at pH 7 and is governed by the electrostatic nature of adsorption, and the mean activation energy estimation suggested it to be the ion exchange phenomenon (Dhillon et al. 2015).

The decline of adsorption was not uniform in the removal of fluoride with Mg-Al-Zr triple metal composite (Wang, Yu, et al. 2017). The adsorption of fluoride declined with increase in the pH value, but after pH 5, the decline was not significant. This is attributed to the electrostatic interaction of adsorption at lower pH and ion exchange mechanism at higher pH. In the case of adsorption of fluoride with alumina-modified expanded graphite oxide composite, the removal at pH lower than pH_{zpc} is governed by electrostatic and ligand exchange, and at $pH > pH_{zpc}$, the mechanism was only governed by ligand exchange reaction (Jin et al. 2015). The variation in optimum pH values for the batch adsorption process of fluoride along with nitrate and perchlorate is presented in Figure 6.1.

TABLE 6.1
Summary of Parameters and Optimized Conditions for Batch Adsorption of Fluoride

Adsorbent	Adsorbate	Surface Area (m^2/g), Pore Volume (cm^3/g), Pore Size (nm)	pHzpc	Experimental Conditions	Adsorption Capacity (mg/g)	Thermodynamic Parameters	Kinetic Model and Curve Fitting	Isotherm Model and Curve Fitting	Maximum Adsorption Conditions	References
Activated carbon	F^-	Surface area = 927	9.97	pH = 7 Dose = 0.33 g/l Concentration = 0.1–60 mg/l Contact time = up to 220 min Temperature = 25°C	2.25			Langmuir and Freundlich	Contact time = 50 min	Velazquez-Jimenez et al. (2014)
Zr-modified activated carbon	F^-	Surface area = 546	3.28	Do	1.85			Langmuir and Freundlich	Do	Velazquez-Jimenez et al. (2014)
Zr and oxalic acid-modified activated carbon	F^-	Surface area = 867	11.18	Do	7.40			Langmuir and Freundlich	Do	Velazquez-Jimenez et al. (2014)
Alumina nanofibers	F^-			pH = 3–9 Dose = 0.5–5 g/l Concentration = 10–100 mg/l Contact time = up to 3 h Temperature = room temperature	1.2		Pseudo-second order model Linear	Freundlich model Linear	pH = 7 Dose = 5 g/l	Mahapatra et al. (2013)
Chitosan–sodium alginate–aluminum (biomaterial scaffold, i.e. BMS) Column study	F^-			pH = 2–9.5	168 at pH 4 and 60 at pH 7		Pseudo-second order model linear	Langmuir	pH = 4 Column study	Kumar et al. (2016)
Calcined Li-Al LDH prepared by precipitation	F^-	Surface area = 37.24 Pore volume = 0.148 Pore size = 24.12		pH = 2–12.5 Dose = 0.01–0.5 g/l Concentration = 20 mg/l Contact time = up to 24 h Temperature = 10°C, 25°C and 40°C			Pseudo-second order model Linear	Freundlich	pH = 2 Dose = 0.1 g/l Temperature = 40°C	Zhang, Li, et al. (2012)

(continued)

TABLE 6.1 (*Continued*)

Summary of Parameters and Optimized Conditions for Batch Adsorption of Fluoride

Adsorbent	Adsorbate	Surface Area (m^2/g), Pore Volume (cm^3/g), Pore Size (nm)	pHzpc	Experimental Conditions	Adsorption Capacity (mg/g)	Thermodynamic Parameters	Kinetic Model and Curve Fitting	Isotherm Model and Curve Fitting	Maximum Adsorption Conditions	References
Calcined Li-Al LDH prepared by homogenous precipitation	F^-	Surface area = 51.27 Pore volume = 0.266 Pore size = 24.18		pH = 2–12.5 Dose = 0.01–0.5 g/l Concentration = 20–60 mg/l Contact time = up to 60 min Temperature = 10°C, 25°C and 40°C			Pseudo-second order Linear	Freundlich	pH = 2 Temperature = 40°C Contact time = 20 min at 20 mg/l and > 50 min at 60 mg/l	Zhang, Li, et al. (2012)
Pine wood biochar	F^-	Surface area = 2.73		pH = 2–10 Dose = 10 g/l Concentration = 10 and 25 mg/l and kinetic studies have concentration range of 20–80 mg/l Contact time = 48 h Temperature = 25°C, 35°C, 45°C	7.66		Pseudo-second order model Linear	Langmuir and Redlich-Peterson nonlinear	pH = 2 Temperature = 25°C	Mohan et al. (2012)
Pine bark biochar	F^-	Surface area = 1.88		Do	10.53		Pseudo-second order model Linear	Langmuir and Redlich–Peterson nonlinear	pH = 2 Temperature = 35°C	Mohan et al. (2012)
Mesoporous alumina	F^-	Surface area = 238 Pore volume = 0.83 Pore size = 11.40 nm (diameter)		pH = 3–11 Dose = 0.5 g/l Concentration = 20 mg/l Contact time = 3 h Temperature = 303–313 K	47.2	ΔG° = negative	Pseudo-second order model Linear	Freundlich model Linear	pH = 3–4 Temperature = 323–333 K	Kundu et al. (2017)

(*continued*)

TABLE 6.1 (*Continued*)
Summary of Parameters and Optimized Conditions for Batch Adsorption of Fluoride

Adsorbent	Adsorbate	Surface Area (m^2/g), Pore Volume (cm^3/g), Pore Size (nm)	pHzpc	Experimental Conditions	Adsorption Capacity (mg/g)	Thermodynamic Parameters	Kinetic Model and Curve Fitting	Isotherm Model and Curve Fitting	Maximum Adsorption Conditions	References
Mesoporous alumina prepared by its sol calcination in the presence of malic acid	F^-	Surface area = 247 Pore volume = 0.83 Pore size = 10.08		Do	49	Do	Do	Do	pH = 3–4 Temperature = 333 K	Kundu et al. (2017)
Mesoporous alumina by its sol calcination in the presence of tartaric acid	F^-	Surface area = 357 Pore volume = 0.88 Pore size = 8.15		Do	51.2	Do	Do	Do	Do	Kundu et al. (2017)
Mesoporous alumina by its sol calcination in the presence of citric acid	F^-	Surface area = 424 Pore volume = 0.118 Pore size = 7.42		Do	62.5	Do	Do	Do	Do	Kundu et al. (2017)
Iron-doped titanium oxide	F^-			pH = 2–12 Dose = 0.5 g/l Agitation speed = 150 rpm Concentration = mg/l Contact time = up to 24 h Temperature = 298–308 K	37.66	$\Delta H^\circ = -13.9$ kJ/mol $\Delta S^\circ = -29.38$ J/mol K ΔG° = negative	Pseudo-second order model	Langmuir model Linear	pH = nonsignificant effect of pH on adsorption capacity, but maximum at pH c.a. 5.2 Contact time = 12 h Temperature = 298 K	Chen et al. (2012)

(*continued*)

TABLE 6.1 (*Continued*)
Summary of Parameters and Optimized Conditions for Batch Adsorption of Fluoride

Adsorbent	Adsorbate	Surface Area (m^2/g), Pore Volume (cm^3/g), Pore Size (nm)	pHzpc	Experimental Conditions	Adsorption Capacity (mg/g)	Thermodynamic Parameters	Kinetic Model and Curve Fitting	Isotherm Model and Curve Fitting	Maximum Adsorption Conditions	References
Fe-doped graphene oxide (SAR-Fe-700)	F^-	Surface area = 220.73 Pore volume = 0.063 Pore size = 2.8 nm (diameter)		pH = 2–12 Dose = 0.5 g/l Concentration = 4–100 mg/l (isotherm study) Contact time = up to 150 min Temperature = room temperature	79			Langmuir model	pH = 2 Contact time = 120 min	Sharma et al. (2017)
$[Ce(L1)_{0.5}(NO_3)(H_2O)_2]\cdot 2DMF$	F^-			pH = 2–10 Dose = 2 g/l Concentration = 12.5 mg/l Contact time = up to 120 min Temperature = 298–318 K	103.95	$\Delta H° = 18.52$ kJ/mol $\Delta S° = 76.56$ J/mol K $\Delta G°$ = negative			pH = 3–7 Contact time = 5 min Temperature = 318 K	Ma et al. (2017)
$[Eu_3(L2)_2(OH)(DMF)_{0.22}(H_2O)_{5.78}]$.guest	F^-			Do	57.01	$\Delta H° = 25.32$ kJ/mol $\Delta S° = 87.55$ J/mol K $\Delta G°$ = negative			Do	Ma et al. (2017)
Iron-aluminum oxide/graphene oxide	F^-	Surface area = 349		pH = 2–11 Dose = 0.1–0.6 g/l Agitation speed = 200 rpm Concentration = 2–50 mg/l (isotherm study) Contact time = up to 20 h Temperature = 25°C	64.72		Pseudo-second order model Linear	Langmuir model	pH = 4.22 Dose = 0.25 g/l as residual concentration below the recommended level of 1.5 mg/l Contact time = 3 h	Liu, Cui, et al. (2016)

(*continued*)

TABLE 6.1 (*Continued*)
Summary of Parameters and Optimized Conditions for Batch Adsorption of Fluoride

Adsorbent	Adsorbate	Surface Area (m^2/g), Pore Volume (cm^3/g), Pore Size (nm)	pHzpc	Experimental Conditions	Adsorption Capacity (mg/g)	Thermodynamic Parameters	Kinetic Model and Curve Fitting	Isotherm Model and Curve Fitting	Maximum Adsorption Conditions	References
Iron-aluminum oxide	F^-	Surface area = 303		Do	46.54		Pseudo-second order model Linear	Langmuir model	pH = 3.04 Dose = not able to remove below the recommended level in the dose range i.e. 1.5 mg/l at pH 6.5 Contact time = 3 h	Liu, Cui, et al. (2016)
Pectin/Al_2O_3-ZrO_2 core/shell beads	F^-			pH = 2–10 Dose = 1 g/l Agitation speed = 140 rpm Concentration = 40–120 mg/l Contact time = up to 8 h Temperature = 15°C–35°C	98.07	$\Delta H° = -11.201$ $\Delta S° = -0.030$ $\Delta G°$ = negative	Pseudo-second order model Nonlinear	Langmuir and Freundlich model Nonlinear	pH = 4 Contact time = 8 h Temperature = nonsignificant effect on adsorption capacity	Zhu et al. (2016)
La–Zr composite	F^-		3	pH = 1–10 Dose = 1 g/l Agitation speed = 150 rpm Concentration = 20–300 mg/l (isotherm study) Contact time = up to 180 min Temperature = 25°C	88.5		Pseudo-second order model Linear	Lan-Fre model	pH = 3 Contact time > 60 min (experiments conducted till 24 h)	Chen, Zhang, Li, et al. (2016)

(*continued*)

TABLE 6.1 (*Continued*)
Summary of Parameters and Optimized Conditions for Batch Adsorption of Fluoride

Adsorbent	Adsorbate	Surface Area (m^2/g), Pore Volume (cm^3/g), Pore Size (nm)	pHzpc	Experimental Conditions	Adsorption Capacity (mg/g)	Thermodynamic Parameters	Kinetic Model and Curve Fitting	Isotherm Model and Curve Fitting	Maximum Adsorption Conditions	References
Calcined lithium/aluminum layered double hydroxide	F^-	Surface area = 113 Pore volume = 0.34	c.a. 7.2	pH = 3–13 Dose =2 g/l (isotherm study) Concentration = 50–500 mg/l (isotherm study) Contact time = up to 6h Temperature = 30°C–50°C	158.7	ΔH° = 6.93 kJ/mol ΔS° = 25.1 J/mol K ΔG° = negative	Pseudo-second order model	Freundlich model Linear	pH = 7 Contact time = 2 h Temperature = 50°C	Zhou et al. (2011)
Composite of basic ammonium sulfate and graphene hydrogel	F^-	Surface area = 128.9		pH = 3.2–11.8 Dose = 0.08 g/l Agitation speed = 375 rpm Concentration = 20 mg/l (kinetic study) Contact time = up to 420 min Temperature = 298–328 K	33.4		Pseudo-first order model	Langmuir model	pH = 7.2 Contact time = 180 min Temperature = 328 K	Chen et al. (2013)
Polypyrrole/TiO_2 composite	F^-	Surface area = 95.71 Pore volume = 0.065	10.3	pH = 1–13 Dose = 0.5–3 g/l Agitation speed = 200 rpm Concentration = 10–200 mg/l (isotherm study) Contact time = up to 180 min Temperature = 25°C–45°C	33.17	ΔH° = −21.39 kJ/mol ΔS° = −62.80 J/K mol ΔG° = negative	Pseudo-first order model	Langmuir model	pH = 3–10 Dose = 2 g/l (to meet the requirement of WHO) Contact time = 60 min Temperature = 25°C	Chen, Shu, et al. (2017)

(continued)

TABLE 6.1 (*Continued*)
Summary of Parameters and Optimized Conditions for Batch Adsorption of Fluoride

Adsorbent	Adsorbate	Surface Area (m^2/g), Pore Volume (cm^3/g), Pore Size (nm)	pHzpc	Experimental Conditions	Adsorption Capacity (mg/g)	Thermodynamic Parameters	Kinetic Model and Curve Fitting	Isotherm Model and Curve Fitting	Maximum Adsorption Conditions	References
Zirconium phosphate nanoflakes	F^-		3.1	pH = 1–13 Dose = 0.1 g/l Agitation speed = 200 rpm Concentration = 10 mg/l (kinetic study) Contact time = up to 240 min Temperature = 298 K	55.8		Pseudo-second order model Linear	Langmuir model Linear	pH = 4 Contact time c.a. 5 min	Zhang, Li, et al. (2017)
Cetyltrimethyl ammonium bromide-coated hydroxyapatite	F^-		6.9	pH = 3–11 Dose = 1 g/l Agitation speed = 200 rpm Concentration = 8–14 mg/l Contact time = 10–60 min Temperature = 303–323 K	9.39	$\Delta H°$ = 13.29 kJ/mol $\Delta S°$ = 0.04 kJ/mol K $\Delta G°$ = negative	Pseudo-second order model intra-particle diffusion model Linear	Freundlich model	pH = 3 Contact time = 30 min Temperature = 323 K	Prabhu and Meenakshi (2014)
Dodecyltrimethyl ammonium bromide-coated hydroxyapatite	F^-		5.8	Do	7.52	$\Delta H°$ = 13.22 kJ/mol $\Delta S°$ = 0.05 kJ/mol K $\Delta G°$ = negative	Pseudo-second order model intra-particle diffusion model Linear	Freundlich model	pH = 3 Contact time = 30 min Temperature = 323 K	Prabhu and Meenakshi (2014)

(continued)

TABLE 6.1 (*Continued*)
Summary of Parameters and Optimized Conditions for Batch Adsorption of Fluoride

Adsorbent	Adsorbate	Surface Area (m^2/g), Pore Volume (cm^3/g), Pore Size (nm)	pHzpc	Experimental Conditions	Adsorption Capacity (mg/g)	Thermodynamic Parameters	Kinetic Model and Curve Fitting	Isotherm Model and Curve Fitting	Maximum Adsorption Conditions	References
Akaganeite-anchored graphene oxide (FeOOH+Ac/GO)	F^-	Surface area = 255.24 Pore size = 2.4	1.8	pH = c.a. 2.11–13 Dose = 2.5 g/l Agitation speed = 160 rpm Concentration = 10–150 mg/l (isotherm study) Contact time = up to 240 min Temperature = 25°C	17.65		Pseudo-second order model Linear	Langmuir and Freundlich model Linear	pH = 3–10 Contact time = 60 min	Kuang et al. (2017)
Goethite-anchored graphene oxide (FeOOH/GO)	F^-	Surface area = 202.60 Pore size = 7.1	3.2	Do	19.82		Pseudo-second order model Linear	Langmuir and Freundlich model Linear	pH = 2.75–10.3 Contact time = 60 min	Kuang et al. (2017)
Fe–Al–La trimetal hydroxide via Fe and Al leaching from red mud	F^-	Surface area = 371.47 Pore size = 10.2		pH = 2–12 Dose = 0.1 g/l Agitation speed = 130 rpm Concentration = 4–16 mg/l (isotherm study) Contact time = up to 240 min Temperature = 25°C–35°C	74.07		Pseudo-second order model linear	Langmuir model linear	pH = 3–9, with maximum at pH 7 Temperature = 25°C	Li, Zhu, et al. (2017)

(*continued*)

TABLE 6.1 (*Continued*)
Summary of Parameters and Optimized Conditions for Batch Adsorption of Fluoride

Adsorbent	Adsorbate	Surface Area (m^2/g), Pore Volume (cm^3/g), Pore Size (nm)	pHzpc	Experimental Conditions	Adsorption Capacity (mg/g)	Thermodynamic Parameters	Kinetic Model and Curve Fitting	Isotherm Model and Curve Fitting	Maximum Adsorption Conditions	References
Dicalcium phosphate inlaid in chitosan bead	F^-			pH = 3–12 Dose = 5 g/l Agitation speed = 100 rpm Concentration = 10 mg/l Contact time = 4 h Temperature = 25°C	50.1		Pseudo-second order model Nonlinear	Langmuir model Nonlinear	pH = 3–12	Shen et al. (2016)
Sulfate-doped hydroxyapatite	F^-	Surface area = 84.2 Pore size = 12.8		pH = 3–11 Dose = 0.5 g/l Agitation speed = 180 rpm Concentration = 2–100 mg/l (isotherm study) Contact time = up to 6 h Temperature = 25°C	28.3		Pseudo-second order model Nonlinear	Freundlich model Linear	pH = 3–10, optimum pH suggested in the article is 3 Contact time = 2 h	Chen, Zhang, He, et al. (2016)
Fe-Ca-Zr hybrid metal oxide nanomaterial	F^-	Surface area = 446.8 Pore volume = 0.511 Pore size = 1.44	7.5	pH = 2–12 Dose = 0.04–0.5 g/l Agitation speed = rpm Concentration = 5–45 mg/l Contact time = 5–50 min Temperature = 303–323 K	250		Pseudo-second order model Nonlinear	Freundlich model Linear	pH = 7 Concentration = 5 mg/l Contact time = 4 min Temperature = 323 K	Dhillon et al. (2015)
Mg-Al-Zr triple metal composite	F^-	Surface area = 200.7 Pore volume = 0.1847 Pore size = 3.82		pH = 3–10 Dose = 1 g/l Agitation speed = 200 rpm Concentration = 10–105 mg/l (concentration) Contact time = up to 24 h Temperature = 25°C	22.9	$\Delta H°$ = 4.75 kJ/mol $\Delta S°$ = 59.61 J/mol K $\Delta G°$ = negative	Pseudo-second order model Nonlinear	Langmuir and Freundlich model Linear	pH = 7 Contact time = 40 min	Wang, Yu, et al. (2017)

(*continued*)

TABLE 6.1 (*Continued*)
Summary of Parameters and Optimized Conditions for Batch Adsorption of Fluoride

Adsorbent	Adsorbate	Surface Area (m^2/g), Pore Volume (cm^3/g), Pore Size (nm)	pHzpc	Experimental Conditions	Adsorption Capacity (mg/g)	Thermodynamic Parameters	Kinetic Model and Curve Fitting	Isotherm Model and Curve Fitting	Maximum Adsorption Conditions	References
Ce–chitosan composite	F^-	Surface area = 16.6	5.3	pH = 2–11 Dose = 0.3 g/l Agitation speed = 150 rpm Concentration = 5–100 mg/l Contact time = up to 720 min Temperature = 283–313 K	153	$\Delta H° = 57.38$ kJ/mol $\Delta S° = 0.24$ kJ/ mol K $\Delta G°$ = negative	Pseudo-second order model Nonlinear	Langmuir model Linear	pH = 3 Contact time = 400 min	Zhu et al. (2017)
Cellulose@ hydroxyapatite nanocomposite	F^-	Surface area = 76.25		pH = 4.1–9 Dose = 1–4 g/l Agitation speed = 200 rpm Concentration = 20 mg/l (effect of pH study) Contact time = up to 720 min Temperature = 25°C	4.2		Pseudo-second order model	Langmuir and Freundlich model	pH = optimum at 6.5 (max at pH 4) Dose = 3 g/l Contact time = 360 min	Yu et al. (2013)
Ce–Fe bimetal oxides	F^-	Surface area = 164.9		pH = 2.9–10.1 Dose = 0.3–1.5 g/l Agitation speed = 180 rpm Concentration = 10 mg/l Contact time = up to 90 min Temperature = 293–313 K	60.97		Pseudo-second order model Linear	Langmuir model Linear	pH = 2.9 –6 Dose = 0.5 g/l Contact time = 40 min Temperature = 293 K	Tang and Zhang (2016)

(continued)

TABLE 6.1 (*Continued*)
Summary of Parameters and Optimized Conditions for Batch Adsorption of Fluoride

Adsorbent	Adsorbate	Surface Area (m^2/g), Pore Volume (cm^3/g), Pore Size (nm)	pHzpc	Experimental Conditions	Adsorption Capacity (mg/g)	Thermodynamic Parameters	Kinetic Model and Curve Fitting	Isotherm Model and Curve Fitting	Maximum Adsorption Conditions	References
UiO-66NH$_2$	F^-	Surface area = 905 Pore volume = 0.43		pH = 3–11 Dose = 0.5 g/l Agitation speed = 300 rpm Concentration = 20 mg/l Contact time = 30 min Temperature = 293–333 K	55	ΔH° = −28.21 kJ/mol ΔS° = 0.023 kJ/mol ΔG° = negative	Pseudo-second order model Linear	No superior isotherm suggested in the study	Maximum is at pH 3, but pH range of 3–7 is recommended Temperature = 293 K	Lin et al. (2016)
Hydroxyl aluminum oxalate	F^-	Surface area = 68.34 Pore size = 7		pH = 2–12 Dose = 1 g/l Agitation speed = 200 rpm Concentration = 20 mg/l Contact time = 24 h Temperature = 298 K	400	ΔH° = 11.55 kJ/mol ΔS° = 93.36 J/mol ΔG° = negative	Pseudo-second order model Linear	Freundlich model Nonlinear	pH = 2 Contact time = 4 h	Wu et al. (2016)
Polypyrrole/hydrous tin oxide	F^-	Surface area = 65.75	7.6	pH = 2.5–10.5 Dose = 0.1–10 g/l Agitation speed = 200 rpm Concentration = 5–20 mg/l Contact time = up to 300 min Temperature = 298–328 K	26.16–28.99	ΔH° = 10.86 kJ/mol ΔS° = 0.098 kJ/mol ΔG° = negative	Pseudo-second order model Linear and nonlinear	Langmuir model Linear and non-linear	pH = 3.5–8.5 Dose = 5 g/l Contact time = 10–45 min Temperature = 328 K	Parashar et al. (2016)
Aluminum fumarate	F^-	Surface area = 1156 Pore size = 1.7	8.1	pH = 2–9 Dose = 0.025–3 g/l Agitation speed = 150 rpm Concentration = 10–1400 mg/l (isotherm study) and 30 mg/l in kinetic study Contact time = up to 360 h Temperature = 293–333 K	600	ΔH° = −32.06 kJ/mol ΔS° = −0.061 kJ/mol ΔG° = negative	Pseudo-second order model Linear	Freundlich model	pH = 2–7 Contact time = 24 h Temperature = 293 K	karmakar et al. (2016)

(continued)

TABLE 6.1 (*Continued*)
Summary of Parameters and Optimized Conditions for Batch Adsorption of Fluoride

Adsorbent	Adsorbate	Surface Area (m^2/g), Pore Volume (cm^3/g), Pore Size (nm)	pHzpc	Experimental Conditions	Adsorption Capacity (mg/g)	Thermodynamic Parameters	Kinetic Model and Curve Fitting	Isotherm Model and Curve Fitting	Maximum Adsorption Conditions	References
MgO nanoplates	F^-	Surface area = 14.7 Pore size = 14.4 and 3.3 (bi-modal)		pH = 2–13 Dose = 1 g/l Agitation speed = 150 rpm Concentration = 20 mg/l (kinetics study) Contact time = up to 24 h Temperature = 25°C	185.5		Pseudo-second order model	Freundlich model	pH = 2–11 Contact time = 60 min	Jin et al. (2016)
Al_2O_3/expanded graphite	F^-		3.88	pH = 2–11 Dose = 1–12 g/l Agitation speed = 200 rpm Concentration = 20 mg/l (kinetics study) Contact time = 150 min Temperature = 303–323 K	1.18	ΔH° = 128.69 kJ/mol ΔS° = 0.4346 kJ/mol ΔG° = negative	Pseudo-second order model Linear	Freundlich model Linear	pH = 4 Dose = 4 g/l (maximum at 10–12 g/l) Contact time = 120 min Temperature = 323 K	Jin et al. (2015)
Zr(IV)-immobilized chitosan	F^-			pH = 3–11 Dose = 1.5–12.5 g/l Agitation speed = 200 rpm Concentration = 50–200 mg/l Contact time = 5–120 min Temperature = 303–323 K	48.26	ΔH° = 3.72 kJ/mol ΔS° = 0.0843 kJ/mol ΔG° = negative	Pseudo-second order model Linear	Langmuir model Linear	pH = 1 Dose = 7.5 g/l Concentration = 50 mg /l Contact time = c.a. 10 min Temperature = 303 K	Liu, Zhang, et al. (2015)

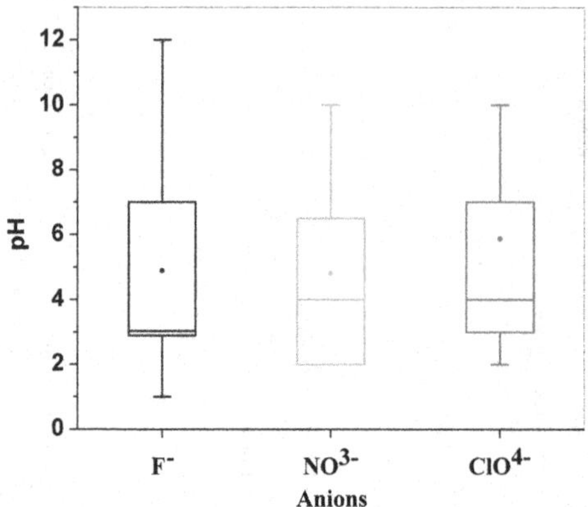

FIGURE 6.1 Box plot of optimum pH for adsorption vs anions graph of literature surveyed (literature surveyed is mentioned in Tables 6.1–6.3. In places where the pH range is specified for optimum pH conditions, both upper and lower limits of pH are taken into consideration. In the box plot, the point and horizontal line inside the box are mean and median values, respectively. Box is set up at 25 and 75 percentile and whiskers at 5 and 95 percentile; the crosses outside the range of whiskers are outliers).

6.1.2 Zeta Potential

The variation in fluoride uptake was in the order of surface charge on the silver nanoparticle-coated biomaterial scaffold (Ag Nps/Al-chitosan-alginate) (Kumar et al. 2016). The higher positive surface charge (c.a. 18.6 mV) at pH 4 coincided with the maximum percentage removal. Also at pH 5 the same surface charge (c.a. 18.6 mV) coincided with the maximum percentage removal, but the fluoride uptake declined. Further increase in pH led to a decline in surface charge, as well as a decline in uptake capacity. Similarly, adsorption capacity with the calcined product of Mg/Fe/La hydrotalcite declined with increase in pH (Wu et al. 2017). The zeta potential also declined with increase in pH. This was due to the electrostatic nature and competition from hydroxyl ions.

6.1.3 Effect of Coexisting Ions

Chloride, sulfate, nitrate, phosphate, hydrogen phosphate, carbonate, bicarbonate (Velazquez-Jimenez et al. 2014; Kumar et al. 2016; Mohan et al. 2012; Kundu et al. 2017; Ma et al. 2017; Liu, Cui, et al. 2016; Chen, Shu, et al. 2017; Wang, Yu, et al. 2017; Wu et al. 2016; Jin et al. 2016; Dhillon et al. 2015; Zhang, Li, et al. 2012; Tang and Zhang 2016), and arsenic (Jing et al. 2012) reduced the uptake of fluoride. Multivalent ions have more effect in the reduction of fluoride adsorption capacity as compared to monovalent anions (Zhang, Li, et al. 2012; Kundu et al. 2017; Wang, Yu, et al. 2017; Karmakar et al. 2016; Chen, Zhang, Li, et al. 2016; Jin et al. 2016). The larger effect of multivalent anions is attributed to the charge density or charge-to-ionic-radii (Liu, Cui, et al. 2016; Tang and Zhang 2016).

The effect of sulfate on the reduction of fluoride's adsorption on akaganeite was more pronounced as compared to chloride and nitrate. The more pronounced effect of sulfate is attributed to the inner sphere complex formation by sulfate ions in comparison to outer sphere complex formation by chloride and nitrate (Kuang et al. 2017). In some cases, sulfate in addition to monovalent anions has also an insignificant effect on fluoride removal (Prabhu and Meenakshi 2014).

The effect of anions on fluoride adsorption capacity is also investigated on the basis of ionic radii in addition to charge-to-ionic radius ratio (Chen, Zhang, He, et al. 2016). The ionic radius of bicarbonate (1.56 Å) was closer to the ionic radius of fluoride (1.33 Å), as compared to that of sulfate (2.30 Å). Hence, bicarbonate easily fits into the mineral arrangement of the adsorbent, i.e. apatite, which leads to a significant decline in adsorption of fluoride as compared to sulfate. Similarly, the chloride and nitrate have higher ionic radius (1.8 Å) and lesser effect on the adsorption of fluoride.

The effect of bicarbonate on fluoride removal capacity is attributed to the increase in pH of the solution, which leads to the increased competition between hydroxide and fluoride ions (Prabhu and Meenakshi 2014). The adsorption capacity of fluoride with groundwater was less as compared to a synthetic solution (Jing et al. 2012). This is attributed to the fact that groundwater sample contained sulfate and carbonate, and competition from the anions led to decrease in adsorption capacity. Sulfate and carbonate reduced fluoride adsorption by competing with fluoride ions for active sites.

6.1.4 Effect of Surface Modification

The modification of the adsorbent led to increase in fluoride adsorption (Prabhu and Meenakshi 2014; Kuang et al. 2017; Chen et al. 2012). The modification of activated carbon with Zr(IV) led to reduction of adsorption capacity. However, the modification of activated carbon with Zr(IV) and oxalic acid led to increase in adsorption capacity (Velazquez-Jimenez et al. 2014). Zr(IV) alone on the surface of activated carbon acts as a Lewis acid center, and oxalic acid's addition led to the conversion of active sites to the basic character. This enhanced the positive charge on Zr(IV) and enhanced the adsorption of fluoride. The unpaired electrons and π bond system of oxalate were the major reasons for conversion of active sites to basic sites. X-ray photoelectron (XPS) spectra of Zr(IV)-modified activated carbon and Zr(IV) and oxalate-modified activated carbon after fluoride adsorption showed an XPS peak at 182.5 and 183.3 eV, which suggests the presence of additional positive surface charge surrounding the Zr atoms.

The doping of titanium dioxide with iron enhanced the formation of hydroxyl groups, as suggested by XPS analysis (Chen et al. 2012). The adsorption of fluoride led to decrease in O1s intensity (530 eV for O^{2-} and 531.5 eV for OH) in XPS spectra. This suggested the exchange of OH^- during fluoride adsorption.

The doping positively affects only up to an optimum ratio, in terms of the adsorption capacity. The adsorption capacity increased with sulfate doping of hydroxyapatite till an S/P atomic ratio of 1:2. On further increase of the S/P atomic ratio (5:6 and 4:3), a new peak in the X-ray diffraction (XRD) spectra emerges, representing the distortion of the crystal structure, which leads to the decline in adsorption capacity.

The addition of cationic surfactants also enhanced the removal of fluoride (Prabhu and Meenakshi 2014). Cationic surfactants, i.e. hexadecylpyridinium chloride, dodecyltrimethyl ammonium bromide, and cetyltrimethyl ammonium bromide, enhanced the hydrophilic character of the hydroxyapatite (adsorbent) and increased the attraction of fluoride ions. The adsorption of hexadecylpyridinium chloride-modified hydroxyapatite was lower than that of dodecyltrimethyl ammonium bromide- and cetyltrimethyl ammonium bromide-modified hydroxyapatite. This is attributed to the restriction of fluoride ions by electron-rich pyridinium groups to tertiary cationic groups.

6.1.5 Effect of Material

Adsorption varied with the synthesis route for preparation of the adsorbent (Zhang, Li, et al. 2012), the ratio of the precursors (Zhou et al. 2011), and terminative pH for precipitation of the precursor for preparation of the adsorbent (Chen et al. 2012). The calcined product of Li-Al layered double hydroxide prepared by the co-precipitation method showed more adsorption capacity than the material prepared by homogenous precipitation (Zhang, Li, et al. 2012). The fluoride ion adsorbs into the interlayer space. The calcination of the adsorbent prepared by homogenous precipitation led to incomplete decomposition of the interlayer, evidenced by Fourier-transform infrared spectroscopy (FT-IR). Hence, there is less adsorption of fluoride by the calcined product of the sample prepared by homogenous precipitation.

The terminative pH during precipitation of the precursors for the adsorbent material led to a change in adsorption capacity (Chen et al. 2012). The optimum terminative pH during synthesis of iron-doped titanium oxide for adsorption of fluoride was 5. In addition, the adsorbent prepared with terminative pH in an acidic pH environment depicted greater adsorption capacity than the adsorbents prepared with terminative pH in an alkaline environment. The XRD suggested the presence of crystalline FeOOH at a terminative pH of 7–9; however, at a terminative pH of 3–6, the crystalline phase of the FeOOH was not depicted.

The higher adsorption capacity of biochar as compared to activated carbon for fluoride is attributed to the opening of pores during the water contact in the adsorption process (Mohan et al. 2012). On contact with water, the swelling phenomenon led to widening of pores. This cannot be determined by the Brunauer–Emmett–Teller (BET) surface area analysis of the dry sample. This can be evidenced by the fact that more water is imbibed by the chars as compared to that measured by BET analysis.

6.1.6 Effect of Temperature

The increase in temperature led to increase (Wang, Yu, et al. 2017; Ma et al. 2017; Zhou et al. 2011; Prabhu and Meenakshi 2014) and decrease (Chen et al. 2012; Chen, Shu, et al. 2017; karmakar et al. 2016) in adsorption capacity. The increase in adsorption capacity of fluoride with mesoporous alumina is attributed to the increase in kinetic energy (Kundu et al. 2017). The XPS analysis for fluoride adsorption on Mg-Al-Zr triple metal composite suggests increase in fluoride adsorbed amount with temperature (Wang, Yu, et al. 2017).

6.1.7 Mechanism

The mechanism of adsorption was examined using different techniques. XRD of the polypyrrole/hydrous tin oxide after fluoride adsorption did not show any change

(Parashar et al. 2016). This suggests that fluoride adsorption is a physical phenomenon involving electrostatic interaction and ion exchange. The Arrhenius activation energy (20.05 kJ/mol) also suggested it to be a physisorption process. XRD of the material after fluoride adsorption on the silver nanoparticle-coated biomaterial scaffold (Ag Nps/Al-chitosan-alginate) explained the formation of a ralstonite-like compound with irreversible adsorption (Kumar et al. 2016).

In addition to XRD, FT-IR is also used to estimate the mechanism of adsorption (Velazquez-Jimenez et al. 2014; Chen, Zhang, He, et al. 2016; Wu et al. 2016). The FT-IR analysis of fluoride adsorption on Zr(IV) and oxalic acid-modified activated carbon suggests the interaction of zirconium ions with carboxylic groups (Velazquez-Jimenez et al. 2014). The sample after adsorption of fluoride with Zr(IV) and oxalate-modified activated carbon led to an FT-IR peak with sharper form at 3400 cm^{-1} and had lower intensity than that of the adsorbent (Velazquez-Jimenez et al. 2014). This depicts involvement of the hydroxyl group in adsorption process. Vibrational changes also appeared in the range of 400–500 cm^{-1}. This is attributed to the fluoride complex with Zr-O groups. Fluoride adsorption occurs with the displacement of the hydroxide group from the Zr-oxalate complex. The chemical rearrangement of the COOH group leads to the formation of zirconium oxyfluoride. The XPS analysis showed that adsorption happened via ion exchange of hydroxide ions.

The XPS analysis for fluoride removal with zirconium phosphate suggested a stronger interaction between them, i.e. inner sphere complex formation (Zhang, Li, et al. 2017). The F1s of pure NaF in XPS spectra showed a peak at 684.9 eV as compared to 685.7 eV in the case of F loaded in ZrP. This suggests the stronger adsorption efficiency between the sample and fluoride ions. In addition, the primitive zirconium phosphate contains two satellite peaks at 183.3 eV (Zr $3d_{5/2}$) and 185.6 eV (Zr $3d_{3/2}$), corresponding to charge transfer from the valence band of the ligand atom to the 4f orbital of the Zr atom. These peaks disappeared, and the appearance of new peaks at ~184.4 and ~186.8 eV corresponded to a newly formed complex. The large energy band shift of 1.1 eV suggests the formation of an inner sphere complex and formation of a new Zr-F complex.

The significant adsorption over the isoelectric point suggests the pH-independent mechanism (Kuang et al. 2017). The adsorption of fluoride with akaganeite-anchored graphene oxide declined with increase in chloride desorption. This suggests that ion exchange can be responsible for adsorption of fluoride on akaganeite-anchored graphene oxide. The goethite-anchored graphene oxide, however, showed similar ion exchange behavior with acetate ions rather than with chloride ions.

The ion exchange behavior in the case of fluoride adsorption with cetyltrimethyl ammonium bromide-modified hydroxyapatite is estimated on the basis of energy dispersive X-ray (EDX) analysis. The EDX analysis of the adsorbent after surface modification of hydroxyapatite with cetyltrimethyl ammonium bromide showed that there were no bromide peaks, which suggested the adsorption of fluoride with ion exchange, i.e. fluoride replacement with bromide (Prabhu and Meenakshi 2014).

The ion exchange behavior is also estimated by FT-IR analysis in the case of fluoride adsorption with hydroxyl aluminum oxalate (Wu et al. 2016). The FT-IR spectra depicted the interchange of fluoride with hydroxyl and oxalate groups (Wu et al. 2016). The FT-IR spectra after adsorption of fluoride showed disappearance of peaks at 3675 and 1715 cm^{-1}. Along with this, there was emergence of a peak at 600 cm^{-1} attributed to Al-F stretching vibration. This suggests ion exchange with hydroxide and oxalate.

The mechanism of adsorption of fluoride onto the Li-Al layered double hydroxide was based on the memory effect (Zhang, Li, et al. 2012). The interlayer anion carbonate decomposed in Li-Al layered double hydroxide synthesized by the co-precipitation method. This is evidenced by the disappearance of FT-IR peaks at 1385 cm^{-1}. However, the interlayer anions in the Li-Al layered double hydroxide prepared by the precipitation method did not disappear completely. Two small peaks were present at 1450–1350 cm^{-1}.

The adsorption of fluoride with cerium-immobilized chitosan is governed by different mechanisms. The maximum adsorption capacity for fluoride with cerium-immobilized chitosan was at pH 3 (Zhu et al. 2017). At pH lower than 3, formation of the HF occurred, which led to the reduction of adsorption of fluoride ions. pH lower than 5.3 led to protonation of hydroxyl and amino groups. Alkaline pH led to decrease in adsorption. The fluoride 1s peak in the XPS spectrum for cerium-immobilized chitosan after fluoride adsorption was higher (685.2 eV) than for NaF (684.5 eV). In addition, the increase in the Ce^{3+} state and decrease in the Ce^{4+} state explained the complexation of fluoride with cerium. The adsorption mechanism was electrostatic attraction, ion exchange with nitrate ions, and complexation with cerium ions. At pH higher than 5.3, only ion exchange and complexation played a major role in the removal of fluoride. q_e (adsorption capacity at equilibrium) at pH =7 was more than half of q_e at pH =5.3; hence, the major role was attributed to ion exchange rather than complexation.

6.1.8 Desorption

The desorption of fluoride was achieved with sodium hydroxide (Chen et al. 2012; Chen, Shu, et al. 2017; Zhu et al. 2017; Dhillon et al. 2015; Zhang and Huang 2019), sodium carbonate (Wu et al. 2017), and sodium aluminate (Wang, Yu, et al. 2017). The high percentage of desorption suggests the physical nature of adsorption e.g. ion exchange (Chen, Shu, et al. 2017; Parashar et al. 2016). Adsorption capacity declined after adsorption in some cases (Wang, Yu, et al. 2017; Dhillon et al. 2015). The non-desorbed amount of the adsorbate is attributed to the chemically bound adsorbate species. Adsorption capacity increases with increase in sodium hydroxide strength in the case of regeneration of aluminum fumarate metal organic framework (karmakar et al. 2016).

6.2 NITRATE

Nitrate is found naturally in the environment and can act as an important plant nutrient (WHO 2017). Leaching from natural vegetation acts as a source of groundwater contamination by nitrate. The major source of nitrate to humans is through vegetables and drinking water. The guideline value for nitrate is 50 mg/l as nitrate ions. Nitrate can cause methemoglobinemia (Johnson 2019). An overview of the experimental parameters and optimized conditions from batch adsorption experiments for nitrate is presented in Table 6.2.

6.2.1 Effect of pH

The adsorption of nitrate is achieved at lower pH with chitosan-modified microsphere (Zhao and Feng 2016), quaternized melamine-formaldehyde resin (Banu and Meenakshi 2017b), Fe_3O_4/ZrO_2/chitosan composite (Jiang et al. 2013), MgO-biochar, FeO-biochar (Usman et al. 2016), and potassium carbonate and ammonium chloride activated carbon (Nunell et al. 2015).

The high removal is attributed to the increase in the electrostatic force of attraction and increase in competition from hydroxide ions on increase in pH (Banu and Meenakshi 2017b; Jiang et al. 2013; Usman et al. 2016) and in some cases, lack of ionization of surface groups at higher pH (Zhao and Feng 2016). However, the maximum removal of nitrate was not achieved at the lowest pH as studied in some cases (Zhao and Feng 2016; Banu and Meenakshi 2017a, b). This phenomenon is attributed to the instability of the adsorbent at the lowest pH studied (Zhao and Feng 2016).

In some cases, the adsorption capacity after attaining the maximum adsorption capacity at a certain pH plateaued up to a pH (Banu and Meenakshi 2017a, b; Hu et al. 2015). The adsorption capacity was also not maximum at low pH in the sorption of nitrate with amino-functionalized MCM-41 (Ebrahimi-Gatkash et al. 2017) and graphene (Ganesan et al. 2013). The maximum adsorption was achieved under near-neutral conditions. The reason was attributed to the electrostatic phenomenon as a major factor governing the adsorption process.

6.2.2 Effect of Coexisting Ions

A number of anions such as bicarbonate, carbonate, sulfate, chloride, and fluoride affect the adsorption of nitrate (Srivastav et al. 2014; Banu and Meenakshi 2017b; Wan et al. 2012; Bagherifam et al. 2014). The effect of sulfate was more pronounced in the case of adsorption with hydrous bismuth oxide (Srivastav et al. 2014), chitosan quaternized resin (Banu and Meenakshi 2017a), quaternized form of melamine–formaldehyde (Banu and Meenakshi 2017b), and granular chitosan-Fe^{3+} complex (Hu et al. 2015).

The effect of sulfate is attributed to the lower stability of sulfate (Srivastav et al. 2014) and charge density (Banu and Meenakshi 2017a, b). The effect of bicarbonate in the case of nitrate removal with hydrous bismuth oxide is attributed to increase of the pH, which led to decline of electrostatic attraction (Srivastav et al. 2014).

The effect of chloride and carbonate was more pronounced than sulfate in the case of adsorption with organoclay (Bagherifam et al. 2014) and calcined hydrotalcite (Wan et al. 2012). The more pronounced effect of carbonate than sulfate on nitrate removal with calcined hydrotalcite is attributed to charge density (Wan et al. 2012). The Hofmeister series of anions with respect to their free energy of hydration affected the removal of nitrate with organoclay. Anions with less free energy of hydration were advocated to have the least competitive effect. Therefore, sulfate has less effect as compared to chloride on nitrate removal.

Anions such as chloride, sulfate, and carbonate caused the decline in percentage removal from 93% in control to 72%, 79%, and 88%, respectively, with organoclay (Bagherifam et al. 2014). The results suggested by the author were in the order of

TABLE 6.2
Summary of Parameters and Optimized Conditions for Batch Adsorption of Nitrate

Adsorbent	Adsorbate	Surface Area (m^2/g), Pore Volume (cm^3/g), Pore Size (nm)	pHzpc	Experimental Conditions	Adsorption Capacity (mg/g)	Thermodynamic Parameters	Kinetic Model and Curve Fitting	Isotherm Model and Curve Fitting	Maximum Adsorption Conditions	References
Modified chitosan	NO_3^-		7.8–8.0	pH = 2–9 Dose = 0.2–2 g/l Agitation speed = 120 rpm Concentration = 10–100 mg/l Contact time = up to 360 min Temperature = 303.15 K	32.15		Pseudo-second order Linear	Langmuir model Linear	pH = 3 Dose = 1 g/l Contact time = 90 min	Zhao and Feng (2016)
Hydrous bismuth oxide	NO_3^-		7.8–8.0	pH = 5–10 Dose = 25–100 g/l Concentration = 14–56 mg N/l Contact time = 60–360 min Temperature = 293–323 K	0.512 mg N/g		Pseudo-first order model Linear	Langmuir and Freundlich model Linear	Dose = 50 g/l Concentration = 14 mg N/l Contact time = 180 min Temperature = 323 K	Srivastav et al. (2014)
Al_2O_3/bio-TiO_2 nanocomposite	NO_3^-			pH = neutral Dose = 1–5 g/l Agitation speed = 120 rpm Concentration = 25–125 mg/l Contact time = 200 min Temperature = room temperature	30.3		Pseudo-second order model Linear	Langmuir model No comparison	Dose = 2 g/l Contact time = 60 min	Suriyaraj et al. (2015)

(continued)

TABLE 6.2 (*Continued*)
Summary of Parameters and Optimized Conditions for Batch Adsorption of Nitrate

Adsorbent	Adsorbate	Surface Area (m^2/g), Pore Volume (cm^3/g), Pore Size (nm)	pHzpc	Experimental Conditions	Adsorption Capacity (mg/g)	Thermodynamic Parameters	Kinetic Model and Curve Fitting	Isotherm Model and Curve Fitting	Maximum Adsorption Conditions	References
Quaternized form of melamine–formaldehyde	NO_3^-		6.3	pH = 2–11 Dose = 0.5 –4 g/l Agitation speed = 120 rpm Concentration = 50–300 mg/l Contact time = 5–40 min Temperature = 303–323 K	40.1	ΔH° = −10.77 kJ/mol ΔS° = 0.02 kJ/mol ΔG° = negative	Pseudo-second order	Freundlich	pH = 4–8 Contact time = 30 min Temperature = 303 K	Banu and Meenakshi (2017b)
Chitosan-grafted quaternized resin	NO_3^-			pH = 2–11 Dose = 0.5–3 g/l Agitation speed = 120 rpm Concentration = 50–400 mg/l Contact time = 10–80 min Temperature = 303–323 K	34.5	ΔH° = −29.36 kJ/mol ΔS° = 0.07 kJ/K mol ΔG° = negative	Pseudo-second order model Linear	Freundlich model Linear	pH = 4–8 Dose = 3 g/l Contact time = 60 min Temperature = 303 K	Banu and Meenakshi (2017a)
Fe_3O_4/ZrO2/chitosan	NO_3^-			pH = 3–8 Dose = 0.5 g/l Agitation speed = 150 rpm Concentration = 1–1000 mg/l (isotherm study) Contact time = c.a. 1440 min Temperature = 25°C	89.3		Pseudo-first order model Linear	Langmuir model Linear	pH = 3 Contact time = 840 min	Jiang et al. (2013)

(continued)

TABLE 6.2 (*Continued*)
Summary of Parameters and Optimized Conditions for Batch Adsorption of Nitrate

Adsorbent	Adsorbate	Surface Area (m^2/g), Pore Volume (cm^3/g), Pore Size (nm)	pHzpc	Experimental Conditions	Adsorption Capacity (mg/g)	Thermodynamic Parameters	Kinetic Model and Curve Fitting	Isotherm Model and Curve Fitting	Maximum Adsorption Conditions	References
Montmorillonite modified with hexadecylpyridinium chloride	NO_3^-			Dose = 0.5–4 g/l Concentration = 0.2–1 mM (isotherm study) Contact time = 24 h	0.67 mmol/g		Pseudo-second order model	Langmuir model Linear	Dose = 2 g/l Contact time = 120 min	Bagherifam et al. (2014)
De-Acidite FF-IP resin (batch and column study)	NO_3^-			pH = 2–10 Dose = 5 g/l Agitation speed = 100 rpm Concentration = 50–300 mg/l Contact time = 0–60 min Temperature = 293–323 K	35 and 30	ΔH° = 27.45 to 34.68 kJ/mol ΔS° = 105.27–131.19 J/mol K ΔG° = negative	Pseudo-first order model Nonlinear	Freundlich model nonlinear	pH = 2–6 Contact time = 25 min Temperature = 323 K	Naushad et al. (2014)
Polyacrylic anion exchange resin (batch and column study)	NO_3^-	Surface area = 3.07 Pore size = 8.14		Dose = 2 g/l Agitation speed = 140 rpm Concentration = 0.805–4.83 mmol/l (kinetic study) Contact time = up to 100 min	2.46 mmol/g		Pseudo first order and second order model	Langmuir and Freundlich model	Contact time = 20 min	Song et al. (2014)

(*continued*)

TABLE 6.2 (*Continued*)
Summary of Parameters and Optimized Conditions for Batch Adsorption of Nitrate

Adsorbent	Adsorbate	Surface Area (m^2/g), Pore Volume (cm^3/g), Pore Size (nm)	pHzpc	Experimental Conditions	Adsorption Capacity (mg/g)	Thermodynamic Parameters	Kinetic Model and Curve Fitting	Isotherm Model and Curve Fitting	Maximum Adsorption Conditions	References
Chemically modified pinewood sawdust	NO_3^-			Dose = 3 g/l Concentration = 10, 30 and 50 mg N/l Contact time = 2 h Temperature = 278, 296 and 313 K	32.8 mg N/g	ΔH° = −6.8 kJ/mol ΔS° = 0.11 kJ/mol K ΔG° = negative		Redlich–Peterson model Linear	Temperature = 278 K	Keränen et al. (2015)
Anionic rice husk	NO_3^-			pH = 3–11 Dose = 4 g/l Agitation speed = 400 rpm Concentration = 50 mg/l Contact time = up to 110 min Temperature = 20°C –50°C	8.32	ΔH° = −19.56 kJ/mol ΔS° = 2.21 kJ/Kmol ΔG° = negative	Pseudo-second order model Linear	Langmuir, Freundlich and Dubinin–Radushkevich model	pH = 7 Contact time = 90 min Temperature = 20°C	Katal, Baei, et al. (2012)
Graphene	NO_3^-		Pzc = 5.7	pH = 2–12 Dose = 0.67 g/l Agitation speed = 200 rpm Concentration = 100–500 mg/l Contact time = 12 h Temperature = 303–343 K	202.43	ΔH° = 1.6102 kJ/mol ΔS° = 5.31 J/Kmol ΔG° = negative	Pseudo-second order model Linear	Langmuir model Linear	pH = 6.5–7.5 Contact time = 45 min Temperature = 343 K	Ganesan et al. (2013)

(*continued*)

TABLE 6.2 (*Continued*)
Summary of Parameters and Optimized Conditions for Batch Adsorption of Nitrate

Adsorbent	Adsorbate	Surface Area (m^2/g), Pore Volume (cm^3/g), Pore Size (nm)	pHzpc	Experimental Conditions	Adsorption Capacity (mg/g)	Thermodynamic Parameters	Kinetic Model and Curve Fitting	Isotherm Model and Curve Fitting	Maximum Adsorption Conditions	References
Calcined hydrotalcite (Mg/Al)	NO_3^-		11.50	pH = not controlled Dose = 2 g/l Agitation speed = 130 rpm Concentration = 25 mg N/l Contact time = 18 h Temperature = 25°C	34.36 mg N/g		Pseudo-second order model Linear	Langmuir model Nonlinear		Wan et al. (2012)
Granular chitosan–Fe^{3+} complex	NO_3^-	Surface area = 8.98 Pore volume = 0.019 Pore size = 5.69	5.4	pH = 3–12 Dose = 5–40 g/l Agitation speed = 140 rpm Concentration = 0.9–4.62 mg NO_3-N/g Contact time = 120 min Temperature = 288–328 K	8.35	ΔH° = −5.06 kJ/mol ΔS° = 27.65 J/mol K ΔG° = negative	Pseudo-second order model Nonlinear	Langmuir–Freundlich model Nonlinear	pH = 3–10 Dose = 40 g/l Contact time = 90 min Temperature = 288 K	Hu et al. (2015)
MgO-biochar	NO_3^-	Surface area = 391.8 Pore volume = 0.012 Pore size = 1.856		pH = 2–8 Dose = 10 g/l Agitation speed = 250 rpm Concentration = 1–100 mg/l (isotherm study) Contact time = up to 120 min Temperature = 25°C	45.36 mmol/kg		Pseudo-second order model Linear	Langmuir model Linear	pH = 2 Time = 30–60 min	Usman et al. (2016)

(*continued*)

TABLE 6.2 (*Continued*)
Summary of Parameters and Optimized Conditions for Batch Adsorption of Nitrate

Adsorbent	Adsorbate	Surface Area (m^2/g), Pore Volume (cm^3/g), Pore Size (nm)	pHzpc	Experimental Conditions	Adsorption Capacity (mg/g)	Thermodynamic Parameters	Kinetic Model and Curve Fitting	Isotherm Model and Curve Fitting	Maximum Adsorption Conditions	References
Amino-functionalized mesoporous MCM-41	NO_3^-	Surface area = 733 Pore volume = 0.336 Pore size = 1.86		pH = 4–8 Dose = 0.5–15 g/l Agitation speed = 150 rpm Concentration = 30–250 mg/l Contact time = 2 h Temperature = 25°C	38.81	Langmuir	pseudo-second order model Linear		pH = 6–7 Dose = 10 g/l Contact time = equilibrium in 10–15 min and but adsorption process proceeded for 120 min	Ebrahimi-Gatkash et al. (2017)
K_2CO_3-activated carbon	NO_3^-	Surface area = 777 cm^2/g Pore volume = 0.35 Pore size = 1.81		pH = 2–7 Dose = up to 1.5 g/l Agitation speed = 300 rpm Concentration = 0.1–6 mmol/l Contact time = 400 min Temperature = 25°C	0.34 mmol/g		Pseudo-second order model	Langmuir model	pH = 2 Dose = 0.5 g/l Contact time = 300 min	Nunell et al. (2015)
NH_4OH-activated carbon	NO_3^-	Surface area = 58 cm^2/g Pore volume = 0.03 Pore size = 1.91		pH = 2–7 Dose = up to 1.5 g/l Agitation speed = 300 rpm Concentration = 0.1–6 mmol/l Contact time = 30 min Temperature = 25°C	0.4 mmol/g		Pseudo-second order model	Freundlich model	pH = 2 Dose = 0.5 g/l Contact time = 30 min	Nunell et al. (2015)

Hofmeister series of anions. Anions with less free energy of hydration have the least competitive effect.

6.2.3 Effect of the Material and Surface Modification

The increase in the Mg/Al ratio in precursor for the formation of calcined hydrocalcite influenced the adsorption capacity of calcined hydrotalcite (Wan et al. 2012). The increase of magnesium to aluminum ratio from 2 to 4 led to increase in adsorption capacity. The increase in the Mg/Al ratio decreased the electric charge density between layers, which led to the increase in the interlayer spacing (Wan et al. 2012). The adsorption of nitrate into the interlayer spacing was favored by the large size of the interlayer space.

The $LaCl_3$ treatment of biochar increased the adsorption capacity by 11.2-fold (Wang, Guo, et al. 2015). This is attributed to the increase in basic functional groups that are positively correlated with nitrate removal efficiency.

The change in the number of amine groups via the surface modifying agent on the surface of MCM-41 led to the change in adsorption capacity (Ebrahimi-Gatkash et al. 2017). The change of the diamine group to the triamine group functionalized on the surface of the adsorbent led to the increase in amine groups. This lead to the increase in the adsorbed nitrate anion.

6.2.4 Mechanism

The mechanism of adsorption of nitrate is estimated by a number of methods. The value of mean adsorption energy was used to estimate the mode of adsorption, i.e. whether physical, chemical, or ion exchange (Srivastav et al. 2014; Banu and Meenakshi 2017a, b). To further support the ion exchange nature of adsorption, the authors used XRD, in which corresponding peaks at two theta 31.91° and 34.09° broadened (Banu and Meenakshi 2017b). The author suggested it to be due to replacement of the chloride ion with the nitrate ion, whereas the same authors in another study used EDX analysis to further prove the ion exchange adsorption process (Banu and Meenakshi 2017a). The role of the surface oxide group was investigated by use of FT-IR (Suriyaraj et al. 2015). The mechanism of adsorption can also be investigated on the basis of adsorption behavior at different pH values. The maximum adsorption of nitrate on graphene occurred at pH above the point of zero charge (PZC) of the adsorbent, i.e. 5.7 (Ganesan et al. 2013). At the point of maximum adsorption, the surface of the adsorbent was negative, and this suggests that the adsorption also occurred through a process other than the electrostatic force of attraction.

The mechanism of adsorption on calcined hydrotalcite occurred through the process of memory effect (Wan et al. 2012). The anion in the interlayer space is vaporized on calcination, and its place is taken by the nitrate anion after the adsorption process. This is supported by the disappearance of XRD peaks at two theta 43° and 62°. The peaks reappeared after adsorption with nitrate. This is also supported by weakening of the FT-IR peak corresponding to hydroxide and carbonate anions after calcination. The FT-IR peak at $1384\,cm^{-1}$ emerges after adsorption of nitrate.

6.2.5 Desorption

Sodium hydroxide (Zhao and Feng 2016; Berkessa et al. 2019), sodium chloride (Banu and Meenakshi 2017b), hydrochloric acid (Banu and Meenakshi 2017a), and alkaline pH (Jiang et al. 2013) have been used for the regeneration of the adsorbent. The adsorption efficiency declined after desorption (Zhao and Feng 2016; Banu and Meenakshi 2017b; Hu et al. 2015). The decline of adsorption capacity after the desorption cycle is attributed to irreversible adsorption sites on the surface of the adsorbent.

6.3 PERCHLORATE

Perchlorate can form naturally in the atmosphere, with trace levels of perchlorate in precipitation (ATSDR 2008). Five perchlorates can be formed in large amounts, i.e. magnesium, potassium, ammonium, sodium, and lithium perchlorate. Perchlorate finds its application in explosives, batteries, adhesives, and bleach. Perchlorate primarily affects the thyroid organ and interferes in the uptake of iodine (ATSDR 2008; Colinas et al. 2016). An overview of the experimental parameters and optimized conditions from batch adsorption experiments for perchlorate is presented in Table 6.3.

6.3.1 Effect of pH

The effect of pH with adsorption capacity of perchlorate varies with the adsorbent. The adsorption was maximum at pH 2 with biochar (Fang et al. 2013) and pH 4 with Mg/(Al + Fe) hydrotalcite (Yang, Gao, Chu, et al. 2012), independent of pH in the range of 4–10 with the calcined product of Mg-$AlCO_3$ layered double hydroxide (Lin et al. 2014) and near isoelectric point with oxidized carbon nanotubes (Fang and Chen 2012).

The perchlorate removal with the nano iron oxide-doped activated carbon was more favorable in acidic medium (max at pH 6; however, increases slightly on rise of pH from 4 to 6) (Xu et al. 2015). This was attributed to the electrostatic attraction. At low pH, the surface of the adsorbent carries more positive charge. On rise of pH (after pH 6), the reduction in adsorption capacity was attributed to competition with hydroxide ions and electrostatic repulsion due to negative charge on the surface of the adsorbent.

Similarly, the perchlorate removal efficiency increased with the calcined product of Mg/(Al + Fe) hydrotalcite-like compound on increasing pH from 2 to 4 (Yang, Gao, Chu, et al. 2012). The low percentage removal at lower pH was due to the destruction of the adsorbent structure. In the pH range of 4–10, the adsorption efficiency plateaued. On increase of pH after 10, the adsorption efficiency declined significantly.

There was little impact of pH on the Fe pillared bentonite, though in the case of Al and Fe-Al pillared bentonite, there was a significant decrease in abatement of the perchlorate uptake above pH 8 (Yang, Xiao, et al. 2013). The increase of the electrostatic repulsion was held as one of the significant reasons for the decline in the percentage removal.

The maximum perchlorate adsorption with biochar (negatively charged at most pH values) occurred at pH near the isoelectric point (Fang et al. 2013). At pH higher

TABLE 6.3
Summary of Parameters and Optimized Conditions for Batch Adsorption of Perchlorate

Adsorbent	Adsorbate	Surface Area (m^2/g), Pore Volume (cm^3/g), Pore Size (nm)	pHzpc	Experimental Conditions	Adsorption Capacity (mg/g)	Thermodynamic Parameters	Kinetic Model and Curve Fitting	Isotherm Model and Curve Fitting	Maximum Adsorption Conditions	References
Calcined LDH	ClO_4^-	Surface area = 124.8 Pore volume = 0.954		pH = 6 Agitation speed = 120 rpm Concentration = 1–450 mg/l (isotherm study) Contact time = 24 h Temperature = 25°C	280			Langmuir Nonlinear (Freundlich under high and low concentration range Linear)	pH = no effect in pH range of 4–9.5	Lin et al. (2014)
Hexadecylpyridinium-modified montmorillonite	ClO_4^-			Concentration = 0.1–0.6 mM (isotherm study) Contact time = 24 h Temperature = 298–318 K	0.081 mmol/g	ΔH° = −49.65 kJ/mol ΔS° = −115. J/mol/K ΔG° = negative	Pseudo-second order Linear	Freundlich and D-R isotherm	Temperature = 298 K	Luo et al. (2016)
Calcined product of Mg(Al-Fe) hydrotalcite	ClO_4^-			pH = 2–12 Dose = up to 0.13–0.33 g/l Agitation speed = 200 rpm Concentration = 200–10,000 μg/l Contact time = 30–2880 min Temperature = 25°C	3605–5001 μg/g		Pseudo-second order Linear	Freundlich model Linear	pH = 4–10 Dose = 1.3 g/l Contact time = 1440 min Temperature = 25°C	Yang, Gao, Chu, et al. (2012)

TABLE 6.3 (*Continued*)
Summary of Parameters and Optimized Conditions for Batch Adsorption of Perchlorate

Adsorbent	Adsorbate	Surface Area (m^2/g), Pore Volume (cm^3/g), Pore Size (nm)	pHzpc	Experimental Conditions	Adsorption Capacity (mg/g)	Thermodynamic Parameters	Kinetic Model and Curve Fitting	Isotherm Model and Curve Fitting	Maximum Adsorption Conditions	References
Oxidized double-walled CNTs	ClO_4^-			pH = 1–10 Agitation speed = 30 rpm Contact time = less than 1 h Temperature = 25°C	2.62 and 3.55 (for contact period of 2h and 8 h oxidation)				pH = neutral pH and pH not adjusted in experiments	Fang and Chen (2012)
Fe-bent Fe-Al-bent Al-bent	ClO_4^-	Surface area = (Fe-bent = 190 Fe-Al-bent = 168 Al-bent = 152) Pore volume = (Fe-bent = 0.223 Fe-Al-bent = 0.191 Al-bent =0.151) Pore size (Diameter) = (Fe-bent = 5.62 Fe-Al-bent = 4.52) Al-Bent = 4.01	zzzzz	pH = 4–10 Dose = 1 g/l Agitation speed = 200 rpm Concentration = 0.05–0.7 mmol/l Contact time = 1–600 min Temperature = 25°C	0.102, 0.107 and 0.117 mmol/g for Al-bent, Fe-Al-bent, Fe-bent	Fe Bent ($\Delta H°$ = 15.684 kJ/mol, $\Delta S°$ = 61.478 J/K mol) For Fe-Al-bent ($\Delta H°$ = 14.048 kJ/mol, $\Delta S°$ = 54.248 J/mol) Al-bent ($\Delta H°$ 15.470 kJ/mol, $\Delta S°$ 58.060 J/mol) $\Delta G°$ negative for all	Pseudo-second order model linear	Langmuir model linear	pH = 4–7 4–10 in case of Fe - bent Contact time = 100 min	Yang, Xiao, et al. (2013)
Iron hydroxide-doped granular activated carbon	ClO_4^-	Surface area = 488–632 Pore volume = 0.251–315 Pore size = 2.13–2.01		pH = 4–10 Dose = 1 g/l Agitation speed = 200 rpm Concentration = 0.1 mmol/l Contact time = 36 h Temperature = 25°C	0.113 mmol/g		Pseudo-second order model only Linear		pH = 6 Time = 10 h but most of experiments conducted till 24 h	Xu et al. (2015)

(continued)

TABLE 6.3 (*Continued*)
Summary of Parameters and Optimized Conditions for Batch Adsorption of Perchlorate

Adsorbent	Adsorbate	Surface Area (m^2/g), Pore Volume (cm^3/g), Pore Size (nm)	pHzpc	Experimental Conditions	Adsorption Capacity (mg/g)	Thermodynamic Parameters	Kinetic Model and Curve Fitting	Isotherm Model and Curve Fitting	Maximum Adsorption Conditions	References
Calcined product of Mg–$FeCO_3$ LDH	ClO_4^-			pH = 2–12 Dose = 0.13–3.33 g/l Agitation speed = 200 rpm Concentration = 2000 μg/l Contact time = 24 h Temperature = 25°C Mg to Fe ratio = 3	2568.4 μg/g		Pseudo-second order model Linear	Freundlich model Linear	pH = 4–10 Dose = 1.33 g/l	Yang, Gao, Deng, et al. (2012)
Nanoiron hydroxide-doped granular activated carbon	ClO_4^-		pH_{IEP} = 8.4–8.8	pH = 2–10 Dose = 1 g/l Agitation speed = 200 rpm Concentration = 0.0816–0.4897 mmol/l (isotherm study) Contact time = 36 h Temperature = 25°C	0.088–0.169 mmol/g		Pseudo-second order model Linear	Langmuir model	pH = 2–3	Xu et al. (2013)

than the isoelectric point, the adsorbent showed a low level of perchlorate adsorption. On the basis of these phenomena, it was suggested that electrostatic attraction was not the major cause of the adsorption.

The change of pH ($pH_{final}-pH_{initial}$) during adsorption of perchlorate with raw carbon nanotubes was positive (Fang and Chen 2012). However, adsorption with oxidized carbon nanotubes led to change of pH, almost nil at adsorption below pH 5 and negative above pH 5. The increase in pH is suggested to be an indication of the ion exchange phenomenon (Xu et al. 2013).

The pH of solution increased with addition of biochar (Fang et al. 2013). The increase of the pyrolysis temperature for biochar led to increase in the pH of the solution. The increase in the pH of the solution by addition of the adsorbent is attributed to ash content.

The pH dependent zeta potential curve of biochar prepared from fir wood chips depicts an S-shaped curve with two inflection points (Fang et al. 2013). The presence of two inflection points suggests that two different forms of the functional group exist on the surface, influencing the dissociation of the functional groups at two different pH values. The coupling of the zeta potential points with the FT-IR led to the identification of two points at hydroxyl and carboxyl groups.

6.3.2 Effect of Coexisting Ions

The presence of anions in the solution declined the adsorption efficiency of perchlorate (Yang, Gao, Chu, et al. 2012; Yang, Xiao, et al. 2013; Xu et al. 2013). Phosphate, sulfate, and nitrate influenced the removal of the perchlorate on Al-, Fe- and Fe-Al pillared bentonite (Yang, Xiao, et al. 2013). The reduction was due to competitive adsorption, and the decline in the electrostatic effect was due to decline in the charge on the surface of the adsorbent.

The effect of anions is in the order of $NO_3^- < H_2PO_4^- < Cl^- < CO_3^{2-} \approx SO_4^{2-}$ for perchlorate adsorption on Mg/(Al-Fe) hydrotalcite (Yang, Gao, Chu, et al. 2012) and $NO_3^- > SO_4^{2-} > Cl^-$ for perchlorate adsorption on nanoiron hydroxide-doped granular activated carbon (Xu et al. 2013). However, adsorption of perchlorate with silver 4,4′-bipyridine nitrate in the presence of carbonate and sulfate did not interfere with perchlorate removal (Colinas et al. 2016). Lower hydration energy of perchlorate and lower solubility of the formed complex, i.e. silver 4,4′-bipyridine perchlorate were held responsible for the noninterference of coexisting anions.

6.3.3 Effect of the Material

The adsorption of perchlorate with uncalcined magnesium aluminum carbonate layered double hydroxide was independent of the Mg/Al ratio (Lin et al. 2014). However, adsorption with calcined magnesium aluminum carbonate layered double hydroxide was dependent on the Mg/Al ratio. Adsorption rises with increase in Mg/Al ratio. This led to the conclusion that adsorption increased with decline in surface charge density. This is in contrary to the general rule. The reason is attributed to the nondependency of adsorption on electrostatic interaction.

The Fe content in Mg/(Al-Fe) hydrotalcite enhanced the adsorption of perchlorate (Yang, Gao, Chu, et al. 2012). The Fe^{3+} content led to the formation of a brucite-like sheet structure. The Al^{3+} substitution with Fe^{3+} was proposed to enhance the positive surface charge. The enhanced positive charge increased the bond strength between layered double hydroxide and anion in the interlayer. However, excessive Fe^{3+} destroyed the brucite-resembling sheet structure. Hence, an optimum ratio of Mg^{2+}/Fe^{3+} is required for maximum charge density.

The removal efficiency of perchlorate also improved with calcination temperature up to 550°C with the calcined product Mg/(Al-Fe) hydrotalcite. The calcination temperature after 600°C declined the adsorption efficiency. The lack of recovery into uncalcined adsorbent configuration (memory effect) for the material prepared at high temperature (> 600°C) contributed to decline in adsorption efficiency. The calcination temperature should be adequate for removal of the interlayer constituent, i.e. carbonate. However, at very elevated calcination temperature, creation of MgO occurred, and adsorbent reconstruction to uncalcined structure state stalled.

The adsorption capacity of double-walled carbon nanotubes enhanced on its oxidation (Fang and Chen 2012). The phenomenon occurred without any significant change of surface area. The enhanced removal was proposed due to the introduction of the oxygen-containing functional groups onto the carbon nanotube surface on oxidation. The change in the surface functional groups was supported by XRD, FT-IR, XPS, and Raman spectroscopy. The FT-IR peaks at 1660, 1724, and 3413 cm^{-1} corresponding to unsaturated ketone, carboxylic acid and saturated ketone, and hydroxyl functional groups intensified after oxidation. This shows that oxidation leads to the introduction of additional functional groups having oxygen atoms. The D band (1338 cm^{-1}) and G band (1580 cm^{-1}) in Raman spectra relate to amorphous carbon and the C-C bond in graphene, respectively. The Raman spectra depict the increase of the D-band to the G-band on oxidation, suggesting the increase in surface defect. The XPS analysis also suggests the increase in carbon with oxygen atoms along with decrease in graphitic carbon. The oxygen-containing sites help in additional H-bonding and electrostatic interaction.

The lower adsorption of perchlorate on sodium bentonite as compared to Al or Fe or Fe-Al pillared bentonite was attributed to the increased number of vacant adsorption sites (Yang, Xiao, et al. 2013).

6.3.4 Mechanism

At different pH values, different mechanisms were working on the adsorption of perchlorate with activated carbon (Xu et al. 2016). At pH 2, the pH which was less than pH_{IEP}, the charge on the surface of the adsorbent was positive and the nitrogen-comprising groups were in the form of $-N^+$ and $-NH_2^+$. This helped in the effective adsorption of perchlorate. In the pH range of 4–8 (larger than the isoelectric point), the surface of the adsorbent was negatively charged, and the nitrogen-comprising groups lost charge and change their form from $-N^+$ and $-NH_2^+$ to –N and –NH, respectively. In addition, –COH and –COOH were the major oxygen-comprising groups, and adsorption is derived by hydrogen bonding. At pH 8–10, a fraction of –COH and –COOH groups were deprotonated to –CO and –COO groups, which led to decrease in adsorption.

The memory effect was proposed for adsorption of perchlorate with calcined magnesium aluminum carbonate layered double hydroxide (Lin et al. 2014). The phenomenon was supported by XRD, SEM, and FT-IR analysis. The peak corresponding to the hydroxide layer structure wiped out after calcination and remerged after perchlorate adsorption. Moreover, the distance between Mg and Al in the parent layered double hydroxide were the same as compared to reconstructed layered double hydroxide. The reconstruction of the hexagonal structure post adsorption was also confirmed by SEM analysis. The FT-IR peak of carbonate at 1062 cm^{-1} in the parent LDH shifted to 1097 cm^{-1} after calcination and reverted back to 1062 cm^{-1} after adsorption.

However, the phenomenon of the memory effect on Mg/Al layered double hydroxide was not ideal. Excessive broadening of the peaks happened at (018) and (015) reflection, leading to the occurrence of stacking faults (Lin et al. 2014). The reconstructed layered double hydroxide showed a strong peak at 936 cm^{-1} in FT-IR spectrum, which was close to perchlorate in solution, i.e. 935 cm^{-1} than to perchlorate of solid state at 954 cm^{-1}. This portrays that perchlorate in the interlayers is in the free state as compared to that in the perchlorate solution.

The memory effect is also observed with calcined Mg/(Al-Fe) hydrotalcite after adsorption with perchlorate (Yang, Gao, Chu, et al. 2012). The mechanism of adsorption on carbon nanotubes of perchlorate also varies with the pH of the solution (Fang and Chen 2012). At very low pH, the surface is protonated, but the enhanced electrostatic attraction prefers smaller ion, i.e. chloride (volume = 0.047 cm^3) than perchlorate (volume = 0.082 nm^3). This leads to the trace amount of perchlorate adsorption. On increase of pH but lower than that of the isoelectric point, the competition from chloride ions declined along with deprotonation of the surface groups. The deprotonation of the surface groups led to increase in the H-bonding. At near-neutral pH, the carbon nanotube surface was slightly negatively charged (pH was +0.85 more than pH_{IEP}) and electrostatic force of attraction was eliminated. The adsorption of ClO_4^- happened by H-bonding rather than by electrostatic attraction. On further increase of pH, perchlorate adsorption declined due to increased electrostatic repulsion.

The mechanism of adsorption of perchlorate on nanoiron hydroxide-doped granular activated carbon follows electrostatic attraction, ion exchange, and surface complexation (Xu et al. 2013). The analogous trend of the zeta potential and pH in the perchlorate adsorption suggests the adsorption follow-up by electrostatic phenomenon. The increase of pH of the solution after adsorption suggests that the ion exchange phenomenon also happens during adsorption. In addition, sulfate ions are also present in the solution, which surged with decline of perchlorate anion in the solution. The relative proportion of the mechanism is estimated by desorption of the adsorbent with an alkaline pH of 11. The perchlorate desorption efficiency of 76% suggests that electrostatic interaction and ion exchange (outer sphere complexation) were dominant mechanisms for perchlorate adsorption, and inner sphere complexation accounts for nearly 24% of the perchlorate removal.

The maximum perchlorate removal with carbon nanotubes happened near the isoelectric point (+0.85) rather than at $pH < pH_{IEP}$ (Fang and Chen 2012). This observation led to the conclusion that electrostatic phenomenon cannot be the principal governing factor at neutral pH, and hydrogen bonding is considered to be the principal governing factor near neutral pH.

6.3.5 Desorption

The regeneration of silver 4,4′-bipyridine perchlorate (SBP) is achieved by increasing the ratio of nitrate to perchlorate and the temperature (Colinas et al. 2016). This led to the shift of the equilibrium and conversion of silver 4,4′-bipyridine perchlorate to individual components. A total of 96% of the material is regenerated and returned to its shape. The regeneration of nanoiron oxide embedded activated carbon can be achieved under alkaline pH (Xu et al. 2015).

6.4 SULFATE

Sulfate occurs naturally and is also produced commercially (WHO 2017). Sulfate is nontoxic, and WHO has also not set up a guideline value for sulfate, due to its level found in water having no health concern. However, high levels of sulfate (600 mg/l) can impact the gastrointestinal system (Silva et al. 2012).

6.4.1 Effect of pH

The removal of sulfate decreases (Gu et al. 2016; Fukushi et al. 2013; Moret and Rubio 2003) with rise in pH. In some cases, the decline in the percentage removal with pH was not linear. The percentage removal increased up to a maximum with rise of pH and declined thereafter, e.g. maximum at pH 4 (Dong et al. 2011) and rise up to pH 3 and remained constant up to pH 9 and declined thereafter (Namasivayam and Sangeetha 2008).

The higher removal at lower pH is attributed to the protonation of the surface, and declination of the removal at higher pH is attributed to the competition from hydroxyl groups (Dong et al. 2011; Moret and Rubio 2003). However, in the case of adsorption with activated carbon, the low removal at low pH was attributed to the competition from chloride ions (Namasivayam and Sangeetha 2008).

The pH affects the formation of inner and outer sphere complex (Gu et al. 2016). The pre-edge intensity in Extended X-ray absorption fine structure (EXAFS) is considered proportional to inner sphere complex formation, and it becomes weaker with increase of pH from 3 to 7 during the adsorption of sulfate on ferrihydrite. Hence, the fraction of the inner sphere complex of adsorbed sulfate on ferrihydrite decreases with rise in pH, while maintaining the ionic strength constant. The increase in pH led to increase in the dominance of $>FeOH$ and $>FeO^-$, and these groups disfavored the inner sphere adsorption. The reason is attributed to stronger and harder Fe-O bonds as compared to $FeOH_2^+$ and decreased electrostatic interaction via decrease in the positive surface charge.

6.4.2 Effect of Coexisting Ions and Surface Modification

The presence of ions in the solution resulted in the decline of adsorption capacity (Howarth et al. 2016; Namasivayam and Sangeetha 2008). The order of influence of anions on the adsorption of sulfate with activated carbon is as follows: molybdate > chlorate > nitrate > chloride > phosphate (Namasivayam and Sangeetha 2008).

The modification of palygorskite (magnesium aluminum phyllosilicate) with octodecyltrimethylammonium chloride (OTMAC) increased the adsorption efficiency of sulfate ten times (Dong et al. 2011). The increased adsorption efficiency is attributed to improved electrostatic interaction. The unmodified palygorskite has a negative charge on the surface of the adsorbent, and there is repulsion between the sulfate and the adsorbent. The modification of the adsorbent with the octodecyltrimethylammonium chloride led to the formation of positive charge on the surface of the adsorbent. The surface modification led to the first layer of the OTMAC grafted on palygoskite. The second layer of the OTMAC interacts with the first layer OTMAC through the hydrophobic force, leading to the formation of the double layer.

The effect of pH on the adsorption of sulfate on activated carbon is the same for both simulated and natural groundwater (Namasivayam and Sangeetha 2008). However, the presence of calcium in natural groundwater declined the adsorbent's dose for treatment of sulfate-contaminated groundwater. Calcium precipitates sulfate and reduces the requirement of the adsorbent's dose.

6.4.3 Mechanism

There are a number of techniques used to determine the mechanism of sulfate adsorption such as EXAFS, differential atomic pair distribution function analysis (d-PDF), (Zhu et al. 2013) and Attenuated total reflection-Fourier transform infrared spectroscopy (ATR-FT-IR) (Johnston and Chrysochoou 2016). The EXAFS spectrum is used for the mechanism of adsorption of sulfate (pH = 4) on ferrihydrite (Zhu et al. 2013). The EXAFS spectra of the sample after adsorption on ferrihydrite showed a weak peak at 8.5 $Å^{-1}$ similar to that of jarosite; this feature was not shown by liquid sulfate solution. This suggests that fewer than three Fe atoms on average surround the sulfur atom of the surface-bound sulfate ion.

The multicurve resolution analysis of ATR-FT-IR indicates the occurrence of the outer and inner sphere sulfate complex on Al-ferrihydrite (Johnston and Chrysochoou 2016). The sulfate species adsorbed at pH higher than 6 depicts two peaks in "multivariate curve resolution ATR-FT-IR" spectra at 1100 and 980 cm^{-1}. The two peaks are attributed to the outer sphere complex. However, on adsorption at lower pH, peaks at 1170, 1120, 1050, and 980 cm^{-1} emerged, which is attributed to the inner sphere adsorption process. The increase in the content of aluminum in ferrihydrite increased the content of the outer-sphere oxyanion. This is attributed to the suppression of the availability of inner-sphere binding sites (Johnston and Chrysochoou 2016).

The EXAFS spectra also decipher the distance between the sulfur and Fe atoms (adsorption with ferrihydrite), which is in the range of 3.18–3.19 Å (Zhu et al. 2013). Differential atomic pair distribution function analysis (d-PDF) along with EXAFS analysis suggests the bond length of sulfur and oxygen as the same. This led to the conclusion that the systematic errors between the two methods were insignificant for S-O bond determination. However, the EXAFS analysis underestimates the Fe-S bond length (3.19 Å), and d-PDF shows that it is in the range of 3.25 ± 0.02 Å.

The surface complexation model was also used to predict the adsorption of sulfate on ferrihydrite. At lower pH, the surface complexation model suggests the existence of inner and outer sphere complex (Gu et al. 2016). The results stimulated by the surface complexation model were supported by results from adsorption experiments and spectroscopic analysis, e.g. the model suggested the decline of the outer sphere complex, increase of sulfate adsorption at lower pH, and absence of the outer sphere complexation maxima at 0.5 M ionic strength.

Mean free energy of adsorption was also used as the parameter for physisorption or chemisorption estimation. The mean free energy for sulfate adsorption with OTMAC-modified palygoskite was calculated as 0.48 kJ/mol. The calculated value was below 8 kJ/mol indicating that the process occurred via physical interaction phenomenon (Dong et al. 2011). The author suggests that the process is controlled by electrostatic interaction rather than a chemical bond.

6.4.4 Desorption

Hydrochloric acid and alkaline pH were used for desorption and reuse of the adsorbent after adsorption with sulfate (Howarth et al. 2016; Namasivayam and Sangeetha 2008). Alkaline pH causes desorption by ion exchange phenomenon (Namasivayam and Sangeetha 2008), and the species adsorbed by chemical adsorption cannot be recovered by desorption via increasing pH value.

7 Impact of Initial Concentration, Adsorbent Dose, and Ionic Strength on Batch Adsorption of Metals and Anions and Elucidation of the Mechanism

Deepak Gusain
Durban University of Technology

Shikha Dubey and Yogesh Chandra Sharma
IIT(BHU), Varanasi

Faizal Bux
Durban University of Technology

CONTENTS

This chapter includes the effects of parameters such as ionic strength and adsorbent dose on the batch adsorption process. These factors usually affect in the same way for most metal anions, e.g. increase in percentage removal with increase of adsorbent dose and decline in initial concentration of the metals and anions. The ionic strength effect depends on the mechanism of adsorption, i.e. whether the adsorption process undergoes inner sphere complex formation or external sphere complex formation.

7.1 EFFECT OF INITIAL CONCENTRATION ON ADSORPTION OF METALS AND ANIONS

The percentage removal declines with increase of initial concentration, e.g. in removal of vanadium (Anirudhan et al. 2009), chromium (Maleki et al. 2015), iron (Al-Anber 2010), copper (Liu, Zhu, et al. 2013), zinc (Zhang, Li, et al. 2010), gallium (Chan 1993), cadmium (Pal and Pal 2017), cesium (Dwivedi et al. 2013; Zong et al. 2017), lead (Venkateswarlu and Yoon 2015a; Moradi et al. 2017; Huang et al. 2011), uranium (Yang, Liu, et al. 2017), fluoride (Zhu et al. 2016; Dhillon et al. 2015; Jin et al. 2016), and nitrate (Zhao and Feng 2016; Banu and Meenakshi 2017a, b; Srivastav et al. 2014; Ganesan, Kamaraj, and Vasudevan 2013; Hu et al. 2015). The decline in percentage removal is attributed to the limited number of active sites, which become saturated after a certain concentration (Huang et al. 2013) or after changing the ratio of surface active sites to total metal ions (Chou et al. 2010).

However, adsorption capacity improved with increase of initial concentration of vanadium with chitosan Zr(IV) composite (Zhang et al. 2014), amine-modified poly(-glycidyl methacrylate)-grafted cellulose (Anirudhan et al. 2009), Ti-doped chitosan bead (Liu and Zhang 2015), manganese on multiwalled carbon nanotubes (Ganesan, Kamaraj, Sozhan, et al. 2013), cobalt (Anirudhan et al. 2016), nickel (Mohammadi et al. 2014; Saleh, Ibrahim, et al. 2017), copper (Liu, Zhu, et al. 2013), zinc (Zhang, Li, et al. 2010), strontium (Chen and Wang 2012; Zhang, Liu, Jiang, et al. 2015), cesium with magnetic Prussian blue (Jang and Lee 2016), copper with a hexacyanoferrate polymer composite (Dwivedi et al. 2013) or Fe_3O_4 @WO_3 (Mu et al. 2017), lead (Mahmoud, Abdou, and Ahmed 2016; Pourbeyram 2016), uranium (Yang, Liu, et al. 2017), fluoride (Zhu et al. 2016; Dhillon et al. 2015; Jin et al. 2016), nitrate (Zhao and Feng 2016; Banu and Meenakshi 2017a, b; Srivastav et al. 2014; Ganesan, Kamaraj, and Vasudevan 2013; Hu et al. 2015), and perchlorate (Yang, Gao, Chu, et al. 2012).

The escalation of adsorption capacity with rise in initial concentration is attributed to increased collision frequency (Verma and Dutta 2015), increased utilization of active sites (Huang et al. 2013), and increased concentration gradient between liquid and solid phase (Anirudhan et al. 2016), which declined the mass transfer resistance (Dwivedi et al. 2013; Zong et al. 2017) and led to enhanced mass transfer (Hamdaoui 2017; Kannamba et al. 2010). However, increase in adsorption capacity is not linear, and the adsorption capacity increased up to a point; afterwards it declined (Zha et al. 2014) or equilibrated (Guo, Jiao, et al. 2017; Zhang, Xia, et al. 2015; Dwivedi et al. 2013; Mu et al. 2017; Dolatyari et al. 2016; Srivastav et al. 2014) (Figure 7.1) or declined insignificantly (Zhou et al. 2013; Yang, Gao, Chu, et al. 2012). The limited number of adsorption sites was held responsible for saturation or declination of adsorption capacity (Zha et al. 2014).

The concentration of gallium or selenium can be increased up to a point, and after that, precipitation or aggregation was observed. Gallium concentration of more than 1 mM causes it to form aggregates in addition to adsorption with poly-γ-glutamate (Hakumai et al. 2016). Similarly, the removal of selenite with increase in initial concentration from 30 to 100–1000 mg/l causes precipitation in addition to the adsorption phenomenon (Zhang, Fu, et al. 2017).

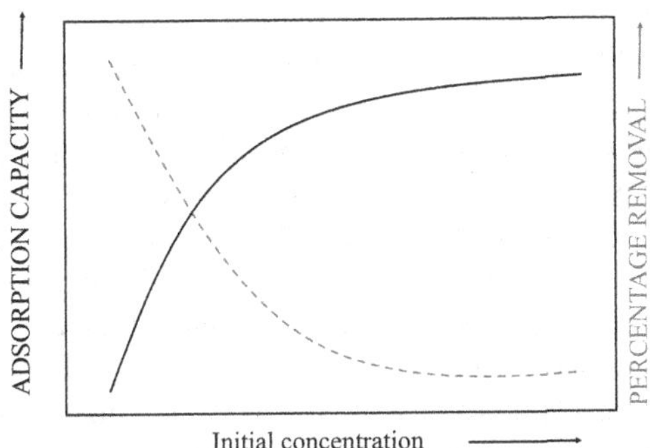

FIGURE 7.1 General effect of initial concentration on the percentage removal and adsorption capacity of inorganic contaminants (solid line curve represents adsorption capacity, and dashed line curve represents percentage removal).

The change in concentration causes a change in the mechanism of adsorption of uranyl ions on carbon nanofibers (Sun et al. 2016). The increase in concentration at pH 4.5 during adsorption of U(VI) on carbon nanofibers led to the transformation of the inner sphere complex to an outer sphere complex. This is evidenced by EXAFS spectra and its corresponding Fourier transform analysis (Sun et al. 2016). This was due to the change in speciation with concentration (Huynh et al. 2017).

The change in concentration of uranium led to a change in the structure of Mg/Al LDH adsorbent after modification (Ma, Huang, et al. 2015). The concentration of less than 50 ppm of uranyl causes no change in basal spacing (d_{basal}) of the adsorbent. However, on changing the concentration from 50 to 120 ppm, a new peak emerged at 0.89 (d_{basal}). On further raising the concentration, the peak becomes more dominant than the peak at 0.82 nm (d_{basal}). However, the peak corresponding to the structure of polysulfide Mg/Al LDH did not change on increasing the concentration. The unchanged peaks in the FT-IR spectrum at 668/669 cm^{-1} (ν (M-O)) and 447 cm^{-1} (δ (M-O-M) indicate the stability of the Mg/Al LDH as an adsorbent after adsorption.

The study of the effect of concentration is also responsible to give better isotherm results; the maximum adsorption capacity was difficult to achieve without saturation and plateau (karmakar et al. 2016). The isotherm study at low concentration gave erroneous results; hence, an initial fluoride concentration of 1400 mg/l was used in the isotherm study for the adsorption of fluoride with aluminum fumarate. The change in the concentration affects the thermodynamic parameters in adsorption of cobalt with multiwalled carbon nanotube/iron oxide composites (Wang et al. 2011). The standard change in entropy and enthalpy declined with the increase in initial concentration of cobalt on multiwalled carbon nanotube/iron oxide composites.

The modification of adsorbent from calcium silicate hydrate to magnetic calcium silicate hydrate affected the variation of percentage removal with uranyl ion concentration (Zhang, Liu, Wang, et al. 2015). The adsorption efficiency of calcium silicate hydrate and magnetic calcium silicate hydrate is the same in the range of concentration of 200–2000 mg/l. The decrease in adsorption capacity with increasing concentration (3000–5000 mg/l) is higher in the case of magnetic calcium silicate hydrate as compared to pristine calcium silicate hydrate.

The percentage removal of selenite in the range of 1–10 mg/l did not change with magnetic graphene oxide (Fu et al. 2014), but in the case of selenate with magnetic graphene oxide, it decreased with an increase in the initial concentration from 1 to 10 mg/l. However, graphene oxide without magnetism depicted different results, and the percentage removal of selenate with graphene oxide remained nearly constant (c.a. 30%) with an increase in the initial concentration from 1 to 10 mg/l.

The increase of adsorption capacity (12.32–47.48 mg/g) of cadmium on the TMU-16-NH_2 metal organic framework occurred with a slight decline of percentage removal (98.6%–95.6%) on increasing the initial concentration (50–200 mg/l) (Roushani et al. 2017). In spite of the increase in adsorption capacity, the optimum concentration was chosen to be 50 mg/l due to higher remainder cadmium concentration at higher initial concentration.

The increase in initial concentration of ferric ions on adsorption with eggshell and hexavalent chromium on aluminum magnesium mixed hydroxide led to a decline in the adsorption rate constant (Yeddou and Bensmaili 2007; Li et al. 2009).

7.2 EFFECT OF ADSORBENT DOSE ON THE ADSORPTION OF METALS AND ANIONS

The percentage removal increased with the increase in adsorbent dose for removal of vanadium with amine-modified poly(glycidyl methacrylate)-grafted cellulose) (Anirudhan et al. 2009), zinc chloride-activated carbon (Namasivayam and Sangeetha 2006), chromium with a metal–organic framework (Maleki et al. 2015), titanium cross-linked chitosan composite (Zhang, Xia, et al. 2015), manganese with sodium dodecyl sulfate-modified alumina (Khobragade and Pal 2016), manganese oxide-coated zeolite (Taffarel and Rubio 2010), cobalt with polyaniline/polypyrrole copolymer nanofibers (Javadian 2014), ferrous ions (Shokry and Hamad 2016), ferric ions (Al-Anber 2010; Bhattacharyya and Gupta 2009), nickel with charcoal ash (Katal, Hasani, et al. 2012), zinc (Zhang, Li, et al. 2010), arsenic (Alijani and Shariatinia 2017; Dehghani et al. 2016; Nashine and Tembhurkar 2016; Mandal et al. 2013; Roy et al. 2014; A. Saleh et al. 2016), selenium (Kameda et al. 2014; Adio et al. 2017), strontium (Wen et al. 2014; Zhao et al. 2014; Yu, Mei, et al. 2015; Zhang, Liu, Jiang, et al. 2015), cadmium (Venkateswarlu and Yoon 2015b; Yan, Zhao, et al. 2015; Roushani et al. 2017; Chen, Shah, et al. 2017; Beyki et al. 2017; Yakout et al. 2016), uranium (Dolatyari et al. 2016; Tan, Wang, Liu, Sun, et al. 2015), fluoride (Zhang, Li, et al. 2012), nitrate (Zhao and Feng 2016; Srivastav et al. 2014; Suriyaraj et al. 2015; Banu and Meenakshi 2017b; Ebrahimi-Gatkash et al. 2017), and perchlorate (Yang, Gao, Chu, et al. 2012). The high percentage removal with the increase in adsorbent dose is attributed to the high rate of superficial adsorption (Hamdaoui 2017) and increased availability of unsaturated sites (SenthilKumar et al. 2011).

However, the increase in adsorbent dose does not lead to a linear increase of the percentage removal, and the percentage removal equilibrated (Figure 7.2.) or slowed

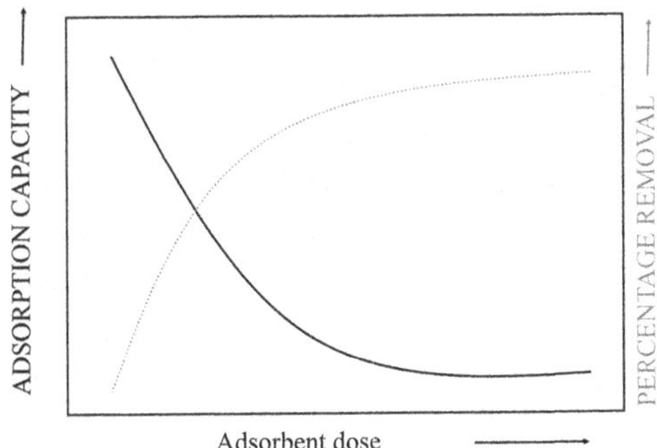

FIGURE 7.2 General effect of adsorbent dose on the percentage removal and adsorption capacity of inorganic contaminants (solid line curve represents adsorption capacity, and dashed line curve represents percentage removal).

down after a particular dose, e.g. in the case of removal of nickel (Fouladgar et al. 2015), fluoride (Zhang, Li, et al. 2012), and nitrate (Zhao and Feng 2016; Suriyaraj et al. 2015; Bagherifam et al. 2014). The percentage removal of nickel with γ -Al_2O_3 (Fouladgar et al. 2015) first increased with the increase in adsorbent dose. However, after a particular adsorbent dose, the percentage removal decreases with increasing adsorbent dose. The decrease in removal is attributed to the screening of active sites by particle interaction or aggregation. The aggregation happened due to hydrogen bonds of alumina with water (Fouladgar et al. 2015). However, the adsorption of nickel with charcoal ash after optimal dose increased slowly; the reason is attributed to the near saturation of the adsorbent (Katal, Hasani, et al. 2012).

Similarly, the percentage removal of cadmium increased on raising the adsorbent dose up to a point, and afterward there was no increase in percentage removal on increasing the adsorbent dose (Venkateswarlu and Yoon 2015b; Yan, Zhao, et al. 2015; Roushani et al. 2017; Chen, Shah, et al. 2017; Beyki et al. 2017; Yakout et al. 2016). The increase in percentage removal is attributed to increase in the number of available sites. The equilibrated state of percentage removal is attributed to adsorbent particle aggregation, which leads to a decrease in surface area and increase in diffusion length (Beyki et al. 2017). The decline in the percentage removal of As(III and V) with the increase in adsorbent dose after optimal dose is attributed to the concentration gradient of arsenic, in addition to aggregation and clumping of the adsorbent. The aggregation and clumping of the adsorbent lead to reduced surface area, decrease in adsorption sites (Yazdani et al. 2016), and increased diffusion path length (A. Saleh et al. 2016).

The percentage removal of fluoride increased with increase in adsorbent dose, and reached equilibrium up to a certain adsorbent dose, and afterward additional fluoride adsorption did not occur (Zhang, Li, et al. 2012) or substantial removal of adsorption did not happen (Mohan et al. 2012; Liu, Cui, et al. 2016). The increase in the removal efficiency is attributed to the increase in the surface area and the high number of unsaturated active sites.

The increase in adsorbent dose led to a decline in the adsorption capacity, e.g. chromium (Maleki et al. 2015; Zhang, Xia, et al. 2015), ferric ions (Bhattacharyya and Gupta 2009), copper (Hamdaoui 2017), zinc (Zhang, Li, et al. 2010), strontium (Wen et al. 2014; Zhao et al. 2014; Yu, Mei, et al. 2015; Zhang, Liu, Jiang, et al. 2015), cesium (Zong et al. 2017; Liu, Xie, et al. 2017; Al Abdullah et al. 2016; Mu et al. 2017), uranium (Dolatyari et al. 2016; Tan, Wang, Liu, Sun, et al. 2015), fluoride (Chen, Shu, et al. 2017; Dhillon et al. 2015), nitrate (Zhao and Feng 2016; Srivastav et al. 2014; Suriyaraj et al. 2015; Banu and Meenakshi 2017b; Ebrahimi-Gatkash et al. 2017), and perchlorate (Yang, Gao, Chu, et al. 2012). The decline in adsorption capacity of chromium (Maleki et al. 2015), copper (Ngah and Fatinathan 2010) and cesium (Zong et al. 2017; Liu, Xie, et al. 2017; Al Abdullah et al. 2016; Mu et al. 2017) with the increase in the adsorbent dose was due to the presence of the unsaturated adsorption sites and aggregation.

The decline in the adsorption capacity of copper on increase in the dose of the adsorbent (Hamdaoui 2017) is also attributed to the decrease in the concentration gradient between the solute concentration in the solution and adsorbed on the surface of the adsorbent. The particle concentration effect is also supposed to be responsible for the decline in adsorption capacity after optimum removal. The higher solid content in the adsorption system blocks the adsorption sites to the adsorbates. The blocking of adsorption sites occurs by blocking or by electrostatic interferences (Sen and Gomez 2011). The increase in adsorbent dose (bentonite) caused increase in pH of the solution. This is attributed to the negative charge on the surface of the adsorbent, which results in the adsorption of hydronium ions on the surface of the adsorbent and leads to increase in pH of the solution.

The decline in adsorption capacity was not linear and not universal with increase in adsorbent dose. In the case of removal of zinc and cadmium with magnetic hydroxyapatite, a maximum adsorbed amount is reached, and after that, the adsorbed amount decreased (Feng et al. 2010). This was due to the increase of vacant sites on the adsorbent. The increase in percentage removal is attributed to the increase in adsorption sites. The adsorbent employed after maximum percentage removal did not further increase the removal. The adsorbent employed after equilibrium is attained between the adsorbent and zinc and remained unused. So, the adsorption capacity declined after optimum capacity since the mass of the adsorbent was not considered in the calculation of the removal capacity (Zhang, Li, et al. 2010).

Similarly, the adsorption capacity of uranyl ions with Fe_3O_4@C@layered double hydroxide composite increased with increase in dose from 0.005 g to 0.01 g/50 ml and then declined afterward (Zhang, Wang, et al. 2013). The increase is attributed to the presence of more binding sites (Tan, Wang, Liu, Wang, et al. 2015; Verma and Dutta 2015).The decline is attributed to agglomeration, which led to decrease in effective surface area (Zhang, Wang, et al. 2013). However, the saturation of percentage removal is achieved at 0.1 g/50 ml at which c.a. 90% adsorption is achieved.

The percentage removal of manganese with iron-impregnated pumice also increased with increase in the adsorbent dose up to 15 g/l, and afterward it saturated (Çifçi and Meriç 2017). However, the adsorption capacity declined, but in the case of pumice (nonimpregnated), the adsorption capacity increased on increase in the adsorbent dose from 5 to 10 g/l and nearly saturated after that.

The nonsignificant increase in adsorption capacity after a particular dose in removal of selenite with nano zero-valent iron is attributed to agglomeration of the adsorbent (Xia et al. 2017). The percentage removal of strontium increased, and adsorption capacity declined with increase in adsorbent dose (Wen et al. 2014; Zhao et al. 2014; Yu, Mei, et al. 2015; Zhang, Liu, Jiang, et al. 2015), attributed to the fact that the surface active sites were exposed fully at low adsorbent dose, and the increase of the adsorbent dose led to particle aggregation and caused the decrease in total surface area and reduction of diffusion path length (Wen et al. 2014).

The adsorbent dose needs to be increased to maintain the same percentage removal with rise in initial concentration, e.g. to maintain the 100% removal efficiency at initial concentrations of 10, 40 and 100 mg/l, the doses of sodium dodecyl sulfate-modified chitosan were 0.45, 0.9, and 1.35 g/l (Pal and Pal 2017). The optimum dose for the composite varied from its individual units. The adsorbent dose required for arsenic removal by hematite was much higher (4 g/l) than that of its composite with multiwalled carbon nanotubes (0.2 g/l) (Alijani and Shariatinia 2017). The percentage removal obtained with hematite was even much lower, i.e. 16%–23%, as compared to the iron and multiwalled carbon nanotube composite, i.e. > 80%.

The adsorbent dose of Na-montmorillonite at less than 1 g/l for strontium removal suggested that the percentage increase was high with amplification of adsorbent dose, but after 1 g/l, the ascent was not steep and near to equilibration (Yu, Mei, et al. 2015). The reason for the large increase was attributed to the increase in functional groups. The decline in amplification of removal was attributed to particle aggregation.

7.3 EFFECT OF IONIC STRENGTH ON THE ADSORPTION OF METALS AND ANIONS

Adsorption declined (Wang et al. 2011; Manohar et al. 2005; Gu et al. 2016), increased (Marco-Lozar et al. 2007), and remained unaffected (Dolatyari et al. 2016; Pan et al. 2017) with increase in ionic strength and varied from case to case (Figure 7.3). Perchlorate (Svecova et al. 2011; Dolatyari et al. 2016; Pan et al. 2017), sodium nitrate (Zhong et al. 2016), potassium nitrate (Guo, Jiao, et al. 2017), sodium chloride, calcium nitrate (Zeng et al. 2015), and calcium chloride (Huang, Yang, et al. 2015) salts have been used to vary the ionic strength of the system. The decrease of percentage removal with ionic strength was taken as macroscopic

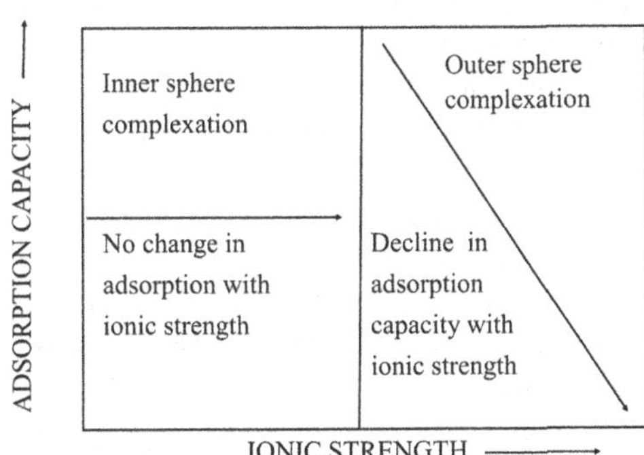

FIGURE 7.3 General effect of ionic strength on percentage removal and adsorption capacity.

evidence for outer sphere complexation (Li, Li, et al. 2012; Min et al. 2015; Guo, Jiao, et al. 2017; Zhong et al. 2016) or the electrostatic nature of adsorption (Zeng et al. 2015), and a process independent of ionic strength was suggested to form the inner sphere complex (Huang, Wu, et al. 2015).

The increase in the ionic strength primarily increases the thickness of the electrical double layer around the adsorbent (Manohar et al. 2005; Wang et al. 2011; Chen and Wang 2006), which leads to a decline in the adsorption capacity. In addition, it also affects the activity coefficient of the metal ions (Wang et al. 2011; Guo, Jiao, et al. 2017; Chen and Wang 2006; Kara et al. 2017), electrostatic interaction (Zhao et al. 2014), accumulation of surface charge (Guo, Jiao, et al. 2017) and competitive sorption (Svecova et al. 2011; Dolatyari et al. 2016; Pan et al. 2017), which, in turn, affect the transfer of metal ions from the solution to the surface of the adsorbent.

The ionic strength influences the thickness and interfacial potential of the double layer, which, in turn, affects the adsorption of the adsorbate. However, the adsorption of copper on multiwalled carbon nanotubes was not affected by ionic strength, as the adsorption is governed by the inner sphere complex formation (Sheng et al. 2010). It is also postulated that the background electrolyte concentration, in turn, affects the ionic strength applied to predict the adsorption. On the basis of the triple-layer model (Hayes and Leckie 1987), β-plane adsorption occurred when ionic strength easily affects the adsorption process; otherwise it follows the o-adsorption process (Sheng et al. 2010). Hence, the adsorption of copper on multiwalled carbon nanotubes participates in o-plane complex reaction. Increase in ionic strength as a result of increasing base in the solution reduced the removal of copper with γ-Al_2O_3 (Fouladgar et al. 2015). The decrease in Gibbs free energy of the hydrated ion solution is postulated to be the reason for this. It decreases the interaction between the cation in the solution and the adsorbent sites having negative charges, in addition to the promotion of formation M^+-OH^- ion pairs (Fouladgar et al. 2015). The reduction in removal is also attributed to the competition of positive ions with the adsorbate, screening of electrostatic interaction, and reduction of activity coefficient of copper ions (Hamdaoui 2017).

The effect of ionic strength varied with the species of selenium on titanate nanotubes (Sheng, Linghu, et al. 2016). The adsorption of selenite on titanate nanotubes is independent of the ionic strength and occurred with decrease in pHzpc; both these phenomena led to the formation of the inner sphere complex, whereas the adsorption of selenate increases with decrease in ionic strength, and the pHzpc of the adsorbent remains unaffected and suggested to form the outer sphere complex.

Similarly, adsorption of As(V) on hydrous cerium oxide increased with increase in ionic strength, but adsorption of As(III) remained unaffected. The increase of ionic strength led to the increase in zeta potential for hydrous cerium oxide, suggesting a decline in negative charge on increase in ionic strength (Li, Li, et al. 2012). The dominant As(III) species was neutral As(III) species, i.e. H_3AsO_3, and the dominant As(V) species was charged As(V) species, i.e. $HAsO_4^{2-}$ at pH 7. The increase or unchanged behavior of arsenic species As(III)) is attributed to inner sphere complex formation. However, this is not the case always; the adsorption of As(III and V) on magnetite remains unaffected by the increase in ionic strength (Liu, Chuang, et al. 2015).

The ionic strength effect on adsorption also depends on pH. The adsorption of cesium with graphene oxide declined the adsorption capacity at pH lower than 6, but its effect was not significant at pH higher than 6 (Tan et al. 2016). This is attributed to the follow-up of the adsorption process by outer sphere complexation at pH lower than 6 and inner sphere complexation at pH higher than 6. Similarly, the adsorption of strontium with sodium rectorite tremendously decreased, was not significantly affected, and was unaffected at pH below 9.5, after 9.5, and after 10.5, respectively, with increase in ionic strength (Zhao et al. 2014). The reduced electrostatic interaction with increasing ionic strength led to the small number of binding sites and, hence, decline in adsorption capacity.

The germanium catechol complex $(Ge(cat)_3)^{2-}$ adsorption on activated carbon increased with increase in ionic strength at pH 10. The increase in adsorption is attributed to the reduction of repulsion between the adsorbate and the adsorbent. The effect is more effective in another adsorbent with high acidic character. However, the adsorption capacity of both the adsorbents reached the same value (Marco-Lozar et al. 2007).

The surface modification of β-zeolite with ethylene diamine changes the mechanism of adsorption by complex formation (Liu, Yuan, et al. 2017). The adsorption of nickel on β-zeolite decreased on increase in ionic strength. However, when the surface of adsorbent is modified with ethylene diamine, then the adsorption process is not affected by the change in ionic strength. The reason stated for this behavior is that the adsorption of nickel on β-zeolite is governed by the outer sphere complex, whereas in the latter case, it is governed by the inner sphere complex.

The adsorption of sulfate on ferrihydrite decreased with increase in ionic strength (Gu et al. 2016). The decline of the outer sphere complex proportion occurred with increasing ionic strength. The decline of the outer sphere complex formation is attributed to the competition from background electrolyte and electrical double layer contraction. The higher ionic strength causes a more pronounced electrical double layer contraction, and this leads to a decline in electric potential (positive) at the adsorbent plane.

In addition, the proportion of outer sphere complex did not remain linear and varied with ionic strength (Gu et al. 2016). The proportion of sulfate as the outer sphere complex via adsorption on ferrihydrite decreases with increase in pH at ionic strength 0.5, but at an ionic strength of 0.02 and 0.1 M, the outer sphere complex formation reached a maximum at pH 5, and afterward it declined. The rise in pH led to two antagonistic conditions promoting and impeding the formation of the outer sphere complex. The rise in pH led to the decline of inner sphere complexation, which causes generation of new sites for outer sphere complexation. In addition, the surface becomes less negatively charged and hence declines the prospects of outer sphere complex formation. At lower ionic strength (0.02 and 0.1 M), both factors are comparable and results in the maximum proportion of outer sphere complexation along with the rise of pH. The high ionic strength (0.5 M) depresses the outer sphere complex formation, and hence, the maximum for outer sphere complexation was absent at high ionic strength. In addition to this, the maximum was not also present at high initial concentration, as most of the active sites are engaged, and the number of active sites that vacated on the decline of inner sphere complexation is inadequate to show any significant increase in the active sites in comparison to the adsorbate. Hence, no significant change in outer sphere complexation is observed.

The decline in adsorption of strontium on modified sawdust on raising the sodium nitrate concentration is attributed to the decrease in strontium species with increase in sodium nitrate concentration estimated by speciation software (Cheng et al. 2012). The decrease in adsorption capacity of strontium with SBA-15 is attributed to the formation of an ion pair between strontium and nitrate (Zhang, Liu, Jiang, et al. 2015). Uranium adsorption with attapulgite depends on ionic strength, but after surface modification with chitosan, it becomes independent of ionic strength (Pan et al. 2017).

8 Kinetic, Isotherm, and Thermodynamic Studies for Batch Adsorption of Metals and Anions, and Management of Adsorbents after the Adsorption Process

Deepak Gusain
Durban University of Technology

Shikha Dubey and Yogesh Chandra Sharma
IIT(BHU), Varanasi

Faizal Bux
Durban University of Technology

CONTENTS

This chapter includes the study of kinetic and isotherm model for metals and anions. The kinetic and isotherm data can be fitted by either linear curve fitting or nonlinear curve fitting. The kinetic and isotherm model can be estimated by the coefficient-of-determination values, chi-square values, and closeness of experimental and theoretical data. Thermodynamic parameters are also evaluated in this chapter, and thermodynamic data depicted the physisorption or chemisorption nature of the adsorption process and the endothermic or exothermic nature of adsorption. In addition, management of an adsorbent post its use is also discussed.

8.1 KINETIC STUDY

The adsorption process is governed by four steps: (a) bulk diffusion or mass transfer, (b) film diffusion or boundary layer diffusion, (c) pore diffusion, and (d) physical or chemical reaction. The first step can be completely ignored or significantly reduced (Wang, Shi, Pan, et al. 2020; Tan and Hameed 2017) in batch adsorption conditions due to stirring. The mass transfer resistance during stirring can be reduced by stirring (Tan and Hameed 2017).

The second step is directly proportional to the concentration of ions and the specific surface area of the adsorbent. The third step, i.e. pore diffusion, is controlled by pore structure, pore size, pore volume, and the size and structure of the adsorbate.

The last step is a physical or chemical reaction, which is quite fast. So, the first and last steps can be ignored in kinetics and the rate is mainly determined by both the second and third steps. If there is a linear relationship between the adsorbate uptake rate and $t^{(1/2)}$, then it can be concluded that only pore diffusion is the rate-controlling step. Therefore, determination of the slowest step or rate-controlling step helps us in determining the rate of adsorbate uptake. The study of adsorbate uptake rate is known as kinetic study in adsorption.

The kinetic study for adsorption was conducted by fitting the kinetic data on the pseudo-first-order model (Panneerselvam et al. 2011; Alijani and Shariatinia 2017; Nashine and Tembhurkar 2016; Yan, Kong, et al. 2015; Kyzas et al. 2016), pseudo-second-order model (Zhang et al. 2014; Liu and Zhang 2015; Namasivayam and Sangeetha 2006; Li et al. 2009), intra-particle diffusion model (Guo et al. 2016; Ngah and Fatinathan 2010), and Weber–Morris model (Sankararamakrishnan et al. 2014).

The best kinetic model is selected on the basis on the coefficient of determination (R^2) obtained during curve fitting analysis of kinetic data. The fitting of the data is analyzed with linear (Ngah and Fatinathan 2010; Sani et al. 2017; Zhu and Li 2015; Liu, Zhu, et al. 2013; Hao et al. 2010) and nonlinear curve fitting analyses (Jian et al. 2015; Qi et al. 2015; Zhang, Gao, et al. 2013; Yu, Ma, et al. 2015; Zhou et al. 2017; Özlem Kocabaş-Ataklı and Yürüm 2013; Hossain et al. 2012). However, in the case of adsorption of cesium on nickel oxide-grafted andic soil, the coefficient of determination was 1, and it was difficult to determine, on the basis of the coefficient of determination, whether the system followed the pseudo-first-order or pseudo-second-order model (Ding et al. 2013). In the case of nickel adsorption with polyvinyl alcohol-based chelating sponge, the system followed both the pseudo-first-order and pseudo-second-order models, and this led to the conclusion in the research article that the system followed the intra-particle diffusion process as the rate-limiting step of adsorption in the solution (Cheng et al. 2014). In addition, the closeness of experimental and theoretical adsorbed amounts is also used for the determination of the most suitable kinetic model (Namasivayam and Sangeetha 2006). Jiang et al. (2013) recommended the use of data on the initial fast rate of adsorption for fitting the kinetic model. It was suggested that the whole kinetic data are not able to fit the pseudo-first-order model, and the model is applicable only for rapid initial stage of adsorption.

The better fit of kinetic data on the pseudo-second-order model follow-up is taken as a proxy for chemisorption (Zhang, Liu, Jiang, et al. 2015; Jin et al. 2017; Zhong et al. 2016; Zha et al. 2014; Venkateswarlu and Yoon 2015b; Huang, Yang, et al. 2015; Roushani et al. 2017; Chen, Shah, et al. 2017; Zeng et al. 2015; Yang, Liu, et al. 2017; Tan, Liu, et al. 2015; Dolatyari et al. 2016; Zhang, Li, et al. 2012; Chen, Zhang, Li, et al. 2016; Chen, Shu, et al. 2017; Wang, Yu, et al. 2017; Zhu et al. 2017; Yu et al. 2013; Tang and Zhang 2016; Wu et al. 2017; Kundu et al. 2017; Zhao and Feng 2016; Usman et al. 2016; Wan et al. 2012; Guo, Su, et al. 2017; Cheng et al. 2014; Liu, Yuan, et al. 2017). The rate of adsorption was fast during the initial stages (Idris 2015; Cheng et al. 2012; Zhao and Feng 2016; Ganesan, Kamaraj, and Vasudevan 2013; Ebrahimi-Gatkash et al. 2017; Hu et al. 2015). The high rate of adsorption was attributed to the high concentration gradient, high availability of the adsorption sites (Anirudhan et al. 2016; Nunell et al. 2015), and passive process of adsorption (Chen, Shu, et al. 2017), e.g. the rate of adsorption increased with an increase in the concentration of uranyl ions on nano-magnesium hydroxide (Chen, Zhuang, et al. 2014). However, in the case of adsorption of manganese with diethylenetriamine-modified silica, the highest initial rate of adsorption was found at the lowest concentration, i.e. 10 ppm (Idris 2015), but in the case of adsorption of nitrate with hydrous bismuth oxide, the adsorption rate was faster in the intermediate concentration range of 28–42 mgN/L as compared to the low (14 mgN/L) and high (56 mgN/L) concentrations. The high rate of adsorption during the initial adsorption of strontium with Na-montmorillonite was attributed to ion exchange or chemical sorption (Yu, Mei, et al. 2015).

The kinetics of U(VI) removal by bovine serum-coated graphene oxide is expected to occur in two parts: the first part is chelation and the second part is intra-particle diffusion (Yang, Liu, et al. 2017). The chelation is confirmed by the merging of a new peak in FTIR spectra at 912 cm^{-1} (O=U=O) and redshift that occurred at 1650 cm^{-1}. This is attributed to chelation between U(VI) and the COOH group. The chelation is also confirmed by the XPS data. The N and O 1s peak shifted to higher energy after adsorption. This suggested the binding of N- and O-containing groups of the adsorbent with uranyl species. The change in binding energy was more in N (1 eV) than in O (0.5 eV). This suggests that the primary group responsible for chemical bond formation is nitrogen and suggests more stability of bond formation with the N-containing groups.

The rate of adsorption varied with temperature and modification of the adsorbent. The rate of adsorption for selenate removal with Mg–Al layered double hydroxide increased with an increase in the temperature (Kameda et al. 2014). Mercury(II) adsorption on the graphene oxide (GO)–iron oxide (Fe_3O_4) magnetic nanoparticle composite (GOMNP) exhibited different mechanisms at varying temperatures. At 20°C, the adsorption mechanism followed pseudo-second-order kinetics, while at higher temperatures, intra-particle diffusion governed the sorption mechanism

(Diagboya et al. 2015). The rate of adsorption was faster after the modification of magnetic mesoporous carbon with polyacrylic acid (Zeng et al. 2015). The faster rate of adsorption was due to the blocking of pores by functionalization. The blocking of pores led to the reduced diffusion path length and decreased diffusion of the adsorbate into the inner surface and pores of the adsorbent.

The concentration also alters the kinetics of the adsorption process. The rate of adsorption of uranyl ion on nano-magnesium hydroxide increased with an increase in the concentration (Chen, Zhuang, et al. 2014). Similarly, an increase in the concentration of manganese led to a decline in the pseudo-second-order rate constant (Idris 2015). The decrease in the adsorption rate constant with an increase in the concentration is attributed to increased occupation of the active sites on the surface of the adsorbent by adsorbates (Idris 2015).

8.2 ISOTHERM STUDY

An adsorption isotherm is a quantitative method to characterize adsorbate equilibrium between the aqueous and solid phases at a constant ambient temperature (Tong et al. 2019). The Langmuir (Zhang et al. 2014; Liu and Zhang 2015; Anirudhan et al. 2009; Li et al. 2009), Freundlich (Guo, Su, et al. 2017; Cheng et al. 2014; Ren, Zhang, et al. 2011; Yazdani et al. 2016), Dubinin–Radushkevich (D-R) (Ngah and Fatinathan 2010; Zhang, Li, et al. 2010; Zhang, Zhu, et al. 2010; Katal, Baei, et al. 2012), Sips (Alatalo et al. 2015; Anirudhan et al. 2016; Reddy and Lee 2013b), and Redlich–Peterson (Kumari et al. 2015; Li, Li, et al. 2012) isotherm models were applied to fit the adsorption isotherm data. In addition, other models were also used, such as the Temkin isotherm model (two-parameter) (Boulaiche et al. 2019; Chakraborty et al. 2019; Bezzina et al. 2020), Flory–Huggins isotherm model (two-parameter) (Arrousse et al. 2020; Jalees et al. 2019), Hill isotherm model (two-parameter) (Lin et al. 2019; Abdelwaheb et al. 2019), Halsey isotherm model (two-parameter) (Ramadoss and Subramaniam 2019; Amin et al. 2019; Shahnaz et al. 2020), and Jovanovic isotherm model (two-parameter) (Karoui et al. 2020; Ghaleh et al. 2020).

Each isotherm model has its own assumption. The Langmuir isotherm is based on the assumption that only one site is available for each adsorbate. The total number of sites is limited, and each adsorbate has the same affinity for each site and does not interfere with the binding of other adsorbates.

The Freundlich isotherm is based on variable kinds of adsorption sites present on the adsorbent (Tong et al. 2019). In this adsorption model, adsorption heat and affinities do not need to be uniformly distributed (opposite to the Langmuir isotherm model) on the heterogeneous surface. There is an exponential decline in adsorption energy with subsequent occupation of adsorption sites (Al-Ghouti and Da'ana 2020).

The Temkin isotherm model is used for the description of adsorption of hydrogen on platinum electrodes, under acidic conditions. It avoids extremely high and low concentrations of the adsorbate (Al-Ghouti and Da'ana 2020). This model assumes that adsorption acts as a function of temperature of all molecules, and adsorption heat of all molecules existing in the layer declines linearly rather than logarithmically with increase in coverage of adsorbent surface.

The Sips isotherm is a combination of both the Langmuir and Freundlich isotherm models and is dependent upon the concentration of the solution where at low concentrations, it favors the Freundlich isotherm, and at the high concentrations, it fits the Langmuir isotherm model (Al-Ghouti and Da'ana 2020).

The Redlich–Peterson model is also a hybrid model of the Langmuir and Freundlich isotherms with three parameters. This isotherm has an exponential function of concentration in the denominator (equilibrium concentration) and varies linearly with equilibrium in the numerator (Al-Ghouti and Da'ana 2020). At high concentrations, it follows mainly the Freundlich isotherm model, and at low concentrations, it follows the Langmuir isotherm model. However, two isotherms, i.e. Langmuir and Freundlich, are predominantly in use.

The coefficient of determination was used to find out the suitable isotherm (Jain et al. 2014; Mohammadi et al. 2014, 2015; Mangaleshwaran et al. 2015; Liu, Yuan, et al. 2017; Kandah and Meunier 2007; Gupta et al. 2014; Saleem et al. 2016). The isotherm data can be fitted with both linear (Sani et al. 2017; Zhu and Li 2015; SenthilKumar et al. 2011; Fouladgar et al. 2015; Pizarro et al. 2015; Lee et al. 2016; Hao et al. 2010; Ren, Zhang, et al. 2011; Alijani and Shariatinia 2017; Vu et al. 2015; Sankararamakrishnan et al. 2014) and nonlinear curve fitting analyses (Hossain et al. 2012; Özlem Kocabaş-Ataklı and Yürüm 2013; Sheng et al. 2010; Zhang, Liu, Wu, et al. 2015; Martinson and Reddy 2009; Zhang, Gao, et al. 2013; Yu, Ma, et al. 2015; Jian et al. 2015; Tang et al. 2013; Zhang, Ren, et al. 2013). The decent fitting of isotherm data on both the Langmuir and Freundlich isotherms makes it difficult to suggest the best adsorption isotherm, e.g. nickel adsorption with nano-alumina (Srivastava et al. 2011) and Fe_3O_4-impregnated tea waste (Panneerselvam et al. 2011). However, the decent fitting of isotherm data on both the Langmuir and Freundlich isotherms for adsorption of uranyl ions on ammonia-modified graphene oxide suggested that it followed both chemisorption and physisorption (Verma and Dutta 2015).

In the case of adsorption of cobalt with nanocellulose/nanobentonite composite, the value of coefficient of determination for the Langmuir, Freundlich, and Sips isotherms was larger than 0.9 (Anirudhan et al. 2016). However, a lower value of χ^2 (chi-square) was used as an additional factor for comparison. The lower value of χ^2 in the Sips isotherm suggests the follow-up of this model by the isotherm data. Hence, the system follows the monolayer adsorption at a lower concentration and multilayer adsorption at a higher concentration.

The isotherm data for chromium removal with magnetite hollow microspheres followed the Freundlich isotherm at a low initial concentration, i.e. 10 mg/l, but at 20 mg/l, it started to deviate from the Langmuir isotherm, and at 40 mg/l, it followed the Langmuir isotherm with a much higher adsorption capacity, i.e. 180 mg/g (Liu et al. 2012). The biphasic behavior is attributed to hollow magnetite microspheres, where the surface complexation of Cr with Fe on the surface hollow structure provided an accessible pathway into the interior of magnetite hollow microspheres.

The Redlich–Peterson isotherm explains the adsorption of Cr(VI) on Fe_3O_4@ poly(m-phenylenediamine) core shell better than the Langmuir and Freundlich. This suggests that the reduction–sorption process was the hybrid process (nonlinear) (Wang, Zhang, et al. 2015).

The mean energy values calculated from the D-R isotherm model were used as theoretical evidence for the mechanism of adsorption (Chen, Zhang, Li, et al. 2016). The mean adsorption energy of less than 8 kJ/mol suggested the physical nature of adsorption (Javadian 2014; Mukhopadhyay et al. 2017; Azari et al. 2015; Elwakeel and Guibal 2015). The mean energy value between 8 and 16 kJ/mol is indicative of the ion exchange process (Duranoğlu et al. 2012; González and Pliego-Cuervo 2014) or physical/chemical adsorption (Zhu et al. 2017), and that more than 16 kJ/mol is indicative of chemisorption (Mukhopadhyay et al. 2017).

The adsorption of chromium on amino-functionalized titanate nanotubes and protonated titanate nanotubes follows the Langmuir isotherm model (Wang, Liu, et al. 2013). The adsorption capacity for amino-functionalized titanate nanotubes is much larger (153.85 mg/g) than for protonated titanate nanotubes. This suggests that the amino groups act as an important factor in the adsorption of chromium.

8.3 THERMODYNAMICS

Adsorption is a temperature-dependent process. The change in temperature led to a change in diffusion of the adsorbate, due to the decline in the viscosity of the solution, which alters the equilibrium status of the adsorption process and hence the change in thermodynamic parameters observed with a change in the temperature (Ray et al. 2020b; Nouri et al. 2007). Thermodynamic parameters provide information about the nature of adsorption such as endothermic or exothermic nature on the basis of standard enthalpy change, the spontaneity of the process on the basis of change in free energy, and the chemical and physical nature of adsorption by the coefficient of standard enthalpy change. So, it is needed to estimate the thermodynamic parameters for the adsorption process.

The change in free energy for manganese (Idris 2015; Ganesan, Kamaraj, Sozhan, et al. 2013; Al-Wakeel et al. 2015), cobalt (Bhatnagar et al. 2010; Ramos et al. 2016; Lingamdinne et al. 2016; Anirudhan et al. 2016; Negm et al. 2015), nickel (Gupta et al. 2014; Panneerselvam et al. 2011; Mohammadi et al. 2015; Saleem et al. 2016; Stojakovic et al. 2016; Liu, Yuan, et al. 2017; Katal, Hasani, et al. 2012), copper (Özlem Kocabaş-Ataklı and Yürüm 2013; SenthilKumar et al. 2011; Fouladgar et al. 2015; Liu, Zhu, et al. 2013; Sheng et al. 2010; Hao et al. 2010), zinc (Afroze et al. 2016; Wang, Yuan, et al. 2013; Sheela et al. 2012; Zhang, Li, et al. 2010; Rashid et al. 2016; Hao et al. 2010), gallium (Zhang, Zhu, et al. 2010; Zhang et al. 2011), arsenite (Mandal et al. 2013; Alijani and Shariatinia 2017; Sankararamakrishnan et al. 2014; A. Saleh et al. 2016), arsenate (Alijani and Shariatinia 2017; Zhou et al. 2017; Sankararamakrishnan et al. 2014), strontium (Zhao et al. 2014; Yu, Mei, et al. 2015; Ghaemi et al. 2011; Kaçan and Kütahyalı 2012), cadmium (Shi et al. 2015; Wang, Liu, et al. 2012; Zhong et al. 2016; Huang, Yang, et al. 2015; Yan, Zhao, et al. 2015; Guo, Jiao, et al. 2017; Roushani et al. 2017; Khan et al. 2015; Huang, Wu, et al. 2015; Yang, Tang, et al. 2014), cesium (Du et al. 2017; Zhang, Wang, and Li 2017; Tan et al. 2016; Zhang, Zhao, et al. 2015; Long et al. 2013; Han, Zhang, and Gu 2013; Zong et al. 2017; Liu, Xie, et al. 2017), mercury (Elhamifar et al. 2016; Liu, Ding, et al. 2016; Guo et al. 2016; Lin and Zou 2017; Luo, Chen, et al. 2015; Patra and Kim 2017; Salamun et al. 2015; Saman et al. 2013), lead (Kumar et al. 2014; Li, Wang, et al.

2017; Yan, Kong, et al. 2015; Hadi Najafabadi et al. 2015; Moradi et al. 2017; Yuan, Zhang, et al. 2017), uranium (Tan, Liu, et al. 2015; Zhang, Liu, Wang, et al. 2015; Verma and Dutta 2015; Zhang, Jing, et al. 2015; Tan, Wang, Liu, Wang, et al. 2015; Bayramoglu and Arica 2016; Dolatyari et al. 2016; Saleh, Tuzen, et al. 2017; Zhang, Wang, et al. 2013), fluoride (Chen et al. 2012; Ma et al. 2017; Zhu et al. 2016, 2017; Chen, Shu, et al. 2017; Prabhu and Meenakshi 2014; Wang, Yu, et al. 2017; ; Lin, Liu, and Chen 2016; Wu et al. 2016; Parashar et al. 2016; karmakar et al. 2016; Jin et al. 2015; Liu, Zhang, et al. 2015), nitrate (Banu and Meenakshi 2017a, b; Naushad et al. 2014; Keränen et al. 2015; Katal, Baei, et al. 2012; Ganesan, Kamaraj, and Vasudevan 2013; Hu et al. 2015), and perchlorate (Luo et al. 2016; Yang, Xiao, et al. 2013) is reported to be negative and spontaneous in nature. However, there were a few cases where the change in free energy was positive and suggested the adsorption process to be nonspontaneous in nature as in the case of copper removal with garden grass (Hossain et al. 2012), arsenite and arsenate removal with air-oxidized zero valent iron-doped multiwalled carbon nanotubes and hematite (Alijani and Shariatinia 2017), arsenate removal with feldspar (Yazdani et al. 2016), strontium removal with potassium tetratitanate and sodium trititanate (Guan et al. 2011), and cadmium removal with a 1,2,4-triazole-3-thiol-modified lignin-based adsorbent at 288–318 K (at 328 and 338 K, the free energy change was negative) (Jin et al. 2017).

The positive free energy change in the case of strontium removal with potassium tetratitanate and sodium trititanate (Guan et al. 2011) may be due to the value of equilibrium constant in the van't Hoff equation as the ratio of q_e to c_e.

Mohamed et al. (2017) have used the magnitude of standard change in free energy to estimate the nature of adsorption, i.e. either physisorption or chemisorption. The change in free energy values between 0 and 20 kJ/mol was used as a macroscopic indicator toward the physisorption, and the values between 80 and 4000 kJ/mol suggested chemisorption. Similarly, other authors have also used it as an indicator of the nature of adsorption (Chen, Shu, et al. 2017; Hadavifar et al. 2014; Azari et al. 2017; Fu et al. 2016).

The positive and negative standard enthalpy changes depict the endothermic and exothermic nature of the adsorption process, respectively. The standard enthalpy change was positive for chromium(VI) (Kumari et al. 2015; Zhou et al. 2016), manganese (Idris 2015; Ganesan, Kamaraj, Sozhan, et al. 2013; Al-Wakeel et al. 2015), cobalt (Negm et al. 2015; Anirudhan et al. 2016; Lingamdinne et al. 2016; Mahmoud, Yakout, et al. 2016; Awual et al. 2015), nickel (Gupta et al. 2014; Panneerselvam et al. 2011; Saleem et al. 2016; Liu, Yuan, et al. 2017; Katal, Hasani, et al. 2012), copper (Özlem Kocabaş-Ataklı and Yürüm 2013; Fouladgar et al. 2015; Liu, Zhu, et al. 2013; Hao et al. 2010; Sheng et al. 2010), zinc (Zhang, Li, et al. 2010; Rashid et al. 2016; Hao et al. 2010), gallium (Zhang, Zhu, et al. 2010; Zhang et al. 2011), arsenite (Mandal et al. 2013; Alijani and Shariatinia 2017; Nashine and Tembhurkar 2016; Sankararamakrishnan et al. 2014; A. Saleh et al. 2016), arsenate (Yazdani et al. 2016; Alijani and Shariatinia 2017; Zhou et al. 2017; Sankararamakrishnan et al. 2014), strontium (Liu, Meng, Luo, et al. 2015; Zhao et al. 2014; Yu, Mei, et al. 2015; Kaçan and Kütahyalı 2012; Guan et al. 2011), cadmium (Jin et al. 2017; Wang, Liu, et al. 2012; Zhong et al. 2016; Huang, Yang, et al. 2015; Yan, Zhao, et al. 2015; Guo, Jiao, et al. 2017; Roushani et al. 2017; Khan et al. 2015; Huang, Wu, et al. 2015; Yang,

Tang, et al. 2014), cesium (Du et al. 2017; Zhang, Wang, and Li 2017; Tan et al. 2016; Zhang, Zhao, et al. 2015; Han, Zhang, and Gu 2013; Zong et al. 2017; Liu, Xie, et al. 2017), mercury (Elhamifar et al. 2016; Guo et al. 2016; Lin and Zou 2017; Wang, Lv, et al. 2016; Thakur et al. 2013), lead (Kumar et al. 2014; Li, Wang, et al. 2017; Yan, Kong, et al. 2015; Hadi Najafabadi et al. 2015; Moradi et al. 2017; Yuan, Zhang, et al. 2017), uranium (Tan, Liu, et al. 2015; Zhang, Liu, Wang, et al. 2015; Verma and Dutta 2015; Zhang, Jing, et al. 2015; Tan, Wang, Liu, Wang, et al. 2015; Bayramoglu and Arica 2016; Dolatyari et al. 2016; Zhang, Wang, et al. 2013), fluoride (Ma et al. 2017; Zhou et al. 2011; Prabhu and Meenakshi 2014; Wang, Yu, et al. 2017; Zhu et al. 2017; Chen et al. 2012), and nitrate (Naushad et al. 2014; Ganesan, Kamaraj, and Vasudevan 2013) and negative for chromium(VI) (Wu et al. 2013; Albadarin et al. 2012; Li, Xie, et al. 2016), cobalt (Bhatnagar et al. 2010; Fang et al. 2014), nickel (Mohammadi et al. 2015; Stojakovic et al. 2016), copper (Hossain et al. 2012; SenthilKumar et al. 2011), zinc (Afroze et al. 2016; Sen and Gomez 2011; Sheela et al. 2012; Wang, Yuan, et al. 2013), strontium (Ghaemi et al. 2011), cadmium (Shi et al. 2015), cesium (Long et al. 2013; Ma et al. 2011), uranium (Saleh, Tuzen, et al. 2017), fluoride (Chen et al. 2012; Zhu et al. 2016; Chen, Shu, et al. 2017; Lin et al. 2016), and nitrate (Banu and Meenakshi 2017b; Keränen et al. 2015; Katal, Baei, et al. 2012; Hu et al. 2015). The endothermic nature of the adsorption process was attributed to the need of energy for breaking the hydration sheath (Gupta et al. 2014; Zhao et al. 2014) and the weakening of the interaction between the solute and solvent in comparison to the solute and adsorbent with an increase in temperature (Ganesan, Kamaraj, and Vasudevan 2013). The positive enthalpy for adsorption of nickel on activated carbon was attributed to breakage of the hydration sheath of metal ion (Gupta et al. 2014). The continuation of the adsorption process needs the breakage of the hydration sheath. The increase in temperature provides the requisite energy and leads to an increase in the adsorption process. Similarly, the positive enthalpy for the adsorption of strontium on Na-rectorite was attributed to breaking of hydration sheath on the adsorbent surface (Zhao et al. 2014). Hence, variation in adsorption with temperature is also used as a factor in the determination of exothermic or endothermic nature of adsorption, e.g. adsorption of nickel on chitosan immobilized on bentonite declined with an increase in temperature, which suggests that the adsorption is an exothermic process (Futalan et al. 2011).

The magnitude of enthalpy change was used to estimate the strength of interaction between the adsorbate and the adsorbent (Zhao et al. 2014; Khan et al. 2015; Yang, Liu, et al. 2017). The standard enthalpy change of less than ca. 20.9 kJ/mol (Yang, Liu, et al. 2017; Khan et al. 2015; Husein 2013; Li, Ye, et al. 2012) or 40 kJ/mol (Saman et al. 2017) (Figure 8.1) is considered as the indicator of the physisorption process, e.g. the standard enthalpy change of ca. 1.31 kJ/mol suggests a weak interaction between the adsorbate and adsorbent (Zhao et al. 2014), and enthalpy change of Cr(VI) adsorption of −8.59 and −10.57 kJ/mol postulates that the mechanism follows either the electrostatic interaction or ion exchange (Duranoğlu et al. 2012).

The change in enthalpy varied with pH, e.g. enthalpy change for nickel adsorption on beta zeolite and ethylenediamine-modified beta zeolite varied along with the change in the pH of the solution (Liu, Yuan, et al. 2017). The thermodynamic parameters, i.e. standard enthalpy change and standard entropy change, declined in the

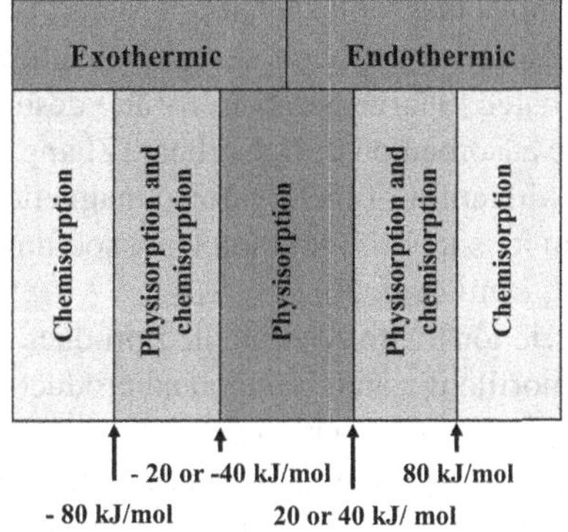

FIGURE 8.1 Enthalpy values as an indicator for chemical or physical adsorption.

presence of magnesium and sodium for adsorption of cesium on poly(b-cyclodextrin)/bentonite (Liu, Xie, et al. 2017). The standard entropy change was positive and negative. The standard entropy change was positive for manganese (Idris 2015; Ganesan, Kamaraj, Sozhan, et al. 2013), cobalt (Anirudhan et al. 2016; Lingamdinne et al. 2016; Mahmoud, Yakout, et al. 2016; Awual et al. 2015), nickel (Gupta et al. 2014; Panneerselvam et al. 2011; Mohammadi et al. 2015; Saleem et al. 2016; Stojakovic et al. 2016; Liu, Yuan, et al. 2017; Katal, Hasani, et al. 2012), copper (Hossain et al. 2012; Özlem Kocabaş-Ataklı and Yürüm 2013; Fouladgar et al. 2015; Liu, Zhu, et al. 2013; Sheng et al. 2010; Hao et al. 2010), arsenite (Alijani and Shariatinia 2017; Nashine and Tembhurkar 2016; Sankararamakrishnan et al. 2014), arsenate (Yazdani et al. 2016; Alijani and Shariatinia 2017; Zhou et al. 2017; Sankararamakrishnan et al. 2014), selenite (Larimi et al. 2013), strontium (Zhao et al. 2014; Yu, Mei, et al. 2015), cadmium (Roushani et al. 2017; Yan, Zhao, et al. 2015), mercury (Elhamifar et al. 2016; Liu, Ding, et al. 2016; Guo et al. 2016; Lin and Zou 2017; Luo, Chen, et al. 2015; Wang, Lv, et al. 2016), lead (Li, Wang, et al. 2017; Yan, Kong, et al. 2015; Hadi Najafabadi et al. 2015; Moradi et al. 2017; Yuan, Zhang, et al. 2017; Venkateswarlu and Yoon 2015a; Wang, Cheng, Yang, et al. 2013), and uranium (Zhang, Liu, Wang, et al. 2015) and negative for manganese (Al-Wakeel et al. 2015), cobalt (Negm et al. 2015), copper (SenthilKumar et al. 2011), arsenite (Alijani and Shariatinia 2017; A. Saleh et al. 2016), and a mixture of arsenite and arsenate (Alijani and Shariatinia 2017).

The positive values of entropy changes for nickel with polyvinyl alcohol-based chelating sponge showed the randomness increased at the solid–liquid interface during the adsorption process (Cheng et al. 2014). Roushani et al. (2017) suggested the positive entropy change as a favorable factor in the adsorption of cadmium on TMU-16-NH_2 metal–organic framework. The positive change in entropy was due to different reasons. The increase in entropy in adsorption of nickel with dolomite was attributed to liberation of hydrated water from the nickel metal ion (Mohammadi et al. 2015). Similarly, in the adsorption of strontium, the positive entropy change was attributed to the disturbance of the hydration layer of the adsorbent and adsorbate (Zhao et al. 2014; Yu, Mei, et al. 2015; Wang et al. 2011). The positive entropy change in cadmium adsorption on graphene oxide–Al_{13} composite was attributed to the substitution of water molecules with chelating groups (Yan, Zhao, et al. 2015). In the case of uranyl ions adsorption on magnetic calcium silicate hydrate, the positive entropy change was attributed to large number of water desorbed from the adsorbent as compared to adsorbate adsorbed on the adsorbent (Zhang, Liu, Wang, et al. 2015). The removal of selenite with amino-functionalized magnetic silica was attributed to the release of two sodium ions into the solution from sodium selenite on the adsorption of one selenite ion (Larimi et al. 2013).

The adsorption of ferric ions with the calcined product of tetrabutylammonium bromide-modified montmorillonite and the calcined product of tetrabutylammonium bromide-modified kaolinite occurred with a decline in entropy (Bhattacharyya and Gupta 2009). The decline in entropy is associated with the more free nature of the ions as compared to the adsorbed amount or immobilized states.

The Arrhenius activation energy of less than 20 kJ/mol for uranyl ion adsorption with Fe_3O_4@C@layered double hydroxide was attributed to the diffusion process (Zhang, Wang, et al. 2013), and the positive values of the activation energy for the adsorption of vanadium were used to depict the endothermic nature of the adsorption process (Liu and Zhang 2015). Similarly, activation energy of −13.40 kJ/mol indicated the physical adsorption nature of adsorption of ferric ions on egg shell (Yeddou and Bensmaili 2007).

8.4 MANAGEMENT OF ADSORBENT AFTER THE ADSORPTION PROCESS

The adsorbent after its use can be managed by various means such as regeneration, reuse, and safe disposal (Figure 8.2). The regeneration can be done by various methods such as acid desorption agent, chelating desorbing agent, alkali desorbing agent, alkali desorbing agent, salt desorbing agent (Vakili et al. 2019), thermal regeneration (Yang et al. 2020), and electrochemical regeneration (Ding et al. 2020). The organic pollutants can also be regenerated in addition to the earlier methods, such as ultrasonic regeneration (Naghizadeh et al. 2017), microbial regeneration, microwave-assisted regeneration, thermal regeneration, chemical regeneration, ozonation, photoassisted oxidation, and electrochemical oxidation (Omorogie et al. 2016).

The effectiveness of the adsorbent reduces subsequently after multiple adsorption–regeneration cycles (Mirza and Ahmad 2018; Zhang et al. 2020; Zhang, Zeng, et al. 2016; Saiz et al. 2014). The procedure makes the adsorbent to be redundant after multiple adsorption–regeneration cycles for the same contaminant. The spent adsorbent can be either disposed in landfill or incinerated (Mohan and Pittman 2007) or put to alternative uses. Spent adsorbents containing hazardous materials are stabilized/solidified before being dumped into landfill (Verbinnen et al. 2015; Paudyal et al. 2020), which leads to an increase in the cost of the life cycle assessment of the adsorbent.

Enhancement of sustainability of the spent adsorbent can be achieved by its proper disposal and economical use in other applications. The spent adsorbent can be applied for a number of alternative uses, for example, as a catalyst (He et al. 2018; Ballav et al. 2018), brick formulation (Devi and Saroha 2016; Avinash and Murugesan 2019; Mukherjee and Halder 2018; Rathore and Mondal 2017), road construction (Mukherjee

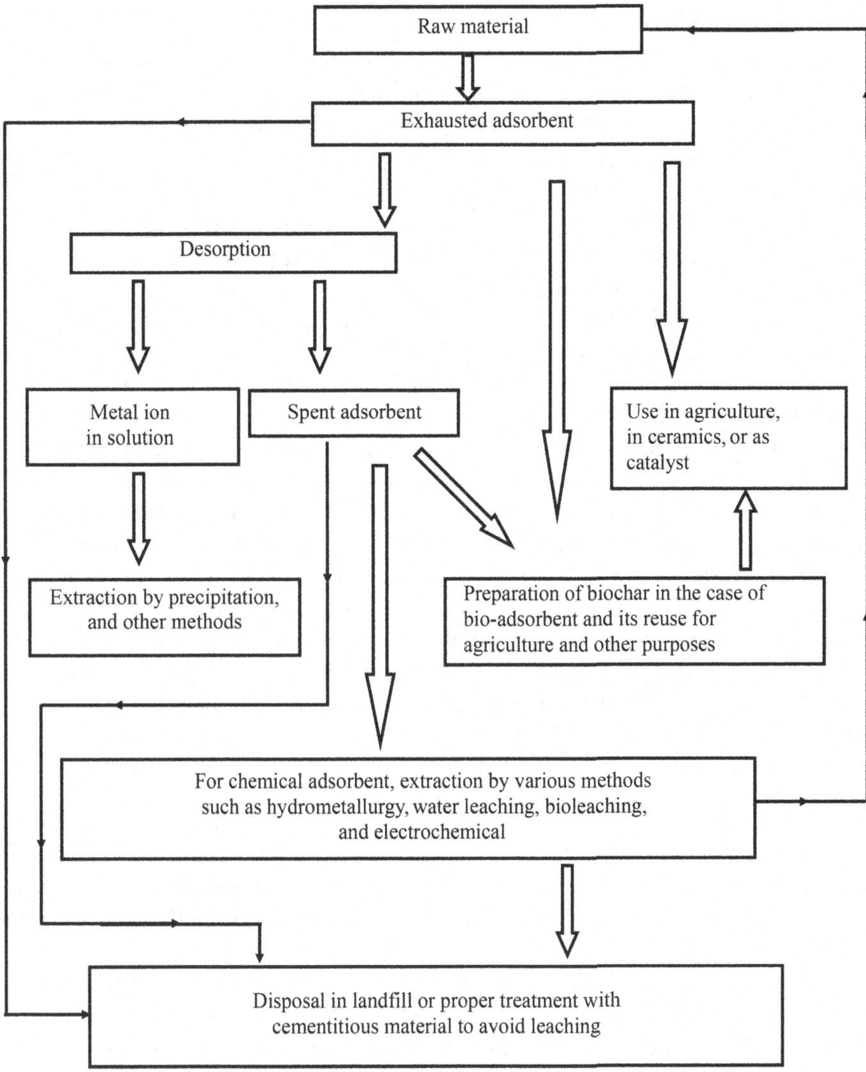

FIGURE 8.2 Disposal management of adsorbent after adsorption.

and Halder 2018), partial cement replacement (Mukherjee and Halder 2018), cement clinkers (Saikia and Goswamee 2019), fertilizer and soil conditioners (Reddy et al. 2017), soil amendment (Kasiuliene et al. 2019), soil, biofuel, and cement (Hossain et al. 2020), glass formation (Majumder et al. 2019), high-temperature dielectric material (Lin et al. 2017), or adsorption of alternative contaminants (Singh and Singhal 2018). The use of the spent adsorbent are broadly classified into three categories:

1. Use as a catalyst
2. Use for the production of ceramics
3. Use as a fertilizer after its use

8.4.1 Use as a Catalyst

The spent adsorbent was applied as a catalyst in photodegradation (Kolinko et al. 2012), reduction of nitrophenol to aminophenol (Meng et al. 2017), hydrocarbon oxidation (Kolinko et al. 2012; Dutta et al. 2017), xylose and xylan conversion into furfural, and phenylacetylene conversion to acetophenone (Ballav et al. 2018; Fu et al. 2019). The end product can be analyzed with NMR spectroscopy (Ballav et al. 2018; Fu et al. 2019), HPLC (Fu et al. 2019), gas chromatography (He et al. 2018), UV–visible spectroscopy (Meng et al. 2017), or FTIR spectroscopy (Kolinko et al. 2012) depending upon the nature of the pollutant.

The catalytic activity in some cases varied with the position of the metal ion on the adsorbent; for example, conversion and selectivity in the oxidation of cyclohexene and ethyl benzene increased with the concentration of chromium sitting on the surface of the adsorbent as compared to chromium attached electrostatically with the amine group of the adsorbent (Dutta et al. 2017). In spite of the wide potential of the spent adsorbent for catalysis, there are some concerns that need to be addressed (Reddy et al. 2017).

The major among them is the leachability of the pollutant or other material from the adsorbent during its use as catalyst. The leaching test used in most of the studies was the California waste extraction test (Rathore et al. 2016) or TCLP (toxicity characteristic leaching procedure) (Mondal and Garg 2017; Ali et al. 2003). The spent adsorbent in many cases contains hazardous materials, and environmental agencies (e.g. USEPA, USA; CPCB, India; DEFRA, UK) have stricter guidelines for disposal of hazardous waste. This issue can also be addressed by enhancing the use of nontoxic spent adsorbents (Reddy et al. 2017).

8.4.2 Use for the Production of Ceramics

Spent adsorbents after their use also find application in the production of ceramic materials such as filler in cement production. The problem associated with the toxic nature of the adsorbent can be solved to some extent by its use in the production of ceramics and road construction.

The leaching of hazardous material of the spent adsorbent was addressed by using proper conditions of the preparation. Spent adsorbent (zeolite- and perlite-supported magnetite after the adsorption of molybdenum) mixed with sludge in ratio of 3/97, which was in reference to adsorption capacity of loaded adsorbent (Verbinnen et al. 2015). The final product (sludge + spent adsorbent) curbed the leaching of molybdenum. The final ceramic product can also restrict the leaching of other heavy metals (such as Cr, Ni, Cu, Zn, As, Cd, and Pb), which were spiked during the ceramic synthesis procedure. This is helpful in treating the contaminated eluent produced during the desorption and adsorption processes. The spent adsorbent can also be disposed by immobilizing into the phosphate glass matrix. Majumder et al. (2019) used up to 20% of the spent adsorbent containing As that has been incorporated during glass formation.

8.4.3 Use as a Fertilizer

The spent adsorbent can also be blended with soil, where it can act as a fertilizer. The adsorbent needs to have some properties for it to be used as a fertilizer, such as affinity toward cations and anions, stability for longer duration under varied environmental conditions, slow release of nutrients, and adequate water permeability and porosity (Majumder et al. 2019).

Mostly, spent biosorbent in its native form or after its conversion to the biochar can be used as a fertilizer. The algal biomass contains a lot of nutrients such as N, P, Ca, and K. These nutrients are released into the soil, which can enhance the fertility of the soil (Bădescu et al. 2018; Cole et al. 2017). Woody biomass (high cellulose and hemicellulose content)-based biosorbent has less biodegradability than algae and can also be applied as a fertilizer, after its use as adsorbent. A synthesized organic adsorbent with suitable biodegradability can also be used as a fertilizer, such as EDTA-modified chitosan carboxymethyl cellulose. EDTA-modified chitosan carboxymethyl cellulose used for removal of copper started to biodegrade in 20 days (Manzoor et al. 2019).

Pyrolysis of spent biosorbent produces biochar, heat, and gases (fuel and nonfuel fraction) (Bădescu et al. 2018). Each product during the production of biochar can be put for separate economic value. The biochar can be applied for the adsorption of pollutants (Abdallah et al. 2019) or can be applied directly to the soil (Reddy et al. 2017). The addition of biochar in addition to supplement the nutrients also restrain the availability of toxic elements to crop present in the soil (Shu et al. 2016; Xu et al. 2018; Ahmad et al. 2016). Chromium and cadmium concentrations in the plant were reduced by 33.50% and 28.73%, respectively (Bashir et al. 2018), while growing in the respective element-contaminated soil by the application of biochar (15 g/kg). Nitrates and phosphates are one of the primary fertilizers needed for crops. However, biochar is also negatively charged, so it has low efficiency toward the anionic pollutants such as nitrate and phosphate. This property limits its application as fertilizer by a low margin. This can be altered by the incorporation of metal ions such as Ca, Mg, and Al (Li, Wang, Zhou, Awasthi, Ali, Zhang, Lahori, et al. 2016; Chen et al. 2011; Arcibar-Orozco et al. 2012; Yao et al. 2013; Zhang and Gao 2013; Wang, Guo, et al. 2015; Wang, Shen, et al. 2016; Zhang, Gao, et al. 2012; Takaya et al. 2016; Li, Wang, Zhou, Awasthi, Ali, Zhang, Gaston, et al. 2016; Fang et al. 2015; Jung et al. 2015; Yin et al. 2017) in biochar prior to nitrate and phosphate removal from the contaminated areas. These elements enhanced the adsorption of nitrate and phosphate via either formation of H bond or precipitation for phosphate and electrostatic attraction for nitrate (Yin et al. 2017). So, biochar produced by this manner contains nitrate and phosphate along with nutrients such as Ca, Mg, and Al available to plants.

The heat produced during biochar production can also be used as a thermal agent, whereas the fuel fraction gases (H_2, CO, CH_4, other hydrocarbons) can be used to produce biofuels (Volli and Singh 2012; Kan et al. 2016; Guedes et al. 2018; Bădescu et al. 2018) and the nonfuel fraction can be used for the synthesis of various chemical reagents of industrial importance (Gu et al. 2015; Kan et al. 2016).

The application of spent bio-adsorbent as a fertilizer has several advantages in the form of metal sequestration; any requirement of desorbing agent (Bădescu et al. 2018; Wosnitza and Barrantes 2006); improvement of nutritional quality of the soil (Bădescu et al. 2017, 2018; Cole et al. 2017); enhancement of soil organic carbon by the application of biochar, which has been declining globally (Reddy et al. 2017); and enhancement of soil water-holding capacity (Mangrich et al. 2015; Woolf et al.

2010; Cornelissen et al. 2013; Subedi et al. 2016; Chen, Rotaru, et al. 2014; Reddy et al. 2017).

However, there are some issues that need to be resolved before the application of spent bio-adsorbent or biochar to soil amendment, such as the determination of toxic metal and its concentration in the spent bio-adsorbent or biochar made from it. Guidelines for certification of biochar differ from country to country (Germany's Federal Soil Protection Act and Switzerland's Chemical Risk Reduction Act) (Schmidt et al. 2013). European guidelines set threshold for each heavy metal concentration in biochar total biomass for its application as biochar. In the case of the heavy metal lead, the concentration of lead should be less than 120 and 150 g/t for basic and premium biochar, respectively (Schmidt et al. 2013). Some other factors that need to be considered are the requirement of biochar in large amount as compared to the commercial fertilizer, controlled release of nutrients to avoid soil contamination and accumulation of metal ions, and initial capital cost to recover all products from pyrolysis of spent adsorbent such as gases and heat during the production of biochar, which can help in reducing carbon footprint.

8.5 ECONOMIC VIABILITY: DESORPTION vs DISPOSAL

After adsorption, the adsorbent can be desorbed and recycled until it favorably keeps the pollutant concentration in eluent within the permissible limit set up by the regulatory agencies. After this, the spent adsorbent can be repurpose for alternative uses like catalyst, production of ceramic, and use for removal of alternate contaminant or can be disposed.

The desorption of contaminants can be done by alkali or acid reagent, chelating agent, and salt (Vakili et al. 2019), and in the case of organic pollutants, by thermal, chemical, microwave, and other methods (San Miguel et al. 2001; Wang et al. 2010; Guilane and Hamdaoui 2016). Lata et al. (2015) observed that alkali was most efficient for removal of heavy metals from chemical-based adsorbent (Table 8.1).

The use of acid, alkali, chelating compound, or chemical as a desorbing agent is associated with the generation of waste (secondary pollution) in eluent laced with contaminants. So, it faces the same problem in disposal as the spent adsorbent due to environmental and economic reasons. However, there are some instances in which the metal laced with heavy metal can be recovered such as Cr with $BaCl_2$ (Zelmanov and Semiat 2011), Hg from EDTA–Hg complex as $HgCl_2$ or $HgSO_4$ (Jeon and Park

TABLE 8.1
Desirable Desorbing Agents for Various Adsorbents

Adsorbent	Desorbing Agent	Examples of Desorbing Agent	References
Chemical sorbent	Alkali	NaOH	Lata et al. (2015)
Bio-adsorbent	Acid	HCl, HNO_3, H_2SO_4	Lata et al. (2015)
Biomass (algae, fungi)	Complexing agent	EDTA	Lata et al. (2015)

2005), and palladium as palladium chloride (Boricha et al. 2007). The summarized processes are as follows.

The adsorbent, i.e. Fe(III) oxide/hydroxide nanoparticle-based agglomerates separated by filtration followed by desorption with NaOH in the pH range of 9–10 (Zelmanov and Semiat 2011). The concentrated Cr solution was treated with $BaCl_2$, which led to the production of $BaCrO_4$ crystals, and the crystals were removed by filtration with a 0.45-μm filter paper. $BaCrO_4$ produced has a larger market value than $BaCl_2$. Similarly, mercury desorption from novel aminated chitosan bead with EDTA was separated as solid EDTA and metal (mercury) chloride or sulfate by using HCl/H_2SO_4 (Jeon and Park 2005).

In another case, palladium (Boricha et al. 2007) adsorbed on the silica gel as palladium phthalocyanine was first thermally calcined in air to partially burn the organic moiety of the complex. Afterwards, 2 M HCl was added to calcined silica so that palladium dissolution takes place as H_2PdCl_4. The palladium was recovered as $PdCl_2$ from the filtrate (solution containing H_2PdCl_4) by adjusting it's pH to 6 with the help of 0.1–0.5 M NaOH.

In addition, there are some examples of recycling available from the petroleum industry, battery sector, and metallurgical sector where metals can be reclaimed after their use. The spent catalyst can be recovered by hydrometallurgical methods (Akcil et al. 2015), water leaching (Kar et al. 2005; Biswas et al. 1985; Zeng and Cheng 2009; Le and Lee 2020), bioleaching (Yu et al. 2020), precipitation method (Wang and Chen 2019; Paudyal et al. 2020), and vacuum heat decomposition (Liu et al. 2018). Some examples from petroleum industries are as follows.

$CoMo/Al_2O_3$ was treated in sulfuric acid environment under solvothermal conditions followed by treatment with elemental sulfur or H_2S. This led to the precipitation of molybdenum and can be solidified by oxidation to molybdic acid and cobalt sulfides can be converted to cobalt sulfate followed by its extraction via ion exchange. In this way, separation and extraction of Al, Co, and Mo were successfully carried out by Hyatt (Hyatt 1987; Akcil et al. 2015).

The process of recovery of metals similar to the nutrient cycle can be achieved by biohydrometallurgical methods. The bacteria and fungi used for the metal solubilization in biohydrometallurgical methods are positively enhanced by the generation of the acidic media or oxidizing media during their growth in the medium (Akcil et al. 2015). Bioleaching can be achieved by *Acidithiobacillus thiooxidans, Acidithiobacillus ferrooxidans* (Yu et al. 2020), or *Aspergillus nomius* (Liu et al. 2018). The recent rise of LED (light emitting diode) has led to increased gallium, copper, and nickel waste in recent times. These metals can be bioleached with the help of *Acidithiobacillus ferrooxidans* (Pourhossein and Mousavi 2018). The bioleaching process has the advantages of low energy requirement and low operation and maintenance costs. However, the longer time period required for operation and the dependency on atmospheric conditions are certain limitations.

To overcome these, there are other methods available such as the water leaching method for extraction of vanadium and molybdenum with the use of temperature, salt, and water only (Biswas et al. 1985; Zeng and Cheng 2009; Kar et al. 2005; Le and Lee 2020), and the use of organic acids such as gluconic acid and lactic acid for leaching (Roshanfar et al. 2019) and oxalic acid for precipitation (Verma et al. 2019).

In the oxalic acid leaching method for battery waste ($LiCoO_2$), the leaching efficiency of greater than 99% and 96% for Li and Co can be achieved, respectively (Verma et al. 2019; Sohn et al. 2006). The lithium ion can be leached into aqueous phase and cobalt can be precipitated out as cobalt oxalate in this method.

In addition to the chemical, biological, and solvothermal methods, there are electrochemical methods, which are costly but more environmentally friendly. One example is the removal of chromium with a polyaniline–resin composite (resin is a porous polymer XAD-4 adsorbent). The recovery of chromium achieved is nearly 62.9% in the first cycle, followed by 91.63% and 90.33% in the second and third cycles, respectively (Ding et al. 2020). In this case, there is negligible generation of secondary waste as compared to the other methods. Some examples of desirable methods and chemicals for the recovery of metals and anions are presented in Table 8.2.

Industries which have expertise in removing metal from batteries and used catalyst are operational at the moment, for example, Eramet, Treibacher Industrie AG, Moxba-Metrex, GS EcoMetal Co. Ltd., Taiyo Koko, Full Yield Industry Co. Ltd. (Akcil et al. 2015), and Tata Chemicals. The spent adsorbents can be sent to industries, thus avoiding the need for additional separate infrastructure.

The success of the adsorption process for upliftment in standard of living depends on the localized conditions. It also depends on the performance cost and appropriateness (Lata et al. 2015), so in countries with low income, it is not possible to use the chemical adsorbents, and thus, bio-sorbents are preferred in those countries. In the case of biosorption of heavy metal, it is best suggested to look for its conversion to biochar and, if possible, use it for the adsorption of alternate pollutant, and after its use, apply it on the soil to improve the soil fertility. However, we should be aware of the limitation regarding the maximum permissible limit of heavy metals or pollutants by the environmental regulatory agency of the region.

So, the applicability of the adsorbent and desorption is based on the economics of the area and the rarity of the material to be used for adsorption. The another reason to avoid chemical adsorbent in countries with low income is that precursor for the chemical adsorbents have to be extracted it from its ore by mining. The mining industry is already marred by environmental damage to biodiversity, and open cast mining is the most preferable mode of mining, which has a severe impact on the local ecosystem.

Figure 8.3 suggests that, if the cost of replacing the adsorbent is high {(cost of fresh adsorbent + economic loss to biodiversity during mining of material for fresh adsorbent) − (economic cost of desorption and extraction + economic value of end product)} and the adsorbent post adsorption cannot find alternative uses, then it is better to desorb the adsorbent for reuse and treat the effluent or eluent (for metal extraction) generated during desorption using various processes. The process can be followed by precipitation of the metal extracted of metal in the spent adsorbent. The spent chemical adsorbent serves as a source of raw material for the production of goods (Federal Ministry for the Environment 2016). This helps in achieving the circular economy.

If extraction is not possible, then it is suitable to dispose the pollutant by adding binders (such as cement) before its final disposal. Arsenic-laden waste disposal is likely carried out by stabilization of solidification, followed by disposal of treated waste. In the absence of proper guidelines, the adsorptive filter media and regenerative waste are dumped into the small brick-lined pits (Mondal and Garg 2017; Sullivan et al. 2010; Ali et al. 2003). The pits need to be tested for TCLP (toxicity characteristic leaching procedure). A column leaching test on the arsenic waste was conducted, and the results showed that leaching was nonsignificant and nonhazardous as per the guidelines of the USEPA (Ali et al. 2003).

The adsorbent can also be made up of Mg and Ca (in addition to adsorbent based on carbon) to make the adsorption process sustainable. These elements (Mg and Ca) can serve as nutrients in the soil after its use as adsorbent. So, focus must be on an efficient process and adsorbent must be made up of a material with large abundance in nature. In the case of spent adsorbent that is applicable in alternative uses such as soil fertilizer and abundant to replace, it is economically and environmentally friendly to put it to alternative use rather than desorption.

8.6 CONCLUSION

The current book gives an insight into the factors affecting the adsorption process. The reason for improvement or declination in the adsorption capacity is suggested to be dependent on a number of factors such as initial concentration, adsorbent dose, ionic strength, coexisting ion, surface modification, and speciation of the material. The pH dependence of the adsorption capacity depends on the electrostatic attraction, speciation of the species, and competition from hydronium ions. In the case of materials, to be independent of pH, the lack of electrostatic force of attraction in determining the mechanism of adsorption is suggested. The presence of competitive ions lead to decrease in the adsorption of metal due to the competition for active sites. However, in some cases like adsorption of cadmium on Al_{13}-pillared montmorillonite increased by the presence of phosphate ions. The inference derived from various experiments such as change in pH, assessment of thermodynamic parameters, and the results from various characterization techniques (XRD, XPS, FTIR, XANES, EXAFS) helped in estimating the mechanism of adsorption. XPS aids in the determination of oxidation state of the substance. However, XANES helps in better explaining the adsorption mechanism, wherever oxidation and reduction of the species varied during drying of the sample, as in the case of arsenic adsorption. The change in the structure after modification of the adsorbent can be examined by XRD and BET. XRD aids in determining the structural integrity of the structure, and BET provides information about the change in pore structure and volume.

Regeneration of adsorbents is carried out with acids, bases, and chelating agent solutions. The proportion of regenerated adsorbent also gives macroscopic evidence for the physisorption or chemisorption nature of the process. The adsorbent success is highly localized and depends on performance appropriateness and per capita income of the area. Spent adsorbent can be put to multiple uses in fields such as agriculture, construction, and catalysis. The better fit of the kinetic model on the pseudo-second order in most studies is taken as a proxy for chemisorption. The best isotherm and kinetic model was estimated on the basis of the values of coefficient of determination and chi-square values or both. The adsorption rate was fast during

TABLE 8.2
Chemicals and Procedures for Recovery of Elements

S. No.	Element or Chemical Species	Starting Material/Adsorbent	Desorbing Agent	Recovery Chemical	References
1	V	Heavy oil desulfurization waste catalyst	Water leaching	NaCl, heat, and H_2O	Biswas et al. (1985); Zeng and Cheng (2009)
2a	Cr	Polyaniline–XAD-4 resin composite	Electrochemical process		Ding et al. (2020)
2b	Cr	Mg–Al LDH	Acid solution and $NaBH_4$		Lv et al. (2019)
3	Ni	Ni-$MoAl_2O_3$	Hydrometallurgical process	Roasting, NH_4OH, $(NH_4)_2CO_3$	Millsap and Reisler (1978)
4a	As	Magnetic biosorbent	NaOH		Baig et al. (2014)
4b	As	$CoFe_2O_4$	NaOH		Liu et al. (2020)
5	Sr	Mucor/Alg-Na	HNO_3		Naeimi and Faghihian (2019)
6	Cd	Iminodiacetic acid-functionalized PP-g-PGMA	Citric acid to form a metal–citrate complex		Lopez et al. (2020)
7	Hg	Synthetic samples representing zinc–air battery waste, fluorescent lamp waste, and chlor-alkali industry waste	Solvent extraction	Polyethylene oxide, sodium citrate, water, HCl, NaOH, and NaOH	da Cunha et al. (2016)
8a	Pb	Pyromellitic dianhydride-modified sugarcane bagasse	HNO_3	Na_3PO_4	Tang et al. (2018)
8b	Pb	Magnetic biosorbent	EDTA		Wang, Tang, et al. (2015)

(Continued)

TABLE 8.2 (*Continued*)
Chemicals and Procedures for Recovery of Elements

S. No.	Element or Chemical Species	Starting Material/Adsorbent	Desorbing Agent	Recovery Chemical	References
9	U	Amidoxime-grafted activated carbon fibers	HNO_3		Lu, Zhang, et al. (2017)
10	Mn	Polymetallic (Al, Mn, Zn, and Cu) ore	Chemical leaching	H_2SO_4 + oxalic acid	El Hazek and Gabr (2016)
11	Co	$CoMo/Al_2O_3$	Hydrometallurgical process	H_2SO_4, H_2S, ion exchange	Akcil et al. (2015); Hyatt (1987)
12	Cu	Copper-, zinc, and iron-containing wastewater	Sequential precipitation	H_2O_2, NaHS, lime	Wang and Chen (2019)
13	Zn	do	do	do	Wang and Chen (2019)
14	Se	Kesterite-based photovoltaic cell	Thermal oxidation and reduction	Zinc, $SnCl_4$, ascorbic acid, 2-phenylacetophenone for the reduction of selenium dioxide if needed	Asensio et al. (2020)
15	Sc	Hydrolyzed sulfuric acid solution of scandium and titanium	Scrubbing and stripping	Na_3PO_4 and H_2O_2 for titanium, followed by NaOH for scandium	Li, Li, et al. (2018)
16	Ti	do	do	do	Li, Li, et al. (2018)
17	Ga	Semiconductor scrap	Vacuum heat decomposition		Chatterjee et al. (2020)
14	Ge	Optical fibers	Hydrometallurgy process	Acid for leaching, followed by the use of trioctylamine for extraction	Chen, Chang, et al. (2017)
15	F	Zirconia-loaded orange waste gel	$Ca(OH)_2$	The desorbing agent $Ca(OH)_2$ also acts as a precipitating agent	Paudyal et al. (2020)
19	PO_4^{3-}	$Fe_3O_4@ZrO_2$	NaOH	$CaCl_2$	Guan et al. (2020)

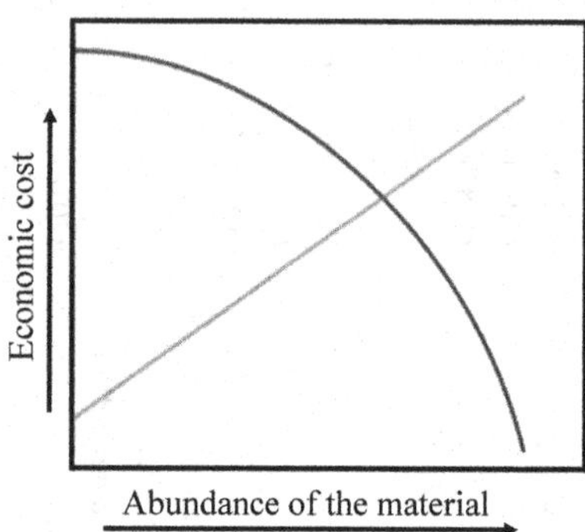

FIGURE 8.3 Economic cost graph for sustainable use of adsorbent; the straight line depicts the economic cost of desorption and extraction and replacement of adsorbent + economic loss to biodiversity during mining of material, and the curved line depicts the cost with respect to the abundance of material (in the case of area above the intersection point, it is suitable to recycle and reuse the adsorbent).

the initial stages in most of the cases. The mean adsorption energy calculated from the D-R isotherm model is also indicative of the nature of adsorption. The mean adsorption energy value of less than 8 kJ/mol is suggested to follow physisorption; 8–16 kJ/mol, to follow ion exchange; and more than 16 kJ/mol, to follow chemisorption. The assessment of thermodynamic parameters gives an insight into spontaneity, endothermic or exothermic nature, and increased or decreased randomness during the adsorption process. The current book helps in further improvement of adsorption processes and also aids in designing better adsorbent for reclamation of material or remediation of water.

References

A. Saleh, T, A Sarı, and M Tuzen. 2016. Chitosan-modified vermiculite for As(III) adsorption from aqueous solution: Equilibrium, thermodynamic and kinetic studies. *Journal of Molecular Liquids* 219:937–945.

Abdallah, MM, MN Ahmad, G Walker, JJ Leahy, and W Kwapinski. 2019. Batch and continuous systems for Zn, Cu, and Pb metal ions adsorption on spent mushroom compost biochar. *Industrial & Engineering Chemistry Research* 58 (17):7296–7307.

Abdelwaheb, M, K Jebali, H Dhaouadi, and S Dridi-Dhaouadi. 2019. Adsorption of nitrate, phosphate, nickel and lead on soils: Risk of groundwater contamination. *Ecotoxicology and Environmental Safety* 179:182–187.

Adeleke, OA, AA Abdul Latiff, MR Saphira, et al. 2019. 2- Locally Derived Activated Carbon From Domestic, Agricultural and Industrial Wastes for the Treatment of Palm Oil Mill Effluent. In *Nanotechnology in Water and Wastewater Treatment*, edited by A Ahsan, and AF Ismail: Elsevier: page no. 35–62.

Adeli, M, Y Yamini, and M Faraji. 2017. Removal of copper, nickel and zinc by sodium dodecyl sulphate coated magnetite nanoparticles from water and wastewater samples. *Arabian Journal of Chemistry* 10 Supplement 1:S514–S521.

Adhikari, A, M Das, S Mondal, et al. 2019. Manganese neurotoxicity: Nano-oxide compensates for ion-damage in mammals. *Biomaterials Science* 7 (11):4491–4502.

Adio, SO, MH Omar, M Asif, and TA Saleh. 2017. Arsenic and selenium removal from water using biosynthesized nanoscale zero-valent iron: A factorial design analysis. *Process Safety and Environmental Protection* 107:518–527.

Afroze, S, TK Sen, and HM Ang. 2016. Adsorption removal of zinc (II) from aqueous phase by raw and base modified Eucalyptus sheathiana bark: Kinetics, mechanism and equilibrium study. *Process Safety and Environmental Protection* 102:336–352.

Agalakova, NI, and OV Nadei. 2020. Inorganic fluoride and functions of brain. *Critical Reviews in Toxicology* 50 (1):28–46.

Ahmad, M, YS Ok, B-Y Kim, et al. 2016. Impact of soybean stover-and pine needle-derived biochars on Pb and As mobility, microbial community, and carbon stability in a contaminated agricultural soil. *Journal of Environmental Management* 166:131–139.

Akcil, A, F Vegliò, F Ferella, MD Okudan, and A Tuncuk. 2015. A review of metal recovery from spent petroleum catalysts and ash. *Waste Management* 45:420–433.

Al-Anber, MA. 2010. Removal of high-level Fe^{3+} from aqueous solution using natural inorganic materials: Bentonite (NB) and quartz (NQ). *Desalination* 250 (3):885–891.

Al-Ghouti, MA, and DA Da'ana. 2020. Guidelines for the use and interpretation of adsorption isotherm models: A review. *Journal of Hazardous Materials*:122383.

Al-Wakeel, KZ, HA El Monem, and MMH Khalil. 2015. Removal of divalent manganese from aqueous solution using glycine modified chitosan resin. *Journal of Environmental Chemical Engineering* 3 (1):179–186.

Al Abdullah, J, AG Al Lafi, Y Amin, and T Alnama. 2016. Adsorption of cesium, cobalt, and lead onto a synthetic nano manganese oxide: Behavior and mechanism. *Water, Air, & Soil Pollution* 227 (7):1–14.

Alatalo, S-M, F Pileidis, E Mäkilä, et al. 2015. Versatile cellulose-based carbon aerogel for the removal of both cationic and anionic metal contaminants from water. *ACS Applied Materials & Interfaces* 7 (46):25875–25883.

Albadarin, AB, C Mangwandi, AH Al-Muhtaseb, GM Walker, SJ Allen, and MNM Ahmad. 2012. Kinetic and thermodynamics of chromium ions adsorption onto low-cost dolomite adsorbent. *Chemical Engineering Journal* 179:193–202.

Ali, I. 2012. New generation adsorbents for water treatment. *Chemical Reviews* 112 (10):5073–5091.

Ali, I, AA Basheer, XY Mbianda, et al. 2019. Graphene based adsorbents for remediation of noxious pollutants from wastewater. *Environment International* 127:160–180.

Ali, M, ABM Badruzzaman, MA Jalil, et al. 2003. Fate of arsenic in wastes generated from arsenic removal units. Paper read at Proceedings of BUET-UNU Symposium on Fate of Arsenic in the Environment, Dhaka.

Ali, RM, HA Hamad, MM Hussein, and GF Malash. 2016. Potential of using green adsorbent of heavy metal removal from aqueous solutions: Adsorption kinetics, isotherm, thermodynamic, mechanism and economic analysis. *Ecological Engineering* 91:317–332.

Alijani, H, and Z Shariatinia. 2017. Effective aqueous arsenic removal using zero valent iron doped MWCNT synthesized by in situ CVD method using natural α-Fe_2O_3 as a precursor. *Chemosphere* 171:502–511.

Alothman, ZA, MA Habila, NH Al-Shalan, et al. 2016. Adsorptive removal of Cu(II) and Pb(II) onto mixed-waste activated carbon: Kinetic, thermodynamic, and competitive studies and application to real wastewater samples. *Arabian Journal of Geosciences* 9 (4):315.

Amin, MT, AA Alazba, and M Shafiq. 2019. Application of the biochar derived from orange peel for effective biosorption of copper and cadmium in batch studies: Isotherm models and kinetic studies. *Arabian Journal of Geosciences* 12 (2):46.

Anirudhan, TS, JR Deepa, and J Christa. 2016. Nanocellulose/nanobentonite composite anchored with multi-carboxyl functional groups as an adsorbent for the effective removal of Cobalt(II) from nuclear industry wastewater samples. *Journal of Colloid and Interface Science* 467:307–320.

Anirudhan, TS, S Jalajamony, and L Divya. 2009. Efficiency of amine-modified poly (glycidyl methacrylate)-grafted cellulose in the removal and recovery of vanadium (V) from aqueous solutions. *Industrial & Engineering Chemistry Research* 48 (4):2118–2124.

Anoop Krishnan, K, KG Sreejalekshmi, V Vimexen, and VV Dev. 2016. Evaluation of adsorption properties of sulphurised activated carbon for the effective and economically viable removal of Zn(II) from aqueous solutions. *Ecotoxicology and Environmental Safety* 124:418–425.

Arancibia-Miranda, N, SE Baltazar, A García, et al. 2016. Nanoscale zero valent supported by Zeolite and Montmorillonite: Template effect of the removal of lead ion from an aqueous solution. *Journal of Hazardous Materials* 301:371–380.

Arcibar-Orozco, JA, M Avalos-Borja, and J Rene Rangel-Mendez. 2012. Effect of phosphate on the particle size of ferric oxyhydroxides anchored onto activated carbon: As(V) removal from water. *Environmental Science & Technology* 46 (17):9577–9583.

Armbrecht, HJ, MA Boltz, S Christakos, and MEH Bruns. 1998. Capacity of 1, 25-dihydroxyvitamin D to stimulate expression of calbindin D changes with age in the rat. *Archives of Biochemistry and Biophysics* 352 (2):159–164.

Arrousse, N, E Mabrouk, R Salim, et al. 2020. Fluorescein as commercial and environmentally friendly inhibitor against corrosion of mild steel in molar hydrochloric acid medium. Materials Today: Proceedings.

Asadi, R, H Abdollahi, M Gharabaghi, and Z Boroumand. 2020. Effective removal of Zn (II) ions from aqueous solution by the magnetic $MnFe_2O_4$ and $CoFe_2O_4$ spinel ferrite nanoparticles with focuses on synthesis, characterization, adsorption, and desorption. *Advanced Powder Technology* 31 (4):1480–1489.

Asensio, MP, E Abas, JL Pinilla, and M Laguna. 2020. High recovery of selenium from kesterite-based photovoltaic cells. *European Journal of Inorganic Chemistry* (22):2203–2209.

Ates, A, and G Akgül. 2016. Modification of natural zeolite with NaOH for removal of manganese in drinking water. *Powder Technology* 287 :285–291.

ATSDR. 1999. Toxicological profile for mercury. In *ATSDR* U.S. Department of Health and Human Services.

ATSDR. 2004. Toxicological profile for cesium. In *ATSDR* U.S. Department of Health and Human Services.

ATSDR. 2007. Toxicological profile for lead. In *ATSDR* U.S. Department of Health and Human Services.

ATSDR. 2008. Toxicological profile for perchlorates. In *ATSDR* U.S. Department of Health and Human Services.

ATSDR. 2012. Toxicological profile for cadmium. In *ATSDR* U.S. Department of Health and Human Services.

ATSDR. 2013. Toxicological profile for uranium. In *ATSDR* U.S. Department of Health and Human Services.

Avinash, A, and A Murugesan. 2019. Judicious recycling of biobased adsorbents for biodiesel purification: A critical review. *Environmental Progress & Sustainable Energy* 38 (3):e13077.

Awual, MR. 2019. A facile composite material for enhanced cadmium (II) ion capturing from wastewater. *Journal of Environmental Chemical Engineering* 7 (5):103378.

Awual, MR, MM Hasan, AM Asiri, and MM Rahman. 2019. Novel optical composite material for efficient vanadium (III) capturing from wastewater. *Journal of Molecular Liquids* 283:704–712.

Awual, MR, T Yaita, H Shiwaku, and S Suzuki. 2015. A sensitive ligand embedded nanoconjugate adsorbent for effective cobalt(II) ions capturing from contaminated water. *Chemical Engineering Journal* 276:1–10.

Awual, MR, T Yaita, SA El-Safty, H Shiwaku, S Suzuki, and Y Okamoto. 2013. Copper(II) ions capturing from water using ligand modified a new type mesoporous adsorbent. *Chemical Engineering Journal* 221:322–330.

Awual, MR, T Yaita, T Kobayashi, H Shiwaku, and S Suzuki. 2020. Improving cesium removal to clean-up the contaminated water using modified conjugate material. *Journal of Environmental Chemical Engineering* 8 (2):103684.

Awual, MR, T Yaita, and Y Okamoto. 2014. A novel ligand based dual conjugate adsorbent for cobalt(II) and copper(II) ions capturing from water. *Sensors and Actuators B: Chemical* 203:71–80.

Azari, A, H Gharibi, Babak K, et al. 2017. Magnetic adsorption separation process: An alternative method of mercury extracting from aqueous solution using modified chitosan coated Fe_3O_4 nanocomposites. *Journal of Chemical Technology and Biotechnology* 92 (1):188–200.

Azari, A, RR Kalantary, G Ghanizadeh, B Kakavandi, M Farzadkia, and E Ahmadi. 2015. Iron–silver oxide nanoadsorbent synthesized by co-precipitation process for fluoride removal from aqueous solution and its adsorption mechanism. *RSC Advances* 5 (106):87377–87391.

Baby, R, B Saifullah, and MZ Hussein. 2019. Carbon nanomaterials for the treatment of heavy metal-contaminated water and environmental remediation. *Nanoscale Research Letters* 14 (1):341.

Bădescu, IS, D Bulgariu, I Ahmad, and L Bulgariu. 2018. Valorisation possibilities of exhausted biosorbents loaded with metal ions–A review. *Journal of Environmental Management* 224:288–297.

Bădescu, IS, D Bulgariu, and L Bulgariu. 2017. Alternative utilization of algal biomass (Ulva sp.) loaded with Zn (II) ions for improving of soil quality. *Journal of Applied Phycology* 29 (2):1069–1079.

Bagherifam, S, S Komarneni, A Lakzian, et al. 2014. Highly selective removal of nitrate and perchlorate by organoclay. *Applied Clay Science* 95:126–132.

Bahramzadeh, A, P Zahedi, and M Abdouss. 2016. Acrylamide-plasma treated electrospun polystyrene nanofibrous adsorbents for cadmium and nickel ions removal from aqueous solutions. *Journal of Applied Polymer Science* 133 (5):42944.

Baig, SA, J Zhu, N Muhammad, T Sheng, and X Xu. 2014. Effect of synthesis methods on magnetic Kans grass biochar for enhanced As (III, V) adsorption from aqueous solutions. *Biomass and Bioenergy* 71:299–310.

Bakhtiari, N, and S Azizian. 2015. Adsorption of copper ion from aqueous solution by nanoporous MOF-5: A kinetic and equilibrium study. *Journal of Molecular Liquids* 206:114–118.

Ballav, N, R Das, S Giri, AM Muliwa, K Pillay, and A Maity. 2018. L-cysteine doped polypyrrole (PPy@ L-Cyst): A super adsorbent for the rapid removal of Hg^{+2} and efficient catalytic activity of the spent adsorbent for reuse. *Chemical Engineering Journal* 345:621–630.

Bankole, MT, AS Abdulkareem, IA Mohammed, et al. 2019. Selected heavy metals removal from electroplating wastewater by purified and polyhydroxylbutyrate functionalized carbon nanotubes adsorbents. *Scientific Reports* 9 (1):4475.

Banu, HT, and S Meenakshi. 2017a. One pot synthesis of chitosan grafted quaternized resin for the removal of nitrate and phosphate from aqueous solution. *International Journal of Biological Macromolecules* 104 (part B):1517–1527.

Banu, HT, and S Meenakshi. 2017b. Synthesis of a novel quaternized form of melamine–formaldehyde resin for the removal of nitrate from water. *Journal of Water Process Engineering* 16:81–89.

Bao, J, Y Fu, and Z Bao. 2013. Thiol-functionalized magnetite/graphene oxide hybrid as a reusable adsorbent for Hg^{2+} removal. *Nanoscale Research Letters* 8 (1):486.

Barakat, MA. 2011. New trends in removing heavy metals from industrial wastewater. *Arabian Journal of Chemistry* 4 (4):361–377.

Bartczak, P, M Norman, Ł Klapiszewski, et al. 2018. Removal of nickel(II) and lead(II) ions from aqueous solution using peat as a low-cost adsorbent: A kinetic and equilibrium study. *Arabian Journal of Chemistry* 11 (8):1209–1222.

Bashir, S, Q Hussain, M Akmal, et al. 2018. Sugarcane bagasse-derived biochar reduces the cadmium and chromium bioavailability to mash bean and enhances the microbial activity in contaminated soil. *Journal of Soils and Sediments* 18 (3):874–886.

Bayramoglu, G, and MY Arica. 2016. MCM-41 silica particles grafted with polyacrylonitrile: Modification in to amidoxime and carboxyl groups for enhanced uranium removal from aqueous medium. *Microporous and Mesoporous Materials* 226:117–124.

Bayuo, J, MA Abukari, and KB Pelig-Ba. 2020. Desorption of chromium (VI) and lead (II) ions and regeneration of the exhausted adsorbent. *Applied Water Science* 10 (7):1–6.

Bedia, J, V Muelas-Ramos, M Peñas-Garzón, A Gómez-Avilés, JJ Rodríguez, and C Belver. 2019. A review on the synthesis and characterization of metal organic frameworks for photocatalytic water purification. *Catalysts* 9 (1):52.

Belova, TP. 2019. Adsorption of heavy metal ions (Cu^{2+}, Ni^{2+}, Co^{2+} and Fe^{2+}) from aqueous solutions by natural zeolite. *Heliyon* 5 (9):e02320.

Berkessa, YW, ST Mereta, and FF Feyisa. 2019. Simultaneous removal of nitrate and phosphate from wastewater using solid waste from factory. *Applied Water Science* 9 (2):28.

Beyki, MH, MH Ghasemi, A Jamali, and F Shemirani. 2017. A novel polylysine–resorcinol base γ-alumina nanotube hybrid material for effective adsorption/preconcentration of cadmium from various matrices. *Journal of Industrial and Engineering Chemistry* 46:165–174.

Bezzina, J, T Robshaw, R Dawson, and MD Ogden. 2020. Single metal isotherm study of the ion exchange removal of Cu (II), Fe (II), Pb (II) and Zn (II) from synthetic acetic acid leachate. *Chemical Engineering Journal*:124862.

Bharath, G, and N Ponpandian. 2015. Hydroxyapatite nanoparticles on dendritic [small alpha]-Fe_2O_3 hierarchical architectures for a heterogeneous photocatalyst and adsorption of Pb(ii) ions from industrial wastewater. *RSC Advances* 5 (103):84685–84693.

Bhatnagar, A, AK Minocha, and M Sillanpää. 2010. Adsorptive removal of cobalt from aqueous solution by utilizing lemon peel as biosorbent. *Biochemical Engineering Journal* 48 (2):181–186.

Bhatnagar, A, E Kumar, and M Sillanpää. 2011. Fluoride removal from water by adsorption—A review. *Chemical Engineering Journal* 171 (3):811–840.

Bhattacharyya, KG, and SS Gupta. 2006. Adsorption of Fe(III) from water by natural and acid activated clays: Studies on equilibrium isotherm, kinetics and thermodynamics of interactions. *Adsorption* 12 (3):185–204.

Bhattacharyya, KG, and SS Gupta. 2009. Calcined tetrabutylammonium kaolinite and montmorillonite and adsorption of Fe(II), Co(II) and Ni(II) from solution. *Applied Clay Science* 46 (2):216–221.

Bhaumik, M, K Setshedi, A Maity, and MS Onyango. 2013. Chromium(VI) removal from water using fixed bed column of polypyrrole/Fe_3O_4 nanocomposite. *Separation and Purification Technology* 110:11–19.

Bian, Y, Z-Y Bian, J-X Zhang, A-Z Ding, S-L Liu, and H Wang. 2015. Effect of the oxygen-containing functional group of graphene oxide on the aqueous cadmium ions removal. *Applied Surface Science* 329:269–275.

Biswas, RK, M Wakihara, and M Taniguchi. 1985. Recovery of vanadium and molybdenum from heavy oil desulphurization waste catalyst. *Hydrometallurgy* 14 (2):219–230.

Bogusz, A, P Oleszczuk, and R Dobrowolski. 2015. Application of laboratory prepared and commercially available biochars to adsorption of cadmium, copper and zinc ions from water. *Bioresource Technology* 196:540–549.

Boland, DD, RN Collins, CJ Miller, CJ Glover, and TD Waite. 2014. Effect of solution and solid-phase conditions on the Fe (II)-accelerated transformation of ferrihydrite to lepidocrocite and goethite. *Environmental Science & Technology* 48 (10):5477–5485.

Borhan, A, NA Abdullah, NA Rashidi, and MF Taha. 2016. Removal of Cu^{2+} and Zn^{2+} from single metal aqueous solution using rubber-seed shell based activated carbon. *Procedia Engineering* 148:694–701.

Boricha, AB, HC Bajaj, PK Ghosh, and RV Jasra. 2007. Recovery of palladium from palladium phthalocyanine complex adsorbed on silica. *Hydrometallurgy* 87 (3–4):140–147.

Bosso, ST, and J Enzweiler. 2002. Evaluation of heavy metal removal from aqueous solution onto scolecite. *Water Research* 36 (19):4795–4800.

Bouhamed, F, Z Elouear, J Bouzid, and B Ouddane. 2016. Multi-component adsorption of copper, nickel and zinc from aqueous solutions onto activated carbon prepared from date stones. *Environmental Science and Pollution Research* 23 (16):15801–15806.

Boulaiche, W, B Hamdi, and M Trari. 2019. Removal of heavy metals by chitin: Equilibrium, kinetic and thermodynamic studies. *Applied Water Science* 9 (2):39.

Cao, C-Y, J Qu, F Wei, H Liu, and W-G Song. 2012. Superb adsorption capacity and mechanism of flowerlike magnesium oxide nanostructures for lead and cadmium ions. *ACS Applied Materials & Interfaces* 4 (8):4283–4287.

Cao, C-Y, J Qu, W-S Yan, J-F Zhu, Z-Y Wu, and W-G Song. 2012. Low-cost synthesis of flowerlike α-Fe_2O_3 nanostructures for heavy metal ion removal: Adsorption property and mechanism. *Langmuir* 28 (9):4573–4579.

Cao, M-l, Y Li, H Yin, and S Shen. 2019. Functionalized graphene nanosheets as absorbent for copper (II) removal from water. *Ecotoxicology and Environmental Safety* 173:28–36.

Cao, X, C Tan, M Sindoro, and H Zhang. 2017. Hybrid micro-/nano-structures derived from metal–organic frameworks: Preparation and applications in energy storage and conversion. *Chemical Society Reviews* 46 (10):2660–2677.

Carmalin Sophia, A, EC Lima, N Allaudeen, and S Rajan. 2016. Application of graphene based materials for adsorption of pharmaceutical traces from water and wastewater–A review. *Desalination and Water Treatment* 57 (57):27573–27586.

Cashin, VB, DS Eldridge, A Yu, and D Zhao. 2018. Surface functionalization and manipulation of mesoporous silica adsorbents for improved removal of pollutants: A review. *Environmental Science: Water Research & Technology* 4 (2):110–128.

Catalano, JG, Y Luo, and B Otemuyiwa. 2011. Effect of aqueous Fe (II) on arsenate sorption on goethite and hematite. *Environmental Science & Technology* 45 (20):8826–8833.

Cerrato, JM, WR Knocke, MF Hochella, AM Dietrich, A Jones, and TF Cromer. 2011. Application of XPS and solution chemistry analyses to investigate soluble manganese removal by MnOx(s)-coated media. *Environmental Science & Technology* 45 (23):10068–10074.

Chakraborty, R, R Verma, A Asthana, S Sree Vidya, and AK Singh. 2019. Adsorption of hazardous chromium (VI) ions from aqueous solutions using modified sawdust: Kinetics, isotherm and thermodynamic modelling. *International Journal of Environmental Analytical Chemistry*:1–18.

Chan, W-C. 1993. Removal and recovery of gallium ion from solution by insoluble amphoteric starches. *Journal of Applied Polymer Science* 50 (10):1733–1738.

Chandra, V, and KS Kim. 2011. Highly selective adsorption of Hg^{2+} by a polypyrrole–reduced graphene oxide composite. *Chemical Communications* 47 (13):3942–3944.

Chatterjee, A, R Das, and J Abraham. 2020. Bioleaching of heavy metals from spent batteries using Aspergillus nomius JAMK1. *International Journal of Environmental Science and Technology* 17 (1):49–66.

Chegrouche, S, and A Bensmaili. 2002. Removal of Ga(III) from aqueous solution by adsorption on activated bentonite using a factorial design. *Water Research* 36 (11):2898–2904.

Chen, B, Z Chen, and S Lv. 2011. A novel magnetic biochar efficiently sorbs organic pollutants and phosphate. *Bioresource Technology* 102 (2):716–723.

Chen, C, and X Wang. 2006. Adsorption of Ni (II) from aqueous solution using oxidized multiwall carbon nanotubes. *Industrial & Engineering Chemistry Research* 45 (26):9144–9149.

Chen, G, KJ Shah, L Shi, and P-C Chiang. 2017. Removal of Cd (II) and Pb (II) ions from aqueous solutions by synthetic mineral adsorbent: Performance and mechanisms. *Applied Surface Science* 409:296–305.

Chen, J, C Shu, N Wang, J Feng, H Ma, and W Yan. 2017. Adsorbent synthesis of polypyrrole/TiO_2 for effective fluoride removal from aqueous solution for drinking water purification: Adsorbent characterization and adsorption mechanism. *Journal of Colloid and Interface Science* 495:44–52.

Chen, K, Y-S Kang, Y Zhao, J-M Yang, Y Lu, and W-Y Sun. 2014. Cucurbit[6]uril-based supramolecular assemblies: Possible application in radioactive cesium cation capture. *Journal of the American Chemical Society* 136 (48):16744–16747.

Chen, L, K-S Zhang, J-Y He, W-H Xu, X-J Huang, and J-H Liu. 2016. Enhanced fluoride removal from water by sulfate-doped hydroxyapatite hierarchical hollow microspheres. *Chemical Engineering Journal* 285:616–624.

Chen, L, S He, B-Y He, et al. 2012. Synthesis of iron-doped titanium oxide nanoadsorbent and its adsorption characteristics for fluoride in drinking water. *Industrial & Engineering Chemistry Research* 51 (40):13150–13156.

Chen, M-L, and M-I An. 2012. Selenium adsorption and speciation with Mg–$FeCO_3$ layered double hydroxides loaded cellulose fibre. *Talanta* 95:31–35.

Chen, P, W Zhang, M Li, P Ai, L Tian, and H Jiang. 2016. Facile synthesis of magnetic La–Zr composite as high effective adsorbent for fluoride removal. *RSC Advances* 6 (42):35859–35867.

Chen, S, A-E Rotaru, PM Shrestha, et al. 2014. Promoting interspecies electron transfer with biochar. *Scientific Reports* 4:5019.

Chen, W-S, B-C Chang, and K-L Chiu. 2017. Recovery of germanium from waste optical fibers by hydrometallurgical method. *Journal of Environmental Chemical Engineering* 5 (5):5215–5221.

Chen, Y, and J Wang. 2012. Removal of radionuclide Sr^{2+} ions from aqueous solution using synthesized magnetic chitosan beads. *Nuclear Engineering and Design* 242:445–451.

Chen, Y, K Fu, S Zhu, et al. 2016. Reduced graphene oxide films with ultrahigh conductivity as Li-Ion battery current collectors. *Nano Letters* 16 (6):3616–3623.

Chen, Y, Q Zhang, L Chen, H Bai, and L Li. 2013. Basic aluminum sulfate@ graphene hydrogel composites: Preparation and application for removal of fluoride. *Journal of Materials Chemistry A* 1 (42):13101–13110.

Chen, Z, Z Zhuang, Q Cao, X Pan, X Guan, and Z Lin. 2014. Adsorption-induced crystallization of U-rich nanocrystals on nano-Mg (OH) 2 and the aqueous uranyl enrichment. *ACS Applied Materials & Interfaces* 6 (2):1301–1305.

Cheng, C, J Wang, X Yang, A Li, and C Philippe. 2014. Adsorption of Ni (II) and Cd (II) from water by novel chelating sponge and the effect of alkali-earth metal ions on the adsorption. *Journal of Hazardous Materials* 264:332–341.

Cheng, Z, Z Gao, W Ma, Q Sun, B Wang, and X Wang. 2012. Preparation of magnetic Fe_3O_4 particles modified sawdust as the adsorbent to remove strontium ions. *Chemical Engineering Journal* 209:451–457.

Chou, W-L, C-T Wang, and Y-H Huang. 2010. Removal of gallium ions from aqueous solution using tea waste by adsorption. *Fresenius Environmental Bulletin* 19 (12):2848–2856.

Choubisa, SL. 2012. Fluoride in drinking water and its toxicosis in tribals of Rajasthan, India. *Proceedings of the National Academy of Sciences, India Section B: Biological Sciences* 82 (2):325–330.

Çifçi, Dİ, and S Meriç. 2017. Manganese adsorption by iron impregnated pumice composite. *Colloids and Surfaces A: Physicochemical and Engineering Aspects* 522 (Supplement C):279–286.

Ciszewski, A, SM Zuber, and Z Szczudło. 1998. Interaction of hydrogen with ultrathin titanium layers on tungsten. *Applied Surface Science* 134 (1):234–242.

Cole, AJ, NA Paul, R De Nys, and DA Roberts. 2017. Good for sewage treatment and good for agriculture: Algal based compost and biochar. *Journal of Environmental Management* 200:105–113.

Colinas, IR, RC Silva, and SR Oliver. 2016. Reversible, selective trapping of perchlorate from water in record capacity by a cationic metal–organic framework. *Environmental Science & Technology* 50 (4):1949–1954.

Constantino, LV,JN Quirino, AM Monteiro, et al. 2017. Sorption-desorption of selenite and selenate on Mg-Al layered double hydroxide in competition with nitrate, sulfate and phosphate. *Chemosphere* 181:627–634.

Cornelissen, G, V Martinsen, V Shitumbanuma, et al. 2013. Biochar effect on maize yield and soil characteristics in five conservation farming sites in Zambia. *Agronomy* 3 (2):256–274.

Cortés-Arriagada, D, and A Toro-Labbé. 2015. Improving As(iii) adsorption on graphene based surfaces: Impact of chemical doping. *Physical Chemistry Chemical Physics* 17 (18):12056–12064.

Crini, G. 2005. Recent developments in polysaccharide-based materials used as adsorbents in wastewater treatment. *Progress in Polymer Science* 30 (1):38–70.

Crini, G. 2006. Non-conventional low-cost adsorbents for dye removal: A review. *Bioresource Technology* 97 (9):1061–1085.

Crini, G, E Lichtfouse, LD Wilson, and N Morin-Crini. 2018. Adsorption-oriented Processes Using Conventional and Non-conventional Adsorbents for Wastewater Treatment. In *Green Adsorbents for Pollutant Removal*, edited by Gregorio Crini and Eric Lichtfouse. Cham, Switzerland: Springer: page no. 23–71.

Crini, G, and P-M Badot. 2007. Traitement et épuration des eaux industrielles polluées. *PUFC, Besançon* 353.

Crini, G, and P-M Badot. 2008. Application of chitosan, a natural aminopolysaccharide, for dye removal from aqueous solutions by adsorption processes using batch studies: A review of recent literature. *Progress in Polymer Science* 33 (4):399–447.

Cui, W, S Wang, J Peng, L Zhang, and G Zhang. 2016. Catechol-functionalized nanosilica for adsorption of germanium ions from aqueous media. *Journal of Sol-Gel Science and Technology* 77 (3):666–674.

Członkowska, A, T Litwin, P Dusek, et al. 2018. Wilson disease. *Nature Reviews Disease Primers* 4 (1):21.

da Cunha, RC, PR Patrício, SJ Rodriguez Vargas, LHM da Silva, and MCH da Silva. 2016. Green recovery of mercury from domestic and industrial waste. *Journal of Hazardous Materials* 304:417–424.

Dąbrowski, A. 2001. Adsorption—From theory to practice. *Advances in Colloid and Interface Science* 93 (1–3):135–224.

Daneshvar, E, MJ Zarrinmehr, M Kousha, et al. 2019. Hexavalent chromium removal from water by microalgal-based materials: Adsorption, desorption and recovery studies. *Bioresource Technology* 293:122064.

Das, R, SBA Hamid, ME Ali, AF Ismail, MSM Annuar, and S Ramakrishna. 2014. Multifunctional carbon nanotubes in water treatment: The present, past and future. *Desalination* 354:160–179.

Das, S, MBJ Lindsay, J Essilfie-Dughan, and MJ Hendry. 2017. Dissolved selenium(VI) removal by zero-valent iron under oxic conditions: Influence of sulfate and nitrate. *ACS Omega* 2 (4):1513–1522.

De Gisi, S, G Lofrano, M Grassi, and M Notarnicola. 2016. Characteristics and adsorption capacities of low-cost sorbents for wastewater treatment: A review. *Sustainable Materials and Technologies* 9:10–40.

Death, C, G Coulson, U Kierdorf, H Kierdorf, WK Morris, and J Hufschmid. 2015. Dental fluorosis and skeletal fluoride content as biomarkers of excess fluoride exposure in marsupials. *Science of the Total Environment* 533 (Supplement C):528–541.

Dehghani, MH, D Sanaei, I Ali, and A Bhatnagar. 2016. Removal of chromium(VI) from aqueous solution using treated waste newspaper as a low-cost adsorbent: Kinetic modeling and isotherm studies. *Journal of Molecular Liquids* 215:671–679.

Deng, S, G Zhang, S Liang, and P Wang. 2017. Microwave assisted preparation of thio-functionalized polyacrylonitrile fiber for the selective and enhanced adsorption of mercury and cadmium from water. *ACS Sustainable Chemistry & Engineering* 5 (7):6054–6063.

Devi, P, and AK Saroha. 2016. Risk assessment and technical feasibility of usage of paper mill sludge biochar-based exhausted adsorbent for geopolymeric brick formation. *Environmental Science and Pollution Research* 23 (21):21641–21651.

Dhillon, A, M Nair, SK Bhargava, and D Kumar. 2015. Excellent fluoride decontamination and antibacterial efficacy of Fe–Ca–Zr hybrid metal oxide nanomaterial. *Journal of Colloid and Interface Science* 457:289–297.

Diagboya, PN, BI Olu-Owolabi, and KO Adebowale. 2015. Synthesis of covalently bonded graphene oxide–iron magnetic nanoparticles and the kinetics of mercury removal. *RSC Advances* 5 (4):2536–2542.

Ding, D, Z Lei, Y Yang, C Feng, and Z Zhang. 2013. Nickel oxide grafted andic soil for efficient cesium removal from aqueous solution: Adsorption behavior and mechanisms. *ACS Applied Materials & Interfaces* 5 (20):10151–10158.

Ding, J, Y Pan, L Li, et al. 2020. Synergetic adsorption and electrochemical classified recycling of Cr (VI) and dyes in synthetic dyeing wastewater. *Chemical Engineering Journal* 384:123232.

Dolatyari, L, MR Yaftian, and S Rostamnia. 2016. Removal of uranium (VI) ions from aqueous solutions using Schiff base functionalized SBA-15 mesoporous silica materials. *Journal of Environmental Management* 169:8–17.

Dong, C, F Zhang, Z Pang, and G Yang. 2016. Efficient and selective adsorption of multi-metal ions using sulfonated cellulose as adsorbent. *Carbohydrate Polymers* 151:230–236.

Dong, H, Y Chen, G Sheng, et al. 2016. The roles of a pillared bentonite on enhancing Se(VI) removal by ZVI and the influence of co-existing solutes in groundwater. *Journal of Hazardous Materials* 304:306–312.

Dong, L, L Hou, Z Wang, P Gu, G Chen, and R Jiang. 2018. A new function of spent activated carbon in BAC process: Removing heavy metals by ion exchange mechanism. *Journal of Hazardous Materials* 359:76–84.

Dong, R, Y Liu, X Wang, and J Huang. 2011. Adsorption of sulfate ions from aqueous solution by surfactant-modified palygorskite. *Journal of Chemical & Engineering Data* 56 (10):3890–3896.

Dong, W, TK Tokunaga, JA Davis, and J Wan. 2012. Uranium(VI) adsorption and surface complexation modeling onto background sediments from the F-area savannah river site. *Environmental Science & Technology* 46 (3):1565–1571.

Drout, RJ, L Robison, Z Chen, T Islamoglu, and OK Farha. 2019. Zirconium metal–organic frameworks for organic pollutant adsorption. *Trends in Chemistry* 1 (3):304–317.

Du, T, X Fang, Y Wei, J Shang, B Zhang, and L Liu. 2017. Synthesis of nanocontainer chabazites from fly ash with a template-and fluoride-free process for cesium ion adsorption. *Energy & Fuels* 31 (4):4301–4307.

Dubey, S, S Banerjee, SN Upadhyay, and YC Sharma. 2017. Application of common nanomaterials for removal of selected metallic species from water and wastewaters: A critical review. *Journal of Molecular Liquids* 240 (Supplement C):656–677.

Duman, O, and E Ayranci. 2010. Attachment of benzo-crown ethers onto activated carbon cloth to enhance the removal of chromium, cobalt and nickel ions from aqueous solutions by adsorption. *Journal of Hazardous Materials* 176 (1–3):231–238.

Duranoğlu, D, İGB Kaya, U Beker, and BF Şenkal. 2012. Synthesis and adsorption properties of polymeric and polymer-based hybrid adsorbent for hexavalent chromium removal. *Chemical Engineering Journal* 181:103–112.

Dutta, D, SK Roy, and AK Talukdar. 2017. Effective removal of Cr (VI) from aqueous solution by diamino-functionalised mesoporous MCM-48 and selective oxidation of cyclohexene and ethylbenzene over the Cr containing spent adsorbent. *Journal of Environmental Chemical Engineering* 5 (5):4707–4715.

Dwivedi, C, A Kumar, KK Singh, et al. 2013. Copper hexacyanoferrate–polymer composite beads for cesium ion removal: Synthesis, characterization, sorption, and kinetic studies. *Journal of Applied Polymer Science* 129 (1):152–160.

Ebrahimi-Gatkash, M, H Younesi, A Shahbazi, and A Heidari. 2017. Amino-functionalized mesoporous MCM-41 silica as an efficient adsorbent for water treatment: Batch and fixed-bed column adsorption of the nitrate anion. *Applied Water Science* 7 (4):1887–1901.

Efome, JE, D Rana, T Matsuura, and CQ Lan. 2018. Metal–organic frameworks supported on nanofibers to remove heavy metals. *Journal of Materials Chemistry A* 6 (10):4550–4555.

Egodawatte, S, A Datt, EA Burns, and SC Larsen. 2015. Chemical Insight into the adsorption of chromium(III) on iron oxide/mesoporous silica nanocomposites. *Langmuir* 31 (27):7553–7562.

El Hazek, MN, and AA Gabr. 2016. Dissolution of manganese from polymetallic material using sulfuric-oxalic acid medium. *American Journal of Analytical Chemistry* 7 (5):469–477.

Elhamifar, D, F Shojaeipoor, and O Yari. 2016. Thiopropyl-containing ionic liquid based periodic mesoporous organosilica as a novel and efficient adsorbent for the removal of Hg (II) and Pb (II) ions from aqueous solutions. *RSC Advances* 6 (63):58658–58666.

Elwakeel, KZ, and E Guibal. 2015. Selective removal of Hg (II) from aqueous solution by functionalized magnetic-macromolecular hybrid material. *Chemical Engineering Journal* 281:345–359.

Erdem Yayayürük, A, and O Yayayürük. 2017. Adsorptive performance of nanosized zero-valent iron for V (V) removal from aqueous solutions. *Journal of Chemical Technology and Biotechnology* 92 (8):1891–1898.

Ersan, G, OG Apul, F Perreault, and T Karanfil. 2017. Adsorption of organic contaminants by graphene nanosheets: A review. *Water Research* 126:385–398.

Etale, A, H Tutu, and DC Drake. 2016. Application of maghemite nanoparticles as sorbents for the removal of Cu(II), Mn(II) and U(VI) ions from aqueous solution in acid mine drainage conditions. *Applied Water Science* 6 (2):187–197.

Fan, L, C Luo, M Sun, and H Qiu. 2012. Synthesis of graphene oxide decorated with magnetic cyclodextrin for fast chromium removal. *Journal of Materials Chemistry* 22 (47):24577–24583.

Fan, Q, D Shao, Y Lu, W Wu, and X Wang. 2009. Effect of pH, ionic strength, temperature and humic substances on the sorption of Ni(II) to Na–attapulgite. *Chemical Engineering Journal* 150 (1):188–195.

Fang, C, T Zhang, P Li, et al. 2015. Phosphorus recovery from biogas fermentation liquid by Ca–Mg loaded biochar. *Journal of Environmental Sciences* 29:106–114.

Fang, F, L Kong, J Huang, et al. 2014. Removal of cobalt ions from aqueous solution by an amination graphene oxide nanocomposite. *Journal of Hazardous Materials* 270:1–10.

Fang, Q, and B Chen. 2012. Adsorption of perchlorate onto raw and oxidized carbon nanotubes in aqueous solution. *Carbon* 50 (6):2209–2219.

Fang, Q, B Chen, Y Lin, and Y Guan. 2013. Aromatic and hydrophobic surfaces of wood-derived biochar enhance perchlorate adsorption via hydrogen bonding to oxygen-containing organic groups. *Environmental Science & Technology* 48 (1):279–288.

Faroon, O, A Ashizawa, S Wright, et al. 2012. Agency for Toxic Substances and Disease Registry (ATSDR) Toxicological Profiles. In *Toxicological Profile for Cadmium*. Atlanta (GA): Agency for Toxic Substances and Disease Registry (US): page no. 1–430.

Faur-Brasquet, C, Z Reddad, K Kadirvelu, and P Le Cloirec. 2002. Modeling the adsorption of metal ions (Cu^{2+}, Ni^{2+}, Pb^{2+}) onto ACCs using surface complexation models. *Applied Surface Science* 196 (1–4):356–365.

Federal Ministry for the Environment, Nature Conservation, Building and Nuclear Safety. 2016. *German Resource Efficiency Programme II: Programme for the Sustainable Use and Conservation of Natural Resources*. edited by Federal Environment Ministry, Division WR III 1: Rostock: Federal Ministry for the Environment, Nature Conservation, Building and Nuclear Safety,: page no. 1–143.

Fellenz, N, FJ Perez-Alonso, PP Martin, et al. 2017. Chromium (VI) removal from water by means of adsorption-reduction at the surface of amino-functionalized MCM-41 sorbents. *Microporous and Mesoporous Materials* 239:138–146.

Feng, M, P Zhang, H-C Zhou, and VK Sharma. 2018. Water-stable metal-organic frameworks for aqueous removal of heavy metals and radionuclides: A review. *Chemosphere* 209:783–800.

Feng, Y, J-L Gong, G-M Zeng, et al. 2010. Adsorption of Cd (II) and Zn (II) from aqueous solutions using magnetic hydroxyapatite nanoparticles as adsorbents. *Chemical Engineering Journal* 162 (2):487–494.

Feng, Z, S Zhu, DRM de Godoi, ACS Samia, and D Scherson. 2012. Adsorption of Cd^{2+} on carboxyl-terminated superparamagnetic iron oxide nanoparticles. *Analytical chemistry* 84 (8):3764–3770.

Fernández-Reyes, B, K Ortiz-Martínez, JA Lasalde-Ramírez, and AJ Hernández-Maldonado. 2020. Chapter 1- Engineered Adsorbents for the Removal of Contaminants of Emerging Concern from Water. In *Contaminants of Emerging Concern in Water and Wastewater*, edited by AJ Hernández-Maldonado, and L Blaney. Oxford (UK): Butterworth-Heinemann: page no. 3–45.

Fordyce, F. 2007. Selenium geochemistry and health. *AMBIO: A Journal of the Human Environment* 36 (1):94–97.

Fouladgar, M, M Beheshti, and H Sabzyan. 2015. Single and binary adsorption of nickel and copper from aqueous solutions by γ-alumina nanoparticles: Equilibrium and kinetic modeling. *Journal of Molecular Liquids* 211:1060–1073.

Frierdich, AJ, and JG Catalano. 2012. Controls on Fe (II)-activated trace element release from goethite and hematite. *Environmental Science & Technology* 46 (3):1519–1526.

Fu, F, and Q Wang. 2011. Removal of heavy metal ions from wastewaters: A review. *Journal of Environmental Management* 92 (3):407–418.

Fu, R, Y Liu, Z Lou, Z Wang, SA Baig, and X Xu. 2016. Adsorptive removal of Pb (II) by magnetic activated carbon incorporated with amino groups from aqueous solutions. *Journal of the Taiwan Institute of Chemical Engineers* 62:247–258.

Fu, W, and Z Huang. 2018. Magnetic dithiocarbamate functionalized reduced graphene oxide for the removal of Cu(II), Cd(II), Pb(II), and Hg(II) ions from aqueous solution: Synthesis, adsorption, and regeneration. *Chemosphere* 209:449–456.

Fu, Y, J Jiang, Z Chen, S Ying, J Wang, and J Hu. 2019. Rapid and selective removal of Hg (II) ions and high catalytic performance of the spent adsorbent based on functionalized mesoporous silica/poly (m-aminothiophenol) nanocomposite. *Journal of Molecular Liquids* 286:110746.

Fu, Y, J Wang, Q Liu, and H Zeng. 2014. Water-dispersible magnetic nanoparticle–graphene oxide composites for selenium removal. *Carbon* 77:710–721.

Fuge, R. 2013. Anthropogenic Sources. In *Essentials of Medical Geology: Revised Edition*, edited by O. Selinus. Dordrecht, Netherlands: Springer: page no. 59–74.

Fukushi, K, K Aoyama, C Yang, N Kitadai, and S Nakashima. 2013. Surface complexation modeling for sulfate adsorption on ferrihydrite consistent with in situ infrared spectroscopic observations. *Applied Geochemistry* 36:92–103.

Futalan, CM, C-C Kan, ML Dalida, K-J Hsien, C Pascua, and M-W Wan. 2011. Comparative and competitive adsorption of copper, lead, and nickel using chitosan immobilized on bentonite. *Carbohydrate Polymers* 83 (2):528–536.

Gale, JD, CRA Catlow, and MJ Gillan. 1999. A density functional study of Ti/MgCl 2-supported Ziegler–Natta catalysts. *Topics in Catalysis* 9 (3):235–250.

Gallegos, TJ, CC Fuller, SM Webb, and W Betterton. 2013. Uranium(VI) interactions with mackinawite in the presence and absence of bicarbonate and oxygen. *Environmental Science & Technology* 47 (13):7357–7364.

Ganesan, P, R Kamaraj, G Sozhan, and S Vasudevan. 2013. Oxidized multiwalled carbon nanotubes as adsorbent for the removal of manganese from aqueous solution. *Environmental Science and Pollution Research* 20 (2):987–996.

Ganesan, P, R Kamaraj, and S Vasudevan. 2013. Application of isotherm, kinetic and thermodynamic models for the adsorption of nitrate ions on graphene from aqueous solution. *Journal of the Taiwan Institute of Chemical Engineers* 44 (5):808–814.

Gedam, AH, and RS Dongre. 2015. Adsorption characterization of Pb(ii) ions onto iodate doped chitosan composite: Equilibrium and kinetic studies. *RSC Advances* 5 (67):54188–54201.

Gezer, N, M Gülfen, and AO Aydın. 2011. Adsorption of selenite and selenate ions onto thiourea-formaldehyde resin. *Journal of Applied Polymer Science* 122 (2):1134–1141.

Ghaemi, A, M Torab-Mostaedi, and M Ghannadi-Maragheh. 2011. Characterizations of strontium (II) and barium (II) adsorption from aqueous solutions using dolomite powder. *Journal of Hazardous Materials* 190 (1):916–921.

Ghaleh, SP, E Khodapanah, and SA Tabatabaei-Nezhad. 2020. Comprehensive monolayer two-parameter isotherm and kinetic studies of thiamine adsorption on clay minerals: Experimental and modeling approaches. *Journal of Molecular Liquids*:112942.

Ghosh, G, and DK Mukhopadhyay. 2019. Human Health Hazards due to Arsenic and Fluoride Contamination in Drinking Water and Food Chain. In *Groundwater Development and Management*, edited by PK Sikdar. Cham, Switzerland: Springer. Chapt. *15*: page no. 351–369.

Gilchrist, M, PG Winyard, and N Benjamin. 2010. Dietary nitrate – Good or bad? *Nitric Oxide* 22 (2):104–109.

Gloria, B, TB Atherholt, and PD Cohn. 2011. *Water Quality and Treatment a Handbook on Drinking Water*, edited by JK Edzwald. 6 ed. Denver, Colorado: American Water Works Association.

González, PG, and YB Pliego-Cuervo. 2014. Adsorption of Cd (II), Hg (II) and Zn (II) from aqueous solution using mesoporous activated carbon produced from Bambusa vulgaris striata. *Chemical Engineering Research and Design* 92 (11):2715–2724.

Goyer, RA. 1989. Mechanisms of lead and cadmium nephrotoxicity. *Toxicology Letters* 46 (1–3):153–162.

Goyer, RA. 1990. Lead toxicity: From overt to subclinical to subtle health effects. *Environmental Health Perspectives* 86:177.

Grassi, M, G Kaykioglu, V Belgiorno, and G Lofrano. 2012. Removal of Emerging Contaminants from Water and Wastewater by Adsorption Process. In *Emerging Compounds Removal from Wastewater: Natural and Solar Based Treatments*, edited by G Lofrano. Dordrecht, Netherlands: Springer. Chapt. *2*: page no. 15–37.

Greenlee, LF, DF Lawler, BD Freeman, B Marrot, and P Moulin. 2009. Reverse osmosis desalination: Water sources, technology, and today's challenges. *Water Research* 43 (9):2317–2348.

Gu, C, Z Wang, JD Kubicki, X Wang, and M Zhu. 2016. X-ray absorption spectroscopic quantification and speciation modeling of sulfate adsorption on ferrihydrite surfaces. *Environmental Science & Technology* 50 (15):8067–8076.

Gu, S, J Zhou, C Yu, Z Luo, Q Wang, and Z Shi. 2015. A novel two-staged thermal synthesis method of generating nanosilica from rice husk via pre-pyrolysis combined with calcination. *Industrial Crops and Products* 65:1–6.

Gu, Z, W Song, Z Yang, and R Zhou. 2018. Metal–organic framework as an efficient filter for the removal of heavy metal cations in water. *Physical Chemistry Chemical Physics* 20 (48):30384–30391.

Guan, T, Y Kuang, X Li, J Fang, W Fang, and D Wu. 2020. The recovery of phosphorus from source-separated urine by repeatedly usable magnetic Fe_3O_4@ ZrO_2 nanoparticles under acidic conditions. *Environment International* 134:105322.

Guan, W, J Pan, H Ou, et al. 2011. Removal of strontium (II) ions by potassium tetratitanate whisker and sodium trititanate whisker from aqueous solution: Equilibrium, kinetics and thermodynamics. *Chemical Engineering Journal* 167 (1):215–222.

Guedes, RE, AS Luna, and AR Torres. 2018. Operating parameters for bio-oil production in biomass pyrolysis: A review. *Journal of Analytical and Applied Pyrolysis* 129:134–149.

Guilane, S, and O Hamdaoui. 2016. Regeneration of exhausted granular activated carbon by low frequency ultrasound in batch reactor. *Desalination and Water Treatment* 57 (34):15826–15834.

Guo, N, S-j Su, B Liao, S-l Ding, and W-y Sun. 2017. Preparation and properties of a novel macro porous Ni^{2+}-imprinted chitosan foam adsorbents for adsorption of nickel ions from aqueous solution. *Carbohydrate Polymers* 165:376–383.

Guo, S, P Jiao, Z Dan, N Duan, G Chen, and J Zhang. 2017. Preparation of L-arginine modified magnetic adsorbent by one-step method for removal of Zn (II) and Cd (II) from aqueous solution. *Chemical Engineering Journal* 317:999–1011.

Guo, Y, J Deng, J Zhu, X Zhou, and R Bai. 2016. Removal of mercury (II) and methylene blue from a wastewater environment with magnetic graphene oxide: Adsorption kinetics, isotherms and mechanism. *RSC Advances* 6 (86):82523–82536.

Guo, Z, X Zhang, Y Kang, and J Zhang. 2017. Biomass-derived carbon sorbents for Cd (II) removal: Activation and adsorption mechanism. *ACS Sustainable Chemistry & Engineering* 5 (5):4103–4109.

Gupta, V, S Agarwal, and TA Saleh. 2011. Chromium removal by combining the magnetic properties of iron oxide with adsorption properties of carbon nanotubes. *Water Research* 45 (6):2207–2212.

Gupta, VK, A Rastogi, and A Nayak. 2010. Adsorption studies on the removal of hexavalent chromium from aqueous solution using a low cost fertilizer industry waste material. *Journal of Colloid and Interface Science* 342 (1):135–141.

Gupta, VK, Suhas, A Nayak, S Agarwal, M Chaudhary, and I Tyagi. 2014. Removal of Ni (II) ions from water using scrap tire. *Journal of Molecular Liquids* 190:215–222.

Gusain, R, N Kumar, and SS Ray. 2020. Recent advances in carbon nanomaterial-based adsorbents for water purification. *Coordination Chemistry Reviews* 405:213111.

Hadavifar, M, N Bahramifar, H Younesi, and Q Li. 2014. Adsorption of mercury ions from synthetic and real wastewater aqueous solution by functionalized multi-walled carbon nanotube with both amino and thiolated groups. *Chemical Engineering Journal* 237:217–228.

Hadi Najafabadi, H, M Irani, LR Rad, AH Haratameh, and I Haririan. 2015. Removal of Cu^{2+}, Pb^{2+} and Cr^{6+} from aqueous solutions using a chitosan/graphene oxide composite nanofibrous adsorbent. *RSC Advances* 5 (21):16532–16539.

Hagemann, N, H-P Schmidt, R Kägi, et al. 2020. Wood-based activated biochar to eliminate organic micropollutants from biologically treated wastewater. *Science of the Total Environment*:138417.

Hakumai, Y, S Oike, Y Shibata, and M Ashiuchi. 2016. Cooperative adsorption of critical metal ions using archaeal poly-γ-glutamate. *BioMetals* 29 (3):527–534.

Hamdaoui, O. 2017. Adsorption of Cu(II) from aqueous phase by Cedar bark. *Journal of Dispersion Science and Technology* 38 (8):1087–1091.

Hamza, MF, A Gamal, G Hussein, et al. 2019. Uranium (VI) and zirconium (IV) sorption on magnetic chitosan derivatives–effect of different functional groups on separation properties. *Journal of Chemical Technology & Biotechnology* 94 (12):3866–3882.

Hamza, MF, Y Wei, HI Mira, A-H Adel, and E Guibal. 2019. Synthesis and adsorption characteristics of grafted hydrazinyl amine magnetite-chitosan for Ni (II) and Pb (II) recovery. *Chemical Engineering Journal* 362:310–324.

Han, C, H Pu, H Li, et al. 2013. The optimization of As(V) removal over mesoporous alumina by using response surface methodology and adsorption mechanism. *Journal of Hazardous Materials* 254–255:301–309.

Han, DS, B Batchelor, and A Abdel-Wahab. 2013. XPS analysis of sorption of selenium(IV) and selenium(VI) to mackinawite (FeS). *Environmental Progress & Sustainable Energy* 32 (1):84–93.

Han, F, G-H Zhang, and P Gu. 2013. Adsorption kinetics and equilibrium modeling of cesium on copper ferrocyanide. *Journal of Radioanalytical and Nuclear Chemistry* 295 (1):369–377.

Hao, L, H Song, L Zhang, X Wan, Y Tang, and Y Lv. 2012. SiO_2/graphene composite for highly selective adsorption of Pb(II) ion. *Journal of Colloid and Interface Science* 369 (1):381–387.

Hao, Y-M, C Man, and Z-B Hu. 2010. Effective removal of Cu (II) ions from aqueous solution by amino-functionalized magnetic nanoparticles. *Journal of Hazardous Materials* 184 (1–3):392–399.

Hasanzadeh, R, PN Moghadam, N Bahri-Laleh, and EN Zare. 2016. Sulfonated magnetic nanocomposite based on reactive PGMA-MAn copolymer@nanoparticles: Effective removal of Cu(II) ions from aqueous solutions. *International Journal of Polymer Science* 2016:2610541.

Hayes, KF, and JO. Leckie. 1987. Modeling ionic strength effects on cation adsorption at hydrous oxide/solution interfaces. *Journal of Colloid and Interface Science* 115 (2):564–572.

He, D, L Zhang, Y Zhao, et al. 2018. Recycling spent Cr adsorbents as catalyst for eliminating methylmercaptan. *Environmental Science & Technology* 52 (6):3669–3675.

He, M, Y Zhu, Y Yang, B Han, and Y Zhang. 2011. Adsorption of cobalt (II) ions from aqueous solutions by palygorskite. *Applied Clay Science* 54 (3):292–296.

Heidari, A, H Younesi, Z Mehraban, and H Heikkinen. 2013. Selective adsorption of Pb (II), Cd (II), and Ni (II) ions from aqueous solution using chitosan–MAA nanoparticles. *International Journal of Biological Macromolecules* 61:251–263.

Hossain, MA, HH Ngo, WS Guo, and T Setiadi. 2012. Adsorption and desorption of copper(II) ions onto garden grass. *Bioresource Technology* 121:386–395.

Hossain, N, MA Bhuiyan, BK Pramanik, S Nizamuddin, and G Griffin. 2020. Waste materials for wastewater treatment and waste adsorbents for biofuel and cement supplement applications: A critical review. *Journal of Cleaner Production* 255:120261.

Howarth, AJ, MJ Katz, TC Wang, et al. 2015. High efficiency adsorption and removal of selenate and selenite from water using metal–organic frameworks. *Journal of the American Chemical Society* 137 (23):7488–7494.

Howarth, AJ, TC Wang, SS Al-Juaid, SG Aziz, JT Hupp, and OK Farha. 2016. Efficient extraction of sulfate from water using a Zr-metal–organic framework. *Dalton Transactions* 45 (1):93–97.

Hu, J, A Aarts, R Shang, B Heijman, and L Rietveld. 2016. Integrating powdered activated carbon into wastewater tertiary filter for micro-pollutant removal. *Journal of Environmental Management* 177:45–52.

Hu, Q, N Chen, C Feng, and W Hu. 2015. Nitrate adsorption from aqueous solution using granular chitosan-Fe^{3+} complex. *Applied Surface Science* 347:1–9.

Hua, M, S Zhang, B Pan, W Zhang, L Lv, and Q Zhang. 2012. Heavy metal removal from water/wastewater by nanosized metal oxides: A review. *Journal of Hazardous Materials* 211–212:317–331.

Hua, T, RJ Haynes, YF Zhou, A Boullemant, and I Chandrawana. 2015. Potential for use of industrial waste materials as filter media for removal of Al, Mo, As, V and Ga from alkaline drainage in constructed wetlands – Adsorption studies. *Water Research* 71:32–41.

Hualei, Z, L Dongyan, T Yajun, and C Yunfa. 2008. Extraction of scandium from red mud by modified activated carbon and kinetics study. *Rare metals* 27 (3):223–227.

Huang, J, Z Wu, L Chen, and Y Sun. 2015. Surface complexation modeling of adsorption of Cd (II) on graphene oxides. *Journal of Molecular Liquids* 209:753–758.

Huang, L, M He, B Chen, and B Hu. 2015. A designable magnetic MOF composite and facile coordination-based post-synthetic strategy for the enhanced removal of Hg^{2+} from water. *Journal of Materials Chemistry A* 3 (21):11587–11595.

Huang, L, M He, B Chen, and B Hu. 2016. A mercapto functionalized magnetic Zr-MOF by solvent-assisted ligand exchange for Hg^{2+} removal from water. *Journal of Materials Chemistry A* 4 (14):5159–5166.

Huang, L, M He, B-b Chen, Q Cheng, and B Hu. 2017. Highly efficient magnetic nitrogen-doped porous carbon prepared by one-step carbonization strategy for Hg^{2+} removal from water. *ACS Applied Materials & Interfaces* 9 (3):2550–2559.

Huang, R, B Yang, and Q Liu. 2013. Removal of chromium (VI) ions from aqueous solutions with protonated crosslinked chitosan. *Journal of Applied Polymer Science* 129 (2):908–915.

Huang, R, M Wu, T Zhang, D Li, P Tang, and Y Feng. 2017. Template-free synthesis of large-pore-size porous magnesium silicate hierarchical nanostructures for high-efficiency removal of heavy metal ions. *ACS Sustainable Chemistry & Engineering* 5 (3):2774–2780.

Huang, Y, C Yang, Z Sun, G Zeng, and H He. 2015. Removal of cadmium and lead from aqueous solutions using nitrilotriacetic acid anhydride modified ligno-cellulosic material. *RSC Advances* 5 (15):11475–11484.

Huang, Y, D Hu, S Wen, M Shen, M Zhu, and X Shi. 2014. Selective removal of mercury ions using thymine-grafted electrospun polymer nanofibers. *New Journal of Chemistry* 38 (4):1533–1539.

Huang, Z-H, X Zheng, W Lv, M Wang, Q-H Yang, and F Kang. 2011. Adsorption of lead (II) ions from aqueous solution on low-temperature exfoliated graphene nanosheets. *Langmuir* 27 (12):7558–7562.

Husein, DZ. 2013. Adsorption and removal of mercury ions from aqueous solution using raw and chemically modified Egyptian mandarin peel. *Desalination and Water Treatment* 51 (34–36):6761–6769.

Huynh, J, R Palacio, F Safizadeh, et al. 2017. Adsorption of uranium over NH2-functionalized ordered silica in aqueous solutions. *ACS Applied Materials & Interfaces* 9 (18):15672–15684.

Hyatt, DE. 1987. Value recovery from spent alumina-base catalyst. Google Patents.

IARC. 2012. *Monographs on the Evaluation of Carcinogenic Risks to Humans Arsenic and Arsenic Compounds*. Lyon, France: IARC.

Ide, Y, N Ochi, and M Ogawa. 2011. Effective and selective adsorption of Zn^{2+} from seawater on a layered silicate. *Angewandte Chemie* 123 (3):680–682.

Idris, SAM. 2015. Adsorption, kinetic and thermodynamic studies for manganese extraction from aqueous medium using mesoporous silica. *Journal of Colloid and Interface Science* 440 (Supplement C):84–90.

Ihsanullah, AA, AM Al-Amer, T Laoui, et al. 2016. Heavy metal removal from aqueous solution by advanced carbon nanotubes: Critical review of adsorption applications. *Separation and Purification Technology* 157:141–161.

Ihsanullah, FAA-K, B Abu-Sharkh, AM Abulkibash, et al. 2016. Effect of acid modification on adsorption of hexavalent chromium (Cr(VI)) from aqueous solution by activated carbon and carbon nanotubes. *Desalination and Water Treatment* 57 (16):7232–7244.

Ijomone, OM, OM Aluko, COA Okoh, AC Martins Jr, and M Aschner. 2019. Role for calcium signaling in manganese neurotoxicity. *Journal of Trace Elements in Medicine and Biology* 56:146–155.

Inyang, MI, B Gao, Y Yao, et al. 2016. A review of biochar as a low-cost adsorbent for aqueous heavy metal removal. *Critical Reviews in Environmental Science and Technology* 46 (4):406–433.

Iravani, S. 2011. Green synthesis of metal nanoparticles using plants. *Green Chemistry* 13 (10):2638–2650.

Jain, M, VK Garg, and K Kadirvelu. 2014. Removal of Ni(II) from aqueous system by chemically modified sunflower biomass. *Desalination and Water Treatment* 52 (28–30):5681–5695.

Jain, R, N Jordan, D Schild, et al. 2015. Adsorption of zinc by biogenic elemental selenium nanoparticles. *Chemical Engineering Journal* 260:855–863.

Jalees, MI, MU Farooq, S Basheer, and S Asghar. 2019. Removal of heavy metals from drinking water using Chikni Mitti (kaolinite): Isotherm and kinetics. *Arabian Journal for Science and Engineering* 44 (7):6351–6359.

Jang, J, and DS Lee. 2016. Magnetic prussian blue nanocomposites for effective cesium removal from aqueous solution. *Industrial & Engineering Chemistry Research* 55 (13):3852–3860.

Jankovský, O, P Šimek, K Klímová, D Sedmidubský, M Pumera, and Z Sofer. 2015. Highly selective removal of Ga^{3+} ions from Al^{3+}/Ga^{3+} mixtures using graphite oxide. *Carbon* 89:121–129.

Javadian, H. 2014. Application of kinetic, isotherm and thermodynamic models for the adsorption of Co (II) ions on polyaniline/polypyrrole copolymer nanofibers from aqueous solution. *Journal of Industrial and Engineering Chemistry* 20 (6):4233–4241.

Jeon, C, and KH Park. 2005. Adsorption and desorption characteristics of mercury (II) ions using aminated chitosan bead. *Water research* 39 (16):3938–3944.

Jia, J, A Bendounan, HMN Kotresh, et al. 2013. Selenium adsorption on Au(111) and Ag(111) surfaces: Adsorbed selenium and selenide films. *The Journal of Physical Chemistry C* 117 (19):9835–9842.

Jian, M, B Liu, G Zhang, R Liu, and X Zhang. 2015. Adsorptive removal of arsenic from aqueous solution by zeolitic imidazolate framework-8 (ZIF-8) nanoparticles. *Colloids and Surfaces A: Physicochemical and Engineering Aspects* 465:67–76.

Jiang, H, P Chen, S Luo, X Tu, Q Cao, and M Shu. 2013. Synthesis of novel nanocomposite Fe_3O_4/ZrO_2/chitosan and its application for removal of nitrate and phosphate. *Applied Surface Science* 284:942–949.

Jiang, H, W Zhang, P Chen, et al. 2016. Equipping an adsorbent with an indicator: A novel composite to simultaneously detect and remove heavy metals from water. *Journal of Materials Chemistry A* 4 (30):11897–11907.

Jiang, L, and P Liu. 2014. Novel magnetic fly ash/poly (acrylic acid) composite microgel for selective adsorption of Pb (II) ion: Synthesis and evaluation. *Industrial & Engineering Chemistry Research* 53 (8):2924–2931.

Jiang, L, S Xiao, and J Chen. 2015. Removal behavior and mechanism of Co (II) on the surface of Fe–Mn binary oxide adsorbent. *Colloids and Surfaces A: Physicochemical and Engineering Aspects* 479:1–10.

Jiang, S, L Huang, TAH Nguyen, et al. 2016. Copper and zinc adsorption by softwood and hardwood biochars under elevated sulphate-induced salinity and acidic pH conditions. *Chemosphere* 142:64–71.

Jiang, S-D, Q-Z Yao, G-T Zhou, and S-Q Fu. 2012. Fabrication of hydroxyapatite hierarchical hollow microspheres and potential application in water treatment. *The Journal of Physical Chemistry C* 116 (7):4484–4492.

Jin, C, X Zhang, J Xin, et al. 2017. Clickable synthesis of 1, 2, 4-triazole modified lignin-based adsorbent for the selective removal of Cd (II). *ACS Sustainable Chemistry & Engineering* 5 (5):4086–4093.

Jin, H, Z Ji, J Yuan, et al. 2015. Research on removal of fluoride in aqueous solution by alumina-modified expanded graphite composite. *Journal of Alloys and Compounds* 620:361–367.

Jin, Z, Y Jia, K-S Zhang, et al. 2016. Effective removal of fluoride by porous MgO nanoplates and its adsorption mechanism. *Journal of Alloys and Compounds* 675:292–300.

Jing, C, J Cui, Y Huang, and A Li. 2012. Fabrication, characterization, and application of a composite adsorbent for simultaneous removal of arsenic and fluoride. *ACS Applied Materials & Interfaces* 4 (2):714–720.

Johansson, CL, NA Paul, R de Nys, and DA Roberts. 2015. The complexity of biosorption treatments for oxyanions in a multi-element mine effluent. *Journal of Environmental Management* 151:386–392.

Johnson, SF 2019. Methemoglobinemia: Infants at risk. *Current Problems in Pediatric and Adolescent Health Care* 49 (3):57–67.

Johnston, CP, and M Chrysochoou. 2016. Mechanisms of chromate, selenate, and sulfate adsorption on Al-substituted ferrihydrite: Implications for ferrihydrite surface structure and reactivity. *Environmental Science & Technology* 50 (7):3589–3596.

Jones, AM, CA Murphy, TD Waite, and RN Collins. 2017. Fe (II) interactions with smectites: Temporal changes in redox reactivity and the formation of green rust. *Environmental Science & Technology* 51 (21):12573–12582.

Jung, KW, T-U Jeong, M-J Hwang, K Kim, and K-H Ahn. 2015. Phosphate adsorption ability of biochar/Mg–Al assembled nanocomposites prepared by aluminum-electrode based electro-assisted modification method with $MgCl_2$ as electrolyte. *Bioresource Technology* 198:603–610.

Jyothi, MN, BJ Ramaiah, and SM Maliyekkal. 2020. Occurrence, Contamination, Speciation and Analysis of Selenium in the Environment. In *Measurement, Analysis and Remediation of Environmental Pollutants*, edited by Tarun Gupta, Swatantra Pratap Singh,Prashant Rajput, Avinash Kumar Agarwal. Singapore: Springer: page no. 245–269.

Kabiri, S, DN Tran, S Azari, and D Losic. 2015. Graphene-diatom silica aerogels for efficient removal of mercury ions from water. *ACS Applied Materials & Interfaces* 7 (22):11815–11823.

Kaçan, E, and C Kütahyalı. 2012. Adsorption of strontium from aqueous solution using activated carbon produced from textile sewage sludges. *Journal of Analytical and Applied Pyrolysis* 97:149–157.

Kameda, T, E Kondo, and T Yoshioka. 2014. Equilibrium and kinetic studies of Se(vi) removal by Mg-Al layered double hydroxide doped with Fe^{2+}. *RSC Advances* 4 (106):61817–61822.

Kan, T, V Strezov, and TJ Evans. 2016. Lignocellulosic biomass pyrolysis: A review of product properties and effects of pyrolysis parameters. *Renewable and Sustainable Energy Reviews* 57:1126–1140.

Kandah, MI, and JL Meunier. 2007. Removal of nickel ions from water by multi-walled carbon nanotubes. *Journal of Hazardous Materials* 146 (1–2):283–288.

Kannamba, B, KL Reddy, and BV AppaRao. 2010. Removal of Cu(II) from aqueous solutions using chemically modified chitosan. *Journal of Hazardous Materials* 175 (1–3):939–948.

Kar, BB, BVR Murthy, and VN Misra. 2005. Extraction of molybdenum from spent catalyst by salt-roasting. *International Journal of Mineral Processing* 76 (3):143–147.

Kara, İ, D Yilmazer, and ST Akar. 2017. Metakaolin based geopolymer as an effective adsorbent for adsorption of zinc (II) and nickel (II) ions from aqueous solutions. *Applied Clay Science* 139:54–63.

karmakar, S, J Dechnik, C Janiak, and S De. 2016. Aluminium fumarate metal-organic framework: A super adsorbent for fluoride from water. *Journal of Hazardous Materials* 303:10–20.

Karoui, S, RB Arfi, K Mougin, A Ghorbal, AA Assadi, and A Amrane. 2020. Synthesis of novel biocomposite powder for simultaneous removal of hazardous ciprofloxacin and methylene blue: Central composite design, kinetic and isotherm studies using Brouers-Sotolongo family models. *Journal of Hazardous Materials* 387:121675.

Kasiuliene, A, I Carabante, P Bhattacharya, and J Kumpiene. 2019. Hydrothermal carbonisation of peat-based spent sorbents loaded with metal(loid)s. *Environmental Science and Pollution Research* 26 (23):23730–23738.

Katal, R, E Hasani, M Farnam, MS Baei, and MA Ghayyem. 2012. Charcoal ash as an adsorbent for Ni(II) adsorption and its application for wastewater treatment. *Journal of Chemical & Engineering Data* 57 (2):374–383.

Katal, R, MS Baei, HT Rahmati, and H Esfandian. 2012. Kinetic, isotherm and thermodynamic study of nitrate adsorption from aqueous solution using modified rice husk. *Journal of Industrial and Engineering Chemistry* 18 (1):295–302.

Keränen, A, T Leiviskä, O Hormi, and J Tanskanen. 2015. Removal of nitrate by modified pine sawdust: Effects of temperature and co-existing anions. *Journal of Environmental Management* 147:46–54.

Khan, TA, SA Chaudhry, and I Ali. 2015. Equilibrium uptake, isotherm and kinetic studies of Cd (II) adsorption onto iron oxide activated red mud from aqueous solution. *Journal of Molecular Liquids* 202:165–175.

Khobragade, MU, and A Pal. 2016. Adsorptive removal of Mn(II) from water and wastewater by surfactant-modified alumina. *Desalination and Water Treatment* 57 (6):2775–2786.

Kim, H, W Um, W-S Kim, and S Chang. 2017. Synthesis of tributyl phosphate-coated hydroxyapatite for selective uranium removal. *Industrial & Engineering Chemistry Research* 56 (12):3399–3406.

Kim, Y, YK Kim, S Kim, D Harbottle, and JW Lee. 2017. Nanostructured potassium copper hexacyanoferrate-cellulose hydrogel for selective and rapid cesium adsorption. *Chemical Engineering Journal* 313:1042–1050.

Kiser, MA, P Westerhoff, T Benn, Y Wang, J Perez-Rivera, and K Hristovski. 2009. Titanium nanomaterial removal and release from wastewater treatment plants. *Environmental Science & Technology* 43 (17):6757–6763.

Klaassen, C. 2013. *Casarett & Doull's Toxicology: The Basic Science of Poisons, Eighth Edition.* New York:McGraw-Hill Education.

Kobielska, PA, AJ Howarth, OK Farha, and S Nayak. 2018. Metal–organic frameworks for heavy metal removal from water. *Coordination Chemistry Reviews* 358:92–107.

Kolinko, PA, PG Smirniotis, DV Kozlov, and AV Vorontsov. 2012. Cr modified TiO_2-loaded MCM-41 catalysts for UV-light driven photodegradation of diethyl sulfide and ethanol. *Journal of Photochemistry and Photobiology A: Chemistry* 232:1–7.

Kong, L, Y Ruan, Q Zheng, et al. 2020. Uranium extraction using hydroxyapatite recovered from phosphorus containing wastewater. *Journal of Hazardous Materials* 382:120784.

Kousalya, GN, MR Gandhi, N Viswanathan, and S Meenakshi. 2010. Preparation and metal uptake studies of modified forms of chitin. *International Journal of Biological Macromolecules* 47 (5):583–589.

Kuang, L, Y Liu, D Fu, and Y Zhao. 2017. FeOOH-graphene oxide nanocomposites for fluoride removal from water: Acetate mediated nano FeOOH growth and adsorption mechanism. *Journal of Colloid and Interface Science* 490:259–269.

Kucharczyk, R, L Jurczyszyn, and P Hadzel. 2010. Adsorption of thin films of titanium on tungsten (111) surface. *Surface Science* 604 (17):1524–1530.

Kumar, A, P Paul, and SK Nataraj. 2016. Bionanomaterial scaffolds for effective removal of fluoride, chromium, and dye. *ACS Sustainable Chemistry & Engineering* 5 (1):895–903.

Kumar, M, JS Chung, and SH Hur. 2019. Graphene composites for lead ions removal from aqueous solutions. *Applied Sciences* 9 (14):2925.

Kumar, S, RR Nair, PB Pillai, SN Gupta, MAR Iyengar, and AK Sood. 2014. Graphene oxide–$MnFe_2O_4$ magnetic nanohybrids for efficient removal of lead and arsenic from water. *ACS Applied Materials & Interfaces* 6 (20):17426–17436.

Kumari, M, CU Pittman Jr, and D Mohan. 2015. Heavy metals [chromium (VI) and lead (II)] removal from water using mesoporous magnetite (Fe_3O_4) nanospheres. *Journal of Colloid and Interface Science* 442:120–132.

Kundu, S, IH Chowdhury, PK Sinha, and MK Naskar. 2017. Effect of organic acid-modified mesoporous alumina toward fluoride ions removal from water. *Journal of Chemical & Engineering Data* 62 (7):2067–2074.

Kurniawan, TA, GYS Chan, WH Lo, and S Babel. 2006. Physico–chemical treatment techniques for wastewater laden with heavy metals. *Chemical Engineering Journal* 118 (1–2):83–98.

Kuroki, A, M Hiroto, Y Urushihara, T Horikawa, K-I Sotowa, and JRA Avila. 2019. Adsorption mechanism of metal ions on activated carbon. *Adsorption* 25 (6):1251–1258.

Kurwadkar, S, TV Hoang, K Malwade, SR Kanel, WF Harper, and G Struckhoff. 2019. Application of carbon nanotubes for removal of emerging contaminants of concern in engineered water and wastewater treatment systems. *Nanotechnology for Environmental Engineering* 4 (1):12.

Kyzas, GZ, EA Deliyanni, and KA Matis. 2016. Activated carbons produced by pyrolysis of waste potato peels: Cobalt ions removal by adsorption. *Colloids and Surfaces A: Physicochemical and Engineering Aspects* 490:74–83.

Kyzas, GZ, and M Kostoglou. 2014. Green adsorbents for wastewaters: A critical review. *Materials* 7 (1):333–364.

Lakouraj, MM, F Hasanzadeh, and EN Zare. 2014. Nanogel and super-paramagnetic nanocomposite of thiacalix[4]arene functionalized chitosan: Synthesis, characterization and heavy metal sorption. *Iranian Polymer Journal* 23 (12):933–945.

Lakouraj, MM, F Mojerlou, and EN Zare. 2014. Nanogel and superparamagnetic nanocomposite based on sodium alginate for sorption of heavy metal ions. *Carbohydrate Polymers* 106:34–41.

Larimi, YN, MH Mallah, MA Moosavian, and J Safdari. 2013. Fabrication of a magmolecule using nanoparticle and evaluation of its adsorption capacity for selenium ions from nuclear wastewater. *Journal of Radioanalytical and Nuclear Chemistry* 298 (3):1511–1518.

Lata, S, PK Singh, and SR Samadder. 2015. Regeneration of adsorbents and recovery of heavy metals: A review. *International Journal of Environmental Science and Technology* 12 (4):1461–1478.

Le, MN, and MS Lee. 2020. A review on hydrometallurgical processes for the recovery of valuable metals from spent catalysts and life cycle analysis perspective. *Mineral Processing and Extractive Metallurgy Review*:1–20.

Lee, C-G, S Lee, J-A Park, et al. 2017. Removal of copper, nickel and chromium mixtures from metal plating wastewater by adsorption with modified carbon foam. *Chemosphere* 166:203–211.

Lee, J-Y, C-H Chen, S Cheng, and H-Y Li. 2016. Adsorption of Pb(II) and Cu(II) metal ions on functionalized large-pore mesoporous silica. *International Journal of Environmental Science and Technology* 13 (1):65–76.

Lee, S-M, D Tiwari, K-M Choi, J-K Yang, Y-Y Chang, and H-D Lee. 2009. Removal of Mn(II) from aqueous solutions using manganese-coated sand samples. *Journal of Chemical & Engineering Data* 54 (6):1823–1828.

Lei, Y, J-J Guan, W Chen, Q-F Ke, C-Q Zhang, and Y-P Guo. 2015. Fabrication of hydroxyapatite/chitosan porous materials for Pb(ii) removal from aqueous solution. *RSC Advances* 5 (32):25462–25470.

Leiviskä, T, MK Khalid, A Sarpola, and J Tanskanen. 2017. Removal of vanadium from industrial wastewater using iron sorbents in batch and continuous flow pilot systems. *Journal of Environmental Management* 190:231–242.

Li, J, X Wang, G Zhao, et al. 2018. Metal–organic framework-based materials: Superior adsorbents for the capture of toxic and radioactive metal ions. *Chemical Society Reviews* 47 (7):2322–2356.

Li, L, Q Zhu, K Man, and Z Xing. 2017. Fluoride removal from liquid phase by Fe-Al-La trimetal hydroxides adsorbent prepared by iron and aluminum leaching from red mud. *Journal of Molecular Liquids* 237:164–172.

Li, L, Y Li, L Cao, and C Yang. 2015. Enhanced chromium (VI) adsorption using nanosized chitosan fibers tailored by electrospinning. *Carbohydrate Polymers* 125:206–213.

Li, M, C Wang, MJ O'Connell, and CK Chan. 2015. Carbon nanosphere adsorbents for removal of arsenate and selenate from water. *Environmental Science: Nano* 2 (3):245–250.

Li, M, LM Farmen, and CK Chan. 2017. Selenium removal from sulfate-containing groundwater using granular layered double hydroxide materials. *Industrial & Engineering Chemistry Research* 56 (9):2458–2465.

Li, R, JJ Wang, B Zhou, et al. 2016. Enhancing phosphate adsorption by Mg/Al layered double hydroxide functionalized biochar with different Mg/Al ratios. *Science of the Total Environment* 559:121–129.

Li, R, JJ Wang, B Zhou, et al. 2016. Recovery of phosphate from aqueous solution by magnesium oxide decorated magnetic biochar and its potential as phosphate-based fertilizer substitute. *Bioresource Technology* 215:209–214.

Li, R, Q Li, S Gao, and JK Shang. 2012. Exceptional arsenic adsorption performance of hydrous cerium oxide nanoparticles: Part A. Adsorption capacity and mechanism. *Chemical Engineering Journal* 185–186:127–135.

Li, T, D Xie, C He, et al. 2016. Simultaneous adsorption of fluoride and hexavalent chromium by synthetic mesoporous alumina: Performance and interaction mechanism. *RSC Advances* 6 (54):48610–48619.

Li, X, S Wang, Y Liu, et al. 2017. Adsorption of Cu(II), Pb(II), and Cd(II) ions from acidic aqueous solutions by diethylenetriaminepentaacetic acid-modified magnetic graphene oxide. *Journal of Chemical & Engineering Data* 62 (1):407–416.

Li, X, Z Ye, Y Zhang, X Ma, L Yang, and Y Li. 2012. Efficient removal of Hg (II) by polymer-supported hydrated metal oxides from aqueous solution. *Separation Science and Technology* 47 (5):729–741.

Li, Y, B Gao, T Wu, et al. 2009. Hexavalent chromium removal from aqueous solution by adsorption on aluminum magnesium mixed hydroxide. *Water Research* 43 (12):3067–3075.

Li, Y, F Liu, B Xia, et al. 2010. Removal of copper from aqueous solution by carbon nanotube/calcium alginate composites. *Journal of Hazardous Materials* 177 (1–3):876–880.

Li, Y, and J Yu. 2014. New stories of zeolite structures: Their descriptions, determinations, predictions, and evaluations. *Chemical Reviews* 114 (14):7268–7316.

Li, Y, L Li, and J Yu. 2017. Applications of zeolites in sustainable chemistry. *Chem* 3 (6):928–949.

Li, Y, Q Li, G Zhang, et al. 2018. Separation and recovery of scandium and titanium from spent sulfuric acid solution from the titanium dioxide production process. *Hydrometallurgy* 178:1–6.

Li, Z-J, L Wang, L-Y Yuan, et al. 2015. Efficient removal of uranium from aqueous solution by zero-valent iron nanoparticle and its graphene composite. *Journal of Hazardous Materials* 290:26–33.

Liang, J, J Liu, X Yuan, et al. 2015. Facile synthesis of alumina-decorated multi-walled carbon nanotubes for simultaneous adsorption of cadmium ion and trichloroethylene. *Chemical Engineering Journal* 273:101–110.

Liang, L, X Guan, Y Huang, et al. 2015. Efficient selenate removal by zero-valent iron in the presence of weak magnetic field. *Separation and Purification Technology* 156:1064–1072.

Liao, X-P, W Tang, R-Q Zhou, and B Shi. 2008. Adsorption of metal anions of vanadium (V) and chromium (VI) on Zr (IV)-impregnated collagen fiber. *Adsorption* 14 (1):55–64.

Lima, ÉC, MA Adebayo, and FM Machado. 2015. Kinetic and Equilibrium Models of Adsorption. In *Carbon Nanomaterials as Adsorbents for Environmental and Biological Applications*, edited by CP Bergmann, and FM Machado. Cham, Switzerland: Springer International Publishing. Chapt. *3*: page no. 33–69.

Lin, G, C Liu, L Zhang, et al. 2017. High temperature dielectric properties of spent adsorbent with zinc sulfate by cavity perturbation technique. *Journal of Hazardous Materials* 330:36–45.

Lin, G, S Wang, L Zhang, et al. 2019. Enhanced and selective adsorption of Hg^{2+} to a trace level using trithiocyanuric acid-functionalized corn bract. *Environmental Pollution* 244:938–946.

Lin, K-YA, Y-T Liu, and S-Y Chen. 2016. Adsorption of fluoride to UiO-66-NH 2 in water: Stability, kinetic, isotherm and thermodynamic studies. *Journal of Colloid and Interface Science* 461:79–87.

Lin, L, and C Zou. 2017. Kinetic and thermodynamic study of magnetic separable β-cyclodextrin inclusion complex with organic phosphoric acid applied to removal of Hg^{2+}. *Journal of Chemical & Engineering Data* 62 (2):762–772.

Lin, Y, Q Fang, and B Chen. 2014. Perchlorate uptake and molecular mechanisms by magnesium/aluminum carbonate layered double hydroxides and the calcined layered double hydroxides. *Chemical Engineering Journal* 237:38–46.

Ling, L, and W-x Zhang. 2015. Enrichment and encapsulation of uranium with iron nanoparticle. *Journal of the American Chemical Society* 137 (8):2788–2791.

Lingamdinne, LP, JR Koduru, H Roh, Y-L Choi, Y-Y Chang, and Jae-Kyu Yang. 2016. Adsorption removal of Co(II) from waste-water using graphene oxide. *Hydrometallurgy* 165, Part 1:90–96.

Liu, B, W Mu, X Xie, et al. 2015. Enhancing the adsorption capacity of Sr^{2+} and Cs^{+} onto hexagonal tungsten oxide by doped niobium. *RSC Advances* 5 (20):15603–15611.

Liu, C-H, Y-H Chuang, T-Y Chen, et al. 2015. Mechanism of arsenic adsorption on magnetite nanoparticles from water: Thermodynamic and spectroscopic studies. *Environmental Science & Technology* 49 (13):7726–7734.

Liu, D, G Zha, L Hu, and W Jiang. 2018. Recovery of gallium and arsenic from gallium arsenide semiconductor scraps. Paper read at TMS Annual Meeting & Exhibition.

Liu, D, and H Zheng. 2017. Enhanced adsorption of radioactive strontium ions from aqueous solution by H_2O_2-modified attapulgite. *Journal of Radioanalytical and Nuclear Chemistry* 311 (3):1883–1890.

Liu, H, S Xie, T Wang, Y Liu, and T Zeng. 2017. Effect of coexisting cations on the adsorption of cesium onto poly (β-cyclodextrin)/bentonite composite. *Journal of Radioanalytical and Nuclear Chemistry* 312 (3):557–565.

Liu, J, B Li, G Wang, et al. 2020. Facile synthesis of flower-like $CoFe_2O_4$ particles for efficient sorption of aromatic organoarsenicals from aqueous solution. *Journal of Colloid and Interface Science* 568:63–75.

Liu, J, and X Du. 2011. Fast removal of aqueous Hg (II) with quaternary ammonium-functionalized magnetic mesoporous silica and silica regeneration. *Journal of Materials Chemistry* 21 (19):6981–6987.

Liu, K, B Zhu, Q Feng, et al. 2013. Adsorption of Cu(II) ions from aqueous solutions on modified chrysotile: Thermodynamic and kinetic studies. *Applied Clay Science* 80–81:38–45.

Liu, L, L Ding, X Wu, F Deng, R Kang, and X Luo. 2016. Enhancing the Hg (II) removal efficiency from real wastewater by novel thymine-grafted reduced graphene oxide complexes. *Industrial & Engineering Chemistry Research* 55 (24):6845–6853.

Liu, L, Z Cui, Q Ma, W Cui, and X Zhang. 2016. One-step synthesis of magnetic iron–aluminum oxide/graphene oxide nanoparticles as a selective adsorbent for fluoride removal from aqueous solution. *RSC Advances* 6 (13):10783–10791.

Liu, M, Y Wang, L Chen, Y Zhang, and Z Lin. 2015. $Mg(OH)_2$ supported nanoscale zero valent iron enhancing the removal of Pb(II) from aqueous solution. *ACS Applied Materials & Interfaces* 7 (15):7961–7969.

Liu, P, N Yuan, W Xiong, Hu Wu, D Pan, and W Wu. 2017. Removal of nickel (II) from aqueous solutions using synthesized β-zeolite and its ethylenediamine derivative. *Industrial & Engineering Chemistry Research* 56 (11):3067–3076.

Liu, Q, L Zhang, B Yang, and R Huang. 2015. Removal of fluoride from aqueous solution using Zr (IV) immobilized cross-linked chitosan. *International Journal of Biological Macromolecules* 77:15–23.

Liu, T, X Han, Y Wang, et al. 2017. Magnetic chitosan/anaerobic granular sludge composite: Synthesis, characterization and application in heavy metal ions removal. *Journal of Colloid and Interface Science* 508:405–414.

Liu, X, and L Zhang. 2015. Insight into the adsorption mechanisms of vanadium (V) on a high-efficiency biosorbent (Ti-doped chitosan bead). *International Journal of Biological Macromolecules* 79:110–117.

Liu, X, M Wang, S Zhang, and B Pan. 2013. Application potential of carbon nanotubes in water treatment: A review. *Journal of Environmental Sciences* 25 (7):1263–1280.

Liu, X, R Ma, X Wang, et al. 2019. Graphene oxide-based materials for efficient removal of heavy metal ions from aqueous solution: A review. *Environmental Pollution* 252:62–73.

Liu, Y, X Meng, M Luo, et al. 2015. Synthesis of hydrophilic surface ion-imprinted polymer based on graphene oxide for removal of strontium from aqueous solution. *Journal of Materials Chemistry A* 3 (3):1287–1297.

Liu, Y, X Meng, Z Liu, et al. 2015. Preparation of a two-dimensional ion-imprinted polymer based on a graphene oxide/SiO_2 composite for the selective adsorption of nickel ions. *Langmuir* 31 (32):8841–8851.

Liu, Y, Y Wang, S Zhou, et al. 2012. Synthesis of high saturation magnetization superparamagnetic Fe_3O_4 hollow microspheres for swift chromium removal. *ACS Applied Materials & Interfaces* 4 (9):4913–4920.

Long, H, P Wu, and N Zhu. 2013. Evaluation of Cs+ removal from aqueous solution by adsorption on ethylamine-modified montmorillonite. *Chemical Engineering Journal* 225:237–244.

Lopez, GEP, JF Madrid, and LV Abad. 2020. Chromium and cadmium adsorption on radiation-grafted polypropylene copolymers: Regeneration, kinetics, and continuous fixed bed column studies. *SN Applied Sciences* 2 (3):1–10.

Lu, C, and C Liu. 2006. Removal of nickel(II) from aqueous solution by carbon nanotubes. *Journal of Chemical Technology & Biotechnology* 81 (12):1932–1940.

Lu, C, and H Chiu. 2006. Adsorption of zinc(II) from water with purified carbon nanotubes. *Chemical Engineering Science* 61 (4):1138–1145.

Lu, J, F Fu, Z Ding, N Li, and B Tang. 2017. Removal mechanism of selenite by Fe_3O_4-precipitated mesoporous magnetic carbon microspheres. *Journal of Hazardous Materials* 330:93–104.

Lu, X, D Zhang, AT Reda, et al. 2017. Synthesis of amidoxime-grafted activated carbon fibers for efficient recovery of uranium (VI) from aqueous solution. *Industrial & Engineering Chemistry Research* 56 (41):11936–11947.

Lu, Z, J Yu, H Zeng, and Q Liu. 2017. Polyamine-modified magnetic graphene oxide nanocomposite for enhanced selenium removal. *Separation and Purification Technology* 183:249–257.

Lujanienė, G, S Šemčuk, A Lečinskytė, et al. 2017. Magnetic graphene oxide based nanocomposites for removal of radionuclides and metals from contaminated solutions. *Journal of Environmental Radioactivity* 166:166–174.

Luo, F, JL Chen, LL Dang, et al. 2015. High-performance Hg^{2+} removal from ultra-low-concentration aqueous solution using both acylamide- and hydroxyl-functionalized metal–organic framework. *Journal of Materials Chemistry A* 3 (18):9616–9620.

Luo, W, T Hirajima, and K Sasaki. 2016. Optimization of hexadecylpyridinium-modified montmorillonite for removal of perchlorate based on adsorption mechanisms. *Applied Clay Science* 123:29–36.

Luo, X, L Ding, and J Luo. 2015. Adsorptive removal of Pb(II) ions from aqueous samples with amino-functionalization of metal–organic frameworks MIL-101(Cr). *Journal of Chemical & Engineering Data* 60 (6):1732–1743.

Lv, L, N Chen, C Feng, J Zhang, and M Li. 2017. Heavy metal ions removal from aqueous solution by xanthate-modified cross-linked magnetic chitosan/poly(vinyl alcohol) particles. *RSC Advances* 7 (45):27992–28000.

Lv, X, X Qin, K Wang, Y Peng, P Wang, and G Jiang. 2019. Nanoscale zero valent iron supported on MgAl-LDH-decorated reduced graphene oxide: Enhanced performance in Cr (VI) removal, mechanism and regeneration. *Journal of Hazardous Materials* 373:176–186.

Ma, A, F Ke, Jing J, et al. 2017. Two lanthanide-based metal–organic frameworks for highly efficient adsorption and removal of fluoride ions from water. *CrystEngComm* 19 (16):2172–2177.

Ma, B, S Oh, WS Shin, and S-J Choi. 2011. Removal of Co^{2+}, Sr^{2+} and Cs^{+} from aqueous solution by phosphate-modified montmorillonite (PMM). *Desalination* 276 (1):336–346.

Ma, H, BS Hsiao, and B Chu. 2012. Ultrafine cellulose nanofibers as efficient adsorbents for removal of UO^{22+} in water. *ACS Macro Letters* 1 (1):213–216.

Ma, J, Z Wang, Y Shi, and Q Li. 2014. Synthesis and characterization of lysine-modified SBA-15 and its selective adsorption of scandium from a solution of rare earth elements. *RSC Advances* 4 (78):41597–41604.

Ma, L, J Zhu, Y Xi, et al. 2015. Simultaneous adsorption of Cd (II) and phosphate on Al 13 pillared montmorillonite. *RSC Advances* 5 (94):77227–77234.

Ma, S, L Huang, L Ma, et al. 2015. Efficient uranium capture by polysulfide/layered double hydroxide composites. *Journal of the American Chemical Society* 137 (10):3670–3677.

Maccotta, A, C Cosentino, R Coccioni, F Frontalini, G Scopelliti, and A Caruso. 2016. Distribution of Cr and Pb in artificial sea water and their sorption in marine sediments: An example from experimental mesocosms. *Environmental Science and Pollution Research* 23 (23):24068–24080.

Madadrang, CJ, HY Kim, G Gao, et al. 2012. Adsorption behavior of EDTA-graphene oxide for Pb (II) removal. *ACS Applied Materials & Interfaces* 4 (3):1186–1193.

Magnet, C, C Lomenech, C Hurel, et al. 2017. Adsorption of nickel ions by oleate-modified magnetic iron oxide nanoparticles. *Environmental Science and Pollution Research*:1–13.

Mahapatra, A, BG Mishra, and G Hota. 2013. Studies on electrospun alumina nanofibers for the removal of chromium (VI) and fluoride toxic ions from an aqueous system. *Industrial & Engineering Chemistry Research* 52 (4):1554–1561.

Mahmoud, ME, AA Yakout, MI Elserw, and MM Osman. 2016. Adsorption behavior of solvent-free microwave assisted nanosilica-functionalized carboxylic acids for the removal of cobalt (II) from water. *CLEAN–Soil, Air, Water* 44 (8):1011–1022.

Mahmoud, ME, AEH Abdou, and SB Ahmed. 2016. Conversion of waste styrofoam into engineered adsorbents for efficient removal of cadmium, lead and mercury from water. *ACS Sustainable Chemistry & Engineering* 4 (3):819–827.

Majumder, A, L Ramrakhiani, D Mukherjee, et al. 2019. Green synthesis of iron oxide nanoparticles for arsenic remediation in water and sludge utilization. *Clean Technologies and Environmental Policy* 21 (4):795–813.

Maleki, A, B Hayati, M Naghizadeh, and SW Joo. 2015. Adsorption of hexavalent chromium by metal organic frameworks from aqueous solution. *Journal of Industrial and Engineering Chemistry* 28:211–216.

Maleki, A, U Hamesadeghi, H Daraei, et al. 2017. Amine functionalized multi-walled carbon nanotubes: Single and binary systems for high capacity dye removal. *Chemical Engineering Journal* 313:826–835.

Malekpour, A, and M Khodadadi. 2016. Albumin-functionalized magnetic nanoparticles as an efficient sorbent for removal of Pb(ii), Cd(ii), Cu(ii) and Cr(vi) ions from aqueous solutions. *RSC Advances* 6 (18):14705–14711.

Mallakpour, S, and M Madani. 2016. Use of valine amino acid functionalized α-MnO_2/chitosan bionanocomposites as potential sorbents for the removal of lead (II) ions from aqueous solution. *Industrial & Engineering Chemistry Research* 55 (30):8349–8356.

Mandal, S, MK Sahu, and RK Patel. 2013. Adsorption studies of arsenic(III) removal from water by zirconium polyacrylamide hybrid material (ZrPACM-43). *Water Resources and Industry* 4:51–67.

Mangaleshwaran, L, A Thirulogachandar, V Rajasekar, C Muthukumaran, and K Rasappan. 2015. Batch and fixed bed column studies on nickel (II) adsorption from aqueous solution by treated polyurethane foam. *Journal of the Taiwan Institute of Chemical Engineers* 55:112–118.

Mangrich, AS, EMC Cardoso, ME Doumer, et al. 2015. Improving the Water Holding Capacity of Soils of Northeast Brazil by Biochar Augmentation. In *Water Challenges and Solutions on a Global Scale*, edited by Satinder Ahuja. Washington(USA): ACS Publications: page no. 339–354.

Manohar, DM, BF Noeline, and TS Anirudhan. 2005. Removal of vanadium (IV) from aqueous solutions by adsorption process with aluminum-pillared bentonite. *Industrial & Engineering Chemistry Research* 44 (17):6676–6684.

Manyangadze, M, NHM Chikuruwo, TB Narsaiah, CS Chakra, M Radhakumari, and G Danha. 2020. Enhancing adsorption capacity of nano-adsorbents via surface modification: A review. *South African Journal of Chemical Engineering* 31:25–32.

Manzoor, Kaiser, Mudasir Ahmad, Suhail Ahmad, and Saiqa Ikram. 2019. Synthesis, characterization, kinetics, and thermodynamics of EDTA-modified chitosan-carboxymethyl cellulose as Cu (II) ion adsorbent. *ACS omega* 4 (17):17425–17437.

Marco-Lozar, JP, D Cazorla-Amorós, and A Linares-Solano. 2007. A new strategy for germanium adsorption on activated carbon by complex formation. *Carbon* 45 (13):2519–2528.

Margeta, K, NZ Logar, M Šiljeg, and A Farkaš. 2013. Natural Zeolites in Water Treatment–How Effective is Their Use. In *Water Treatment*, edited by W Elshorbagy, and R Chowdhury. London (UK): IntechOpen. Chapt. 5: page no. 81–112.

Martínez, C, and A Corma. 2011. Inorganic molecular sieves: Preparation, modification and industrial application in catalytic processes. *Coordination Chemistry Reviews* 255 (13):1558–1580.

Martinson, CA, and KJ Reddy. 2009. Adsorption of arsenic(III) and arsenic(V) by cupric oxide nanoparticles. *Journal of Colloid and Interface Science* 336 (2):406–411.

Matheickal, JT, Q Yu, and GM Woodburn. 1999. Biosorption of cadmium(II) from aqueous solutions by pre-treated biomass of marine alga Durvillaea potatorum. *Water Research* 33 (2):335–342.

McCarthy, SA, G-L Davies, and YK Gun'ko. 2012. Preparation of multifunctional nanoparticles and their assemblies. *Nature Protocols* 7 (9):1677–1693.

Meitei, MD, and MNV Prasad. 2014. Adsorption of Cu (II), Mn (II) and Zn (II) by Spirodela polyrhiza (L.) Schleiden: Equilibrium, kinetic and thermodynamic studies. *Ecological Engineering* 71:308–317.

Meng, J, F Rao, S Changmei, Q Rongjun, and Z Ying. 2017. Silica gel-based adsorbents prepared via homogeneous and heterogeneous routes: Adsorption properties and recycling as heterogeneous catalysts. *Polymer International* 66 (12):1913–1920.

Mihara, Y, MT Sikder, H Yamagishi, et al. 2016. Adsorption kinetic model of alginate gel beads synthesized micro particle-prussian blue to remove cesium ions from water. *Journal of Water Process Engineering* 10:9–19.

Millsap, WA, and N Reisler. 1978. Cotter's new plant diets on spent catalysts. *Engineering and Mining Journal* 179 (5):105–107.

Min, L-L, Z-H Yuan, L-B Zhong, Q Liu, R-X Wu, and Y-M Zheng. 2015. Preparation of chitosan based electrospun nanofiber membrane and its adsorptive removal of arsenate from aqueous solution. *Chemical Engineering Journal* 267:132–141.

Mirza, A, and R Ahmad. 2018. Novel recyclable (Xanthan gum/montmorillonite) bionanocomposite for the removal of Pb (II) from synthetic and industrial wastewater. *Environmental Technology & Innovation* 11:241–252.

Mohamed, A, WS Nasser, TA Osman, MS Toprak, M Muhammed, and A Uheida. 2017. Removal of chromium (VI) from aqueous solutions using surface modified composite nanofibers. *Journal of Colloid and Interface Science* 505:682–691.

Mohammadi, M, A Ghaemi, M Torab-Mostaedi, M Asadollahzadeh, and A Hemmati. 2015. Adsorption of cadmium (II) and nickel (II) on dolomite powder. *Desalination and Water Treatment* 53 (1):149–157.

Mohammadi, SZ, H Hamidian, and Z Moeinadini. 2014. High surface area-activated carbon from Glycyrrhiza glabra residue by $ZnCl_2$ activation for removal of Pb(II) and Ni(II) from water samples. *Journal of Industrial and Engineering Chemistry* 20 (6):4112–4118.

Mohan, D, and CU Pittman. 2007. Arsenic removal from water/wastewater using adsorbents—A critical review. *Journal of Hazardous Materials* 142 (1):1–53.

Mohan, D, R Sharma, VK Singh, P Steele, and CU Pittman Jr. 2012. Fluoride removal from water using bio-char, a green waste, low-cost adsorbent: Equilibrium uptake and sorption dynamics modeling. *Industrial & Engineering Chemistry Research* 51 (2):900–914.

Mondal, MK, and R Garg. 2017. A comprehensive review on removal of arsenic using activated carbon prepared from easily available waste materials. *Environmental Science and Pollution Research* 24 (15):13295–13306.

Monier, M, and DA Abdel-Latif. 2012. Preparation of cross-linked magnetic chitosan-phenylthiourea resin for adsorption of Hg (II), Cd (II) and Zn (II) ions from aqueous solutions. *Journal of Hazardous Materials* 209:240–249.

Mor, S, K Ravindra, and NR Bishnoi. 2007. Adsorption of chromium from aqueous solution by activated alumina and activated charcoal. *Bioresource Technology* 98 (4):954–957.

Moradi, A, PN Moghadam, R Hasanzadeh, and M Sillanpaa. 2017. Chelating magnetic nanocomposite for the rapid removal of Pb(ii) ions from aqueous solutions: Characterization, kinetic, isotherm and thermodynamic studies. *RSC Advances* 7 (1):433–448.

Moreno-Barbosa, JJ, C López-Velandia, AdP Maldonado, L Giraldo, and JC Moreno-Piraján. 2013. Removal of lead(II) and zinc(II) ions from aqueous solutions by adsorption onto activated carbon synthesized from watermelon shell and walnut shell. *Adsorption* 19 (2):675–685.

Moret, A, and J Rubio. 2003. Sulphate and molybdate ions uptake by chitin-based shrimp shells. *Minerals Engineering* 16 (8):715–722.

Mu, W, Q Yu, X Li, H Wei, and Y Jian. 2017. Efficient removal of Cs^+ and Sr^{2+} from aqueous solution using hierarchically structured hexagonal tungsten trioxide coated Fe_3O_4. *Chemical Engineering Journal* 319:170–178.

Mubarak, NM, JN Sahu, EC Abdullah, and NS Jayakumar. 2014. Removal of heavy metals from wastewater using carbon nanotubes. *Separation & Purification Reviews* 43 (4):311–338.

Mukherjee, S, and G Halder. 2018. A review on the sorptive elimination of fluoride from contaminated wastewater. *Journal of Environmental Chemical Engineering* 6 (1):1257–1270.

Mukhopadhyay, K, A Ghosh, SK Das, B Show, P Sasikumar, and UC Ghosh. 2017. Synthesis and characterisation of cerium (iv)-incorporated hydrous iron (iii) oxide as an adsorbent for fluoride removal from water. *RSC Advances* 7 (42):26037–26051.

Naeem, A, P Westerhoff, and S Mustafa. 2007. Vanadium removal by metal (hydr)oxide adsorbents. *Water Research* 41 (7):1596–1602.

Naeimi, S, and H Faghihian. 2019. Application of novel adsorbent prepared by mucor hiemalis biomass impregnated with calcium alginate for removal of Sr^{2+} from aqueous solutions. *Journal of Polymers and the Environment* 27 (7):1572–1583.

Naghizadeh, A, F Momeni, and E Derakhshani. 2017. Efficiency of ultrasonic process in regeneration of graphene nanoparticles saturated with humic acid. *Desalination and Water Treatment* 70 (2017):290–293.

Najafi, F, O Moradi, M Rajabi, et al. 2015. Thermodynamics of the adsorption of nickel ions from aqueous phase using graphene oxide and glycine functionalized graphene oxide. *Journal of Molecular Liquids* 208:106–113.

Namasivayam, C, and D Sangeetha. 2006. Removal and recovery of vanadium (V) by adsorption onto $ZnCl_2$ activated carbon: Kinetics and isotherms. *Adsorption* 12 (2):103–117.

Namasivayam, C, and D Sangeetha. 2008. Application of coconut coir pith for the removal of sulfate and other anions from water. *Desalination* 219 (1–3):1–13.

Nashine, AL, and AR Tembhurkar. 2016. Equilibrium, kinetic and thermodynamic studies for adsorption of As(III) on coconut (Cocos nucifera L.) fiber. *Journal of Environmental Chemical Engineering* 4 (3):3267–3273.

Naushad, M, MA Khan, ZA ALOthman, and MR Khan. 2014. Adsorptive removal of nitrate from synthetic and commercially available bottled water samples using De-Acidite FF-IP resin. *Journal of Industrial and Engineering Chemistry* 20 (5):3400–3407.

Ndayambaje, G, K Laatikainen, M Laatikainen, et al. 2016. Adsorption of nickel(II) on polyacrylonitrile nanofiber modified with 2-(2′-pyridyl)imidazole. *Chemical Engineering Journal* 284:1106–1116.

Negm, NA, R El Sheikh, AF El-Farargy, HHH Hefni, and M Bekhit. 2015. Treatment of industrial wastewater containing copper and cobalt ions using modified chitosan. *Journal of Industrial and Engineering Chemistry* 21:526–534.

Ng, JC, JP Wang, B Zheng, et al. 2005. Urinary porphyrins as biomarkers for arsenic exposure among susceptible populations in Guizhou province, China. *Toxicology and Applied Pharmacology* 206 (2):176–184.

Ngah, WSW, and S Fatinathan. 2010. Adsorption characterization of Pb(II) and Cu(II) ions onto chitosan-tripolyphosphate beads: Kinetic, equilibrium and thermodynamic studies. *Journal of Environmental Management* 91 (4):958–969.

Nguyen, TC, P Loganathan, TV Nguyen, S Vigneswaran, J Kandasamy, and R Naidu. 2015. Simultaneous adsorption of Cd, Cr, Cu, Pb, and Zn by an iron-coated Australian zeolite in batch and fixed-bed column studies. *Chemical Engineering Journal* 270:393–404.

Nordstrom, DK. 2002. Worldwide occurrences of arsenic in ground water. *Science* 296 (5576):2143–2145.

Nouri, L, I Ghodbane, O Hamdaoui, and M Chiha. 2007. Batch sorption dynamics and equilibrium for the removal of cadmium ions from aqueous phase using wheat bran. *Journal of Hazardous Materials* 149 (1):115–125.

Nunell, GV, ME Fernandez, PR Bonelli, and AL Cukierman. 2015. Nitrate uptake from water by means of tailored adsorbents. *Water, Air, & Soil Pollution* 226 (8):278.

Okoro, HK, SO Ayika, JC Ngila, and AC Tella. 2018. Rising profile on the use of metal–organic frameworks (MOFs) for the removal of heavy metals from the environment: An overview. *Applied Water Science* 8 (6):169.

Omorogie, MO, JO Babalola, and EI Unuabonah. 2016. Regeneration strategies for spent solid matrices used in adsorption of organic pollutants from surface water: A critical review. *Desalination and water Treatment* 57 (2):518–544.

Opoku, F, EM Kiarii, PP Govender, and MA Mamo. 2017. Metal Oxide Polymer Nanocomposites in Water Treatments. In *Descriptive Inorganic Chemistry Researches of Metal Compounds*, edited by Takashiro Akitsu. London (UK): IntechOpen.Chapt. *8*: page no. 173–199.

Özlem Kocabaş-Ataklı, Z, and Y Yürüm. 2013. Synthesis and characterization of anatase nanoadsorbent and application in removal of lead, copper and arsenic from water. *Chemical Engineering Journal* 225:625–635.

Padilla-Rodríguez, A, JA Hernández-Viezcas, JR Peralta-Videa, JL Gardea-Torresdey, O Perales-Pérez, and FR Román-Velázquez. 2015. Synthesis of protonated chitosan flakes for the removal of vanadium (III, IV and V) oxyanions from aqueous solutions. *Microchemical Journal* 118:1–11.

Pakulski, D, W Czepa, S Witomska, et al. 2018. Graphene oxide-branched polyethylenimine foams for efficient removal of toxic cations from water. *Journal of Materials Chemistry A* 6 (20):9384–9390.

Pal, P, and A Pal. 2017. Surfactant-modified chitosan beads for cadmium ion adsorption. *International Journal of Biological Macromolecules* 104:1548–1555.

Pan, D, Q Fan, F Fan, Y Tang, Y Zhang, and W Wu. 2017. Removal of uranium contaminant from aqueous solution by chitosan@ attapulgite composite. *Separation and Purification Technology* 177:86–93.

Panneerselvam, P, N Morad, and KA Tan. 2011. Magnetic nanoparticle (Fe_3O_4) impregnated onto tea waste for the removal of nickel(II) from aqueous solution. *Journal of Hazardous Materials* 186 (1):160–168.

Parashar, K, N Ballav, S Debnath, K Pillay, and A Maity. 2016. Rapid and efficient removal of fluoride ions from aqueous solution using a polypyrrole coated hydrous tin oxide nanocomposite. *Journal of Colloid and Interface Science* 476:103–118.

Patel, H. 2019. Fixed-bed column adsorption study: A comprehensive review. *Applied Water Science* 9 (3):45.

Patil, DS, SM Chavan, and JUK Oubagaranadin. 2016. A review of technologies for manganese removal from wastewaters. *Journal of Environmental Chemical Engineering* 4 (1):468–487.

Patra, AK, and D Kim. 2017. Smart design of self-assembled mesoporous α-FeOOH nanoparticles: High-surface-area sorbent for Hg^{2+} from wastewater. *ACS Sustainable Chemistry & Engineering* 5 (2):1272–1279.

Paudyal, H, K Ohto, H Kawakita, and K Inoue. 2020. Recovery of fluoride from water through adsorption using orange-waste gel, followed by desorption using saturated lime water. *Journal of Material Cycles and Waste Management* 22:1484–1491.

Peacock, CL, and DM Sherman. 2004. Vanadium(V) adsorption onto goethite (α-FeOOH) at pH 1.5 to 12: A surface complexation model based on ab initio molecular geometries and EXAFS spectroscopy. *Geochimica et Cosmochimica Acta* 68 (8):1723–1733.

Peng, Y, H Huang, Y Zhang, et al. 2018. A versatile MOF-based trap for heavy metal ion capture and dispersion. *Nature Communications* 9 (1):187.

Perreault, F, AF de Faria, and M Elimelech. 2015. Environmental applications of graphene-based nanomaterials. *Chemical Society reviews* 44 (16):5861–5896.

Piccin, JS, TRSA Cadaval, LAA de Pinto, and GL Dotto. 2017. Adsorption isotherms in liquid phase: Experimental, Modeling, and Interpretations. In *Adsorption Processes for Water Treatment and Purification*, edited by A Bonilla-Petriciolet, DI Mendoza-Castillo, and HE Reynel-Ávila. Cham, Switzerland: Springer International Publishing. Chapt. 2: page no. 19–51.

Pizarro, J, X Castillo, S Jara, et al. 2015. Adsorption of Cu^{2+} on coal fly ash modified with functionalized mesoporous silica. *Fuel* 156:96–102.

Pourbeyram, S. 2016. Effective removal of heavy metals from aqueous solutions by graphene oxide–zirconium phosphate (GO–Zr-P) nanocomposite. *Industrial & Engineering Chemistry Research* 55 (19):5608–5617.

Pourhossein, F, and SM Mousavi. 2018. Enhancement of copper, nickel, and gallium recovery from LED waste by adaptation of Acidithiobacillus ferrooxidans. *Waste Management* 79:98–108.

Prabhu, SM, and S Meenakshi. 2014. Synthesis of surface coated hydroxyapatite powders for fluoride removal from aqueous solution. *Powder Technology* 268:306–315.

Qi, J, G Zhang, and H Li. 2015. Efficient removal of arsenic from water using a granular adsorbent: Fe–Mn binary oxide impregnated chitosan bead. *Bioresource Technology* 193:243–249.

Qi, X, N Li, Q Xu, D Chen, H Li, and J Lu. 2014. Water-soluble Fe_3O_4 superparamagnetic nanocomposites for the removal of low concentration mercury (II) ions from water. *Rsc Advances* 4 (88):47643–47648.

Qin, H, J Li, H Yang, B Pan, W Zhang, and X Guan. 2017. Coupled effect of ferrous ion and oxygen on the electron selectivity of zerovalent iron for selenate sequestration. *Environmental Science & Technology* 51 (9):5090–5097.

Qing, Y, J Li, B Kang, et al. 2015. Selective sorption mechanism of Cs^+ on potassium nickel hexacyanoferrate (II) compounds. *Journal of Radioanalytical and Nuclear Chemistry* 304 (2):527–533.

Qiu, B, C Xu, D Sun, et al. 2014. Polyaniline coated ethyl cellulose with improved hexavalent chromium removal. *ACS Sustainable Chemistry & Engineering* 2 (8):2070–2080.

Rabiul Awual, M, MM Hasan, T Ihara, and T Yaita. 2014. Mesoporous silica based novel conjugate adsorbent for efficient selenium(IV) detection and removal from water. *Microporous and Mesoporous Materials* 197:331–338.

Rajput, S, LP Singh, CU Pittman Jr, and D Mohan. 2017. Lead (Pb^{2+}) and copper (Cu^{2+}) remediation from water using superparamagnetic maghemite (γ-Fe_2O_3) nanoparticles synthesized by Flame Spray Pyrolysis (FSP). *Journal of Colloid and Interface Science* 492:176–190.

Ramadoss, R, and D Subramaniam. 2019. Removal of divalent nickel from aqueous solution using blue-green marine algae: Adsorption modeling and applicability of various isotherm models. *Separation Science and Technology* 54 (6):943–961.

Ramasamy, DL, V Puhakka, E Repo, S Khan, and M Sillanpää. 2017. Coordination and silica surface chemistry of lanthanides (III), scandium (III) and yttrium (III) sorption on 1-(2-pyridylazo)-2-napththol (PAN) and acetylacetone (acac) immobilized gels. *Chemical Engineering Journal* 324:104–112.

Ramos, SNdC, ALP Xavier, FS Teodoro, LF Gil, and LVA Gurgel. 2016. Removal of cobalt(II), copper(II), and nickel(II) ions from aqueous solutions using phthalate-functionalized sugarcane bagasse: Mono- and multicomponent adsorption in batch mode. *Industrial Crops and Products* 79:116–130.

Rashid, A, HN Bhatti, M Iqbal, and S Noreen. 2016. Fungal biomass composite with bentonite efficiency for nickel and zinc adsorption: A mechanistic study. *Ecological Engineering* 91:459–471.

Rathore, E, P Pal, and K Biswas. 2017. Layered metal chalcophosphate (K-MPS-1) for efficient, selective, and ppb level sequestration of Pb from water. *The Journal of Physical Chemistry C* 121 (14):7959–7966.

Rathore, VK, DK Dohare, and P Mondal. 2016. Competitive adsorption between arsenic and fluoride from binary mixture on chemically treated laterite. *Journal of Environmental Chemical Engineering* 4 (2):2417–2430.

Rathore, VK, and P Mondal. 2017. Competitive adsorption of arsenic and fluoride onto economically prepared aluminum oxide/hydroxide nanoparticles: Multicomponent isotherms and spent adsorbent management. *Industrial & Engineering Chemistry Research* 56 (28):8081–8094.

Ravi, S, and M Selvaraj. 2014. Incessant formation of chain-like mesoporous silica with a superior binding capacity for mercury. *Dalton Transactions* 43 (14):5299–5308.

Ravi, S, M Selvaraj, H Park, H-H Chun, and C-S Ha. 2014. Novel hierarchically dispersed mesoporous silica spheres: Effective adsorbents for mercury from wastewater and a thermodynamic study. *New Journal of Chemistry* 38 (8):3899–3906.

Ray, PZ, and HJ Shipley. 2015. Inorganic nano-adsorbents for the removal of heavy metals and arsenic: A review. *RSC Advances* 5 (38):29885–29907.

Ray, SS, R Gusain, and N Kumar. 2020a. Adsorption in Context of Water Purification. In *Carbon Nanomaterial-Based Adsorbents for Water Purification: Fundamentals and Applications*, edited by SS Ray, R Gusain, and N Kumar. Amsterdam: Elsevier. Chapt. *4*: page no. 67–100.

Ray, SS, R Gusain, and N Kumar. 2020b. *Carbon Nanomaterial-Based Adsorbents for Water Purification: Fundamentals and Applications*. Amsterdam (Netherland): Elsevier.

Reddy, DHK, K Vijayaraghavan, JA Kim, and Y-S Yun. 2017. Valorisation of post-sorption materials: Opportunities, strategies, and challenges. *Advances in Colloid and Interface Science* 242:35–58.

Reddy, DHK, and S-M Lee. 2013. Three-dimensional porous spinel ferrite as an adsorbent for Pb (II) removal from aqueous solutions. *Industrial & Engineering Chemistry Research* 52 (45):15789–15800.

Reinsch, H. 2016. "Green" synthesis of metal-organic frameworks. *European Journal of Inorganic Chemistry* 2016 (27):4290–4299.

Ren, X, C Chen, M Nagatsu, and X Wang. 2011. Carbon nanotubes as adsorbents in environmental pollution management: A review. *Chemical Engineering Journal* 170 (2):395–410.

Ren, Z, G Zhang, and J Paul Chen. 2011. Adsorptive removal of arsenic from water by an iron–zirconium binary oxide adsorbent. *Journal of Colloid and Interface Science* 358 (1):230–237.

Renu, MA, and K Singh. 2016. Heavy metal removal from wastewater using various adsorbents: A review. *Journal of Water Reuse and Desalination* 7 (4):387–419.

Roshanfar, M, R Golmohammadzadeh, and F Rashchi. 2019. An environmentally friendly method for recovery of lithium and cobalt from spent lithium-ion batteries using gluconic and lactic acids. *Journal of Environmental Chemical Engineering* 7 (1):102794.

Roushani, M, Z Saedi, and YM Baghelani. 2017. Removal of cadmium ions from aqueous solutions using TMU-16-NH 2 metal organic framework. *Environmental Nanotechnology, Monitoring & Management* 7:89–96.

Roy, P, NK Mondal, and K Das. 2014. Modeling of the adsorptive removal of arsenic: A statistical approach. *Journal of Environmental Chemical Engineering* 2 (1):585–597.

RSC. 2020. *Arsenic*. Royal society of chemistry 2020a [cited 24 August 2020]. Available from http://www.rsc.org/periodic-table/element/33/arsenic.

RSC. 2020. *Cadmium*. Royal society of chemistry 2020b [cited 25 August 2020]. Available from http://www.rsc.org/periodic-table/element/48/cadmium.

RSC. 2020. *Caesium*. Royal society of chemistry 2020c [cited 24 August 2020]. Available from http://www.rsc.org/periodic-table/element/55/caesium.

RSC. 2020. *Copper*. Royal society of chemistry 2020d [cited 24 August 2020]. Available from http://www.rsc.org/periodic-table/element/29/copper.

RSC. 2020. *Gallium*. Royal society of chemistry 2020e [cited 24 August 2020]. Available from http://www.rsc.org/periodic-table/element/31/Gallium.

RSC. 2020. *Germanium*. RSC 2020f [cited 24 August 2020]. Available from http://www.rsc.org/periodic-table/element/32/Germanium.

RSC. 2020. *Nickel*. Royal society of chemistry 2020g [cited 25 August 2020]. Available from http://www.rsc.org/periodic-table/element/28/nickel.

RSC. 2020. *Scandium*. Royal society of chemistry 2020h [cited 24 August 2020]. Available from http://www.rsc.org/periodic-table/element/21/Scandium.

RSC. 2020. *Strontium*. Royal society of chemistry 2020i [cited 24 August 2020]. Available from http://www.rsc.org/periodic-table/element/38/strontium.

RSC. 2020. *Titanium*. Royal society of chemistry 2020j [cited 24 August 2020]. Available from http://www.rsc.org/periodic-table/element/22/titanium.

Saber-Samandari, S, S Saber-Samandari, H Joneidi-Yekta, and M Mohseni. 2017. Adsorption of anionic and cationic dyes from aqueous solution using gelatin-based magnetic nanocomposite beads comprising carboxylic acid functionalized carbon nanotube. *Chemical Engineering Journal* 308:1133–1144.

Sadegh, H, GAM Ali, VK Gupta, et al. 2017. The role of nanomaterials as effective adsorbents and their applications in wastewater treatment. *Journal of Nanostructure in Chemistry* 7 (1):1–14.

Saha, D, S Barakat, SE Van Bramer, KA Nelson, DK Hensley, and J Chen. 2016. Noncompetitive and competitive adsorption of heavy metals in sulfur-functionalized ordered mesoporous carbon. *ACS Applied Materials & Interfaces* 8 (49):34132–34142.

Saikia, J, and RL Goswamee. 2019. Use of carbon coated ceramic barriers for adsorptive removal of fluoride and permanent immobilization of the spent adsorbent barriers. *SN Applied Sciences* 1 (6):634.

Saiz, J, E Bringas, and I Ortiz. 2014. New functionalized magnetic materials for As^{5+} removal: Adsorbent regeneration and reuse. *Industrial & Engineering Chemistry Research* 53 (49):18928–18934.

Sakulthaew, C, C Chokejaroenrat, A Poapolathep, T Satapanajaru, and S Poapolathep. 2017. Hexavalent chromium adsorption from aqueous solution using carbon nano-onions (CNOs). *Chemosphere* 184:1168–1174.

Salamun, N, S Triwahyono, AA Jalil, T Matsuura, and NFM Salleh. 2015. Acid-vacuo heat treated low cost banana stems fiber for efficient biosorption of Hg (II). *RSC Advances* 5 (19):14129–14137.

Saleem, M, N Wongsrisujarit, and S Boonyarattanakalin. 2016. Removal of nickel (II) ion by adsorption on coconut copra meal biosorbent. *Desalination and Water Treatment* 57 (12):5623–5635.

Saleh, AS, AG Ibrahim, F Abdelhai, EM Elsharma, E Metwally, and T Siyam. 2017. Preparation of poly (chitosan-acrylamide) flocculant using gamma radiation for adsorption of Cu (II) and Ni (II) ions. *Radiation Physics and Chemistry* 134:33–39.

Saleh, TA, M Tuzen, and A Sarı. 2017. Polyethylenimine modified activated carbon as novel magnetic adsorbent for the removal of uranium from aqueous solution. *Chemical Engineering Research and Design* 117:218–227.

Saman, N, K Johari, and H Mat. 2013. Adsorption characteristics of sulfur-functionalized silica microspheres with respect to the removal of Hg (II) from aqueous solutions. *Industrial & Engineering Chemistry Research* 53 (3):1225–1233.

Saman, N, K Johari, S-T Song, H Kong, S-C Cheu, and H Mat. 2017. High removal efficacy of Hg (II) and MeHg (II) ions from aqueous solution by organoalkoxysilane-grafted lignocellulosic waste biomass. *Chemosphere* 171:19–30.

San Miguel, G, SD Lambert, and NJD Graham. 2001. The regeneration of field-spent granular-activated carbons. *Water Research* 35 (11):2740–2748.

Sani, HA, MB Ahmad, MZ Hussein, NA Ibrahim, A Musa, and TA Saleh. 2017. Nanocomposite of ZnO with montmorillonite for removal of lead and copper ions from aqueous solutions. *Process Safety and Environmental Protection* 109:97–105.

Sankararamakrishnan, N, A Gupta, and SR Vidyarthi. 2014. Enhanced arsenic removal at neutral pH using functionalized multiwalled carbon nanotubes. *Journal of Environmental Chemical Engineering* 2 (2):802–810.

Santhosh, C, V Velmurugan, G Jacob, SK Jeong, AN Grace, and A Bhatnagar. 2016. Role of nanomaterials in water treatment applications: A review. *Chemical Engineering Journal* 306:1116–1137.

Sarkar, B, S Mandal, YF Tsang, P Kumar, K-H Kim, and YS Ok. 2018. Designer carbon nanotubes for contaminant removal in water and wastewater: A critical review. *Science of the Total Environment* 612:561–581.

Schedin, F, AK Geim, SV Morozov, et al. 2007. Detection of individual gas molecules adsorbed on graphene. *Nature Materials* 6 (9):652–655.

Schmidt, HP, T Bucheli, C Kammann, B Glaser, S Abiven, and J Leifeld. 2013. EBC (2012)'European biochar certificate-guidelines for a sustainable production of biochar. *European Biochar Foundation (EBC).*

Sen, TK, and D Gomez. 2011. Adsorption of zinc (Zn^{2+}) from aqueous solution on natural bentonite. *Desalination* 267 (2–3):286–294.

SenthilKumar, P, S Ramalingam, V Sathyaselvabala, S Dinesh Kirupha, and S Sivanesan. 2011. Removal of copper(II) ions from aqueous solution by adsorption using cashew nut shell. *Desalination* 266 (1–3):63–71.

Seyed Dorraji, MS, AR Amani-Ghadim, Y Hanifehpour, et al. 2017. Performance of chitosan based nanocomposite hollow fibers in the removal of selenium(IV) from water. *Chemical Engineering Research and Design* 117:309–317.

Shahat, A, MR Awual, and M Naushad. 2015. Functional ligand anchored nanomaterial based facial adsorbent for cobalt(II) detection and removal from water samples. *Chemical Engineering Journal* 271:155–163.

Shahnaz, T, V Sharma, S Subbiah, and S Narayanasamy. 2020. Multivariate optimisation of Cr (VI), Co (III) and Cu (II) adsorption onto nanobentonite incorporated nanocellulose/chitosan aerogel using response surface methodology. *Journal of Water Process Engineering* 36:101283.

Shao, D, G Hou, J Li, T Wen, X Ren, and X Wang. 2014. PANI/GO as a super adsorbent for the selective adsorption of uranium (VI). *Chemical Engineering Journal* 255:604–612.

Sharma, M, D Mondal, Nripat S, et al. 2017. Seaweed-derived nontoxic functionalized graphene sheets as sustainable materials for the efficient removal of fluoride from high fluoride containing drinking water. *ACS Sustainable Chemistry & Engineering* 5 (4):3488–3498.

Shaw, R, R Sharma, S Tiwari, and SK Tiwari. 2016. Surface engineered zeolite: An active interface for rapid adsorption and degradation of toxic contaminants in water. *ACS Applied Materials & Interfaces* 8 (19):12520–12527.

Sheela, T, Y Arthoba Nayaka, R Viswanatha, S Basavanna, and TG Venkatesha. 2012. Kinetics and thermodynamics studies on the adsorption of Zn (II), Cd (II) and Hg (II) from aqueous solution using zinc oxide nanoparticles. *Powder Technology* 217:163–170.

Sheha, RR, and EA El-Shazly. 2010. Kinetics and equilibrium modeling of Se(IV) removal from aqueous solutions using metal oxides. *Chemical Engineering Journal* 160 (1):63–71.

Shen, C, L Wu, Y Chen, et al. 2016. Efficient removal of fluoride from drinking water using well-dispersed monetite bundles inlaid in chitosan beads. *Chemical Engineering Journal* 303:391–400.

Shen, F, X Chen, P Gao, and G Chen. 2003. Electrochemical removal of fluoride ions from industrial wastewater. *Chemical Engineering Science* 58 (3):987–993.

Sheng, G, A Alsaedi, W Shammakh, et al. 2016. Enhanced sequestration of selenite in water by nanoscale zero valent iron immobilization on carbon nanotubes by a combined batch, XPS and XAFS investigation. *Carbon* 99:123–130.

Sheng, G, J Li, D Shao, et al. 2010. Adsorption of copper(II) on multiwalled carbon nanotubes in the absence and presence of humic or fulvic acids. *Journal of Hazardous Materials* 178 (1–3):333–340.

Sheng, G, W Linghu, Z Chen, et al. 2016. Sequestration of selenate and selenite onto titanate nanotube: A combined classical batch and advanced EXAFS approach. *Environmental Nanotechnology, Monitoring & Management* 6:152–158.

Sheng, G, X Shao, Y Li, et al. 2014. Enhanced removal of uranium (VI) by nanoscale zerovalent iron supported on Na–bentonite and an investigation of mechanism. *The Journal of Physical Chemistry A* 118 (16):2952–2958.

Shi, J, H Li, H Lu, and X Zhao. 2015. Use of carboxyl functional magnetite nanoparticles as potential sorbents for the removal of heavy metal ions from aqueous solution. *Journal of Chemical & Engineering Data* 60 (7):2035–2041.

Shokry, H, and H Hamad. 2016. Effect of superparamagnetic nanoparticles on the physicochemical properties of nano hydroxyapatite for groundwater treatment: Adsorption mechanism of Fe (II) and Mn (II). *RSC Advances* 6 (85):82244–82259.

Shu, R, Y Wang, and H Zhong. 2016. Biochar amendment reduced methylmercury accumulation in rice plants. *Journal of Hazardous Materials* 313:1–8.

Silva, AM, RMF Lima, and VA Leão. 2012. Mine water treatment with limestone for sulfate removal. *Journal of Hazardous Materials* 221:45–55.

Singh, D, RK Gautam, R Kumar, BK Shukla, V Shankar, and V Krishna. 2014. Citric acid coated magnetic nanoparticles: Synthesis, characterization and application in removal of Cd (II) ions from aqueous solution. *Journal of Water Process Engineering* 4:233–241.

Singh, J, T Dutta, K-H Kim, M Rawat, P Samddar, and P Kumar. 2018. 'Green' synthesis of metals and their oxide nanoparticles: Applications for environmental remediation. *Journal of Nanobiotechnology* 16 (1):84.

Singh, T, and R Singhal. 2018. Efficient and Economical Application of a Spent Waste Adsorbent Cu^{2+}-Loaded Poly (AAc-AM-SH) Superabsorbent Hydrogels by Reusing It for Adsorption of Phosphate Ion. In *Water Quality Management*, edited by Vijay P singh, Shalini Yadav, Ram N Yadava. Singapore: Springer: page no. 257–267.

Sitko, R, E Turek, Beata Z, et al. 2013. Adsorption of divalent metal ions from aqueous solutions using graphene oxide. *Dalton Transactions* 42 (16):5682–5689.

Sohn, J-S, S-M Shin, D-H Yang, S-K Kim, and C-K Lee. 2006. Comparison of two acidic leaching processes for selecting the effective recycle process of spent lithium ion battery. *Geosystem Engineering* 9 (1):1–6.

Soltermann, D, B Baeyens, MH Bradbury, and MM Fernandes. 2014. Fe (II) uptake on natural montmorillonites. II. Surface complexation modeling. *Environmental Science & Technology* 48 (15):8698–8705.

Song, B, P Xu, G Zeng, et al. 2018. Carbon nanotube-based environmental technologies: The adopted properties, primary mechanisms, and challenges. *Reviews in Environmental Science and Bio/Technology* 17 (3):571–590.

Song, H, Z Yao, C Shuang, and A Li. 2014. Accelerated removal of nitrate from aqueous solution by utilizing polyacrylic anion exchange resin with magnetic separation performance. *Journal of Industrial and Engineering Chemistry* 20 (5):2888–2894.

Song, HS, OS Kwon, J-H Kim, J Conde, and N Artzi. 2017. 3D hydrogel scaffold doped with 2D graphene materials for biosensors and bioelectronics. *Biosensors and Bioelectronics* 89:187–200.

Sreenu, B, K Imran, K Seshaiah, P Sharma, and AP Singh. 2016. Synthesis of new hybrid sorbent 2-mercaptobenzaldehyde SBA-15 and its application in solid phase extraction of Cd (II) from water and food samples. *Analytical Methods* 8 (14):2947–2954.

Srivastav, AL, PK Singh, CH Weng, and YC Sharma. 2014. Novel adsorbent hydrous bismuth oxide for the removal of nitrate from aqueous solutions. *Journal of Hazardous, Toxic, and Radioactive Waste* 19 (2):04014028.

Srivastava, V, CH Weng, VK Singh, and YC Sharma. 2011. Adsorption of nickel ions from aqueous solutions by nano alumina: Kinetic, mass transfer, and equilibrium studies. *Journal of Chemical & Engineering Data* 56 (4):1414–1422.

Staicu, LC, N Morin-Crini, and G Crini. 2017. Desulfurization: Critical step towards enhanced selenium removal from industrial effluents. *Chemosphere* 172:111–119.

Stankovich, S, DA Dikin, RD Piner, et al. 2007. Synthesis of graphene-based nanosheets via chemical reduction of exfoliated graphite oxide. *Carbon (New York)* 45 (7):1558–1565.

Stojakovic, D, J Milenkovic, S Stupar, Z Velickovic, and N Rajic. 2016. Binary adsorption of nickel and zinc from aqueous solutions onto the Serbian natural clinoptilolite. *Desalination and Water Treatment* 57 (40):18748–18754.

Sturgill, JA, JT Swartzbaugh, and PM Randall. 2000. Pollution prevention in the semiconductor industry through recovery and recycling of gallium and arsenic from GaAs polishing wastes. *Clean Products and Processes* 2 (1):18–27.

Subedi, R, N Taupe, I Ikoyi, et al. 2016. Chemically and biologically-mediated fertilizing value of manure-derived biochar. *Science of the Total Environment* 550:924–933.

Sullivan, C, M Tyrer, CR Cheeseman, and NJ Graham. 2010. Disposal of water treatment wastes containing arsenic—A review. *Science of the Total Environment* 408 (8):1770–1778.

Summers, RS, DRU Knappe, and VL Snoeyink. 2011. Adsorption of organic compounds by activated carbon. *Water Quality & Treatment: A Handbook On Drinking Water*:14.1–14.105.

Sun, Q, B Aguila, J Perman, et al. 2017. Postsynthetically modified covalent organic frameworks for efficient and effective mercury removal. *Journal of the American Chemical Society* 139 (7):2786–2793.

Sun, W, W Pan, F Wang, and N Xu. 2015. Removal of Se(IV) and Se(VI) by MFe_2O_4 nanoparticles from aqueous solution. *Chemical Engineering Journal* 273:353–362.

Sun, X, L Yang, Q Li, et al. 2014. Amino-functionalized magnetic cellulose nanocomposite as adsorbent for removal of Cr (VI): Synthesis and adsorption studies. *Chemical Engineering Journal* 241:175–183.

Sun, Y, Z-Y Wu, X Wang, et al. 2016. Macroscopic and microscopic investigation of U(VI) and Eu(III) adsorption on carbonaceous nanofibers. *Environmental Science & Technology* 50 (8):4459–4467.

Suriyaraj, SP, MM Pillai, A Bhattacharyya, and R Selvakumar. 2015. Scavenging of nitrate ions from water using hybrid Al_2O_3/bio-TiO_2 nanocomposite impregnated thermoplastic polyurethane nanofibrous membrane. *RSC Advances* 5 (84):68420–68429.

Svecova, L, M Dossot, S Cremel, et al. 2011. Sorption of selenium oxyanions on TiO_2 (rutile) studied by batch or column experiments and spectroscopic methods. *Journal of Hazardous Materials* 189 (3):764–772.

Swenson, H, and NP Stadie. 2019. Langmuir's theory of adsorption: A centennial review. *Langmuir* 35 (16):5409–5426.

Szczudło, Z, A Ciszewski, and YB Losovyj. 2001. Field electron emission study of Ti and Hf adsorption layers on W. *Applied Surface Science* 174 (2):138–147.

Taffarel, SR, and J Rubio. 2010. Removal of Mn^{2+} from aqueous solution by manganese oxide coated zeolite. *Minerals Engineering* 23 (14):1131–1138.

Takaya, CA, LA Fletcher, S Singh, UC Okwuosa, and AB Ross. 2016. Recovery of phosphate with chemically modified biochars. *Journal of Environmental Chemical Engineering* 4 (1):1156–1165.

Tan, KL, and BH Hameed. 2017. Insight into the adsorption kinetics models for the removal of contaminants from aqueous solutions. *Journal of the Taiwan Institute of Chemical Engineers* 74:25–48.

Tan, L, J Wang, Q Liu, et al. 2015. The synthesis of a manganese dioxide-iron oxide-graphene magnetic nanocomposite for enhanced uranium(vi) removal. *New Journal of Chemistry* 39 (2):868–876.

Tan, L, Q Liu, X Jing, et al. 2015. Removal of uranium (VI) ions from aqueous solution by magnetic cobalt ferrite/multiwalled carbon nanotubes composites. *Chemical Engineering Journal* 273:307–315.

Tan, L, S Wang, W Du, and T Hu. 2016. Effect of water chemistries on adsorption of Cs (I) onto graphene oxide investigated by batch and modeling techniques. *Chemical Engineering Journal* 292:92–97.

Tan, L, Y Wang, Q Liu, et al. 2015. Enhanced adsorption of uranium (VI) using a three-dimensional layered double hydroxide/graphene hybrid material. *Chemical Engineering Journal* 259:752–760.

Tan, Z. 2014. Separation of Particles from a Gas. In *Air Pollution and Greenhouse Gases: From Basic Concepts to Engineering Applications for Air Emission Control*, edited by Z Tan. Singapore: Springer. Chapt. 5: page no. 117–149.

Tang, C, YH Huang, H Zeng, and Z Zhang. 2014. Reductive removal of selenate by zero-valent iron: The roles of aqueous Fe^{2+} and corrosion products, and selenate removal mechanisms. *Water research* 67 166–174.

Tang, D, and G Zhang. 2016. Efficient removal of fluoride by hierarchical Ce–Fe bimetal oxides adsorbent: Thermodynamics, kinetics and mechanism. *Chemical Engineering Journal* 283:721–729.

Tang, J-q, J-b Xi, J-x Yu, R-a Chi, and J-d Chen. 2018. Novel combined method of biosorption and chemical precipitation for recovery of Pb^{2+} from wastewater. *Environmental Science and Pollution Research* 25 (28):28705–28712.

Tang, W, Y Su, Q Li, S Gao, and JK Shang. 2013. Superparamagnetic magnesium ferrite nanoadsorbent for effective arsenic (III, V) removal and easy magnetic separation. *Water Research* 47 (11):3624–3634.

Tesh, SJ, and TB Scott. 2014. Nano-composites for water remediation: A review. *Advanced Materials* 26 (35):6056–6068.

Thakur, S, G Das, PK Raul, and N Karak. 2013. Green one-step approach to prepare sulfur/reduced graphene oxide nanohybrid for effective mercury ions removal. *The Journal of Physical Chemistry C* 117 (15):7636–7642.

Tian, N, Z Zhou, X Tian, C Yang, and Y Li. 2017. Superior capability of $MgAl_2O_4$ for selenite removal from contaminated groundwater during its reconstruction of layered double hydroxides. *Separation and Purification Technology* 176:66–72.

Tien, C. 1994. Macroscopic Description of Adsorption Processes. In *Adsorption Calculations and Modelling*, edited by C Tien. Boston: Butterworth-Heinemann. Chapt. *6*: page no. 87–98.

Tien, C. 2018. Introduction. In *Introduction to Adsorption: Basics, Analysis, and Applications*, edited by C Tien. Amsterdam: Elsevier. Chapt. *1*: page no. 1–6.

Tirtom, VN, A Dinçer, S Becerik, T Aydemir, and A Çelik. 2012. Comparative adsorption of Ni (II) and Cd (II) ions on epichlorohydrin crosslinked chitosan–clay composite beads in aqueous solution. *Chemical Engineering Journal* 197:379–386.

Tofighy, MA, and T Mohammadi. 2011. Adsorption of divalent heavy metal ions from water using carbon nanotube sheets. *Journal of Hazardous Materials* 185 (1):140–147.

Tokar, EJ, WA Boyd, JH Freedman, and MP Waalkes. 2015. Toxic Effects of Metals. In *Toxicology the Basic Science of Poisons*, edited by C Klaassen. New York: McGraw Hill Professional. Chapt. *23*: page no. 981–1030.

Tong, Y, PJ McNamara, and BK Mayer. 2019. Adsorption of organic micropollutants onto biochar: A review of relevant kinetics, mechanisms and equilibrium. *Environmental Science: Water Research & Technology* 5 (5):821–838.

Treviño, S, and A Diaz. 2020. Vanadium and insulin: Partners in metabolic regulation. *Journal of Inorganic Biochemistry*:111094.

Turanov, AN, VK Karandashev, NS Sukhinina, VM Masalov, and GA Emelchenko. 2016. Adsorption of lanthanides and scandium ions by silica sol-gel material doped with novel bifunctional ionic liquid, trioctylmethylammonium 1-phenyl-3-methyl-4-benzoyl-5-onate. *Journal of Environmental Chemical Engineering* 4 (4):3788–3796.

Upadhyay, RK, N Soin, and SS Roy. 2014. Role of graphene/metal oxide composites as photocatalysts, adsorbents and disinfectants in water treatment: A review. *RSC Advances* 4 (8):3823–3851.

Upadhyayula, VKK, S Deng, MC Mitchell, and GB Smith. 2009. Application of carbon nanotube technology for removal of contaminants in drinking water: A review. *Science of the Total Environment* 408 (1):1–13.

USEPA. 2017. *Contaminant candidate list (CCL) and regulatory determination*. USEPA 2017 [cited 18/12/2017 2017]. Available from https://www.epa.gov/ccl/chemical-contaminants-ccl-4.

Usman, AR, M Ahmad, M El-Mahrouky, et al. 2016. Chemically modified biochar produced from conocarpus waste increases NO_3 removal from aqueous solutions. *Environmental Geochemistry and Health* 38 (2):511–521.

Vacca, K, J Komlos, and BM Wadzuk. 2016. Phosphorus removal in constructed stormwater wetland mesocosms amended with water treatment residuals. *Water Environment Research* 88 (9):898–906.

Vakili, M, S Deng, G Cagnetta, et al. 2019. Regeneration of chitosan-based adsorbents used in heavy metal adsorption: A review. *Separation and Purification Technology* 224:373–387.

Velazquez-Jimenez, LH, RH Hurt, J Matos, and JR Rangel-Mendez. 2014. Zirconium–carbon hybrid sorbent for removal of fluoride from water: Oxalic Acid mediated Zr(IV) assembly and adsorption mechanism. *Environmental Science & Technology* 48 (2):1166–1174.

Venkateswarlu, S, and M Yoon. 2015a. Core–shell ferromagnetic nanorod based on amine polymer composite (Fe_3O_4@DAPF) for fast removal of Pb(II) from aqueous solutions. *ACS Applied Materials & Interfaces* 7 (45):25362–25372.

Venkateswarlu, S, and M Yoon. 2015b. Rapid removal of cadmium ions using green-synthesized Fe_3O_4 nanoparticles capped with diethyl-4-(4 amino-5-mercapto-4 H-1, 2, 4-triazol-3-yl) phenyl phosphonate. *RSC Advances* 5 (80):65444–65453.

Verbinnen, B, C Block, J Van Caneghem, and C Vandecasteele. 2015. Recycling of spent adsorbents for oxyanions and heavy metal ions in the production of ceramics. *Waste Management* 45:407–411.

Verbinnen, B, C Block, P Lievens, A Van Brecht, and C Vandecasteele. 2013. Simultaneous removal of molybdenum, antimony and selenium oxyanions from wastewater by adsorption on supported magnetite. *Waste and Biomass Valorization* 4 (3):635–645.

Verma, A, R Kore, DR Corbin, and MB Shiflett. 2019. Metal recovery using oxalate chemistry: A technical review. *Industrial & Engineering Chemistry Research* 58 (34):15381–15393.

Verma, S, HP Mungse, N Kumar, et al. 2011. Graphene oxide: An efficient and reusable carbocatalyst for aza-Michael addition of amines to activated alkenes. *Chemical Communications* 47 (47):12673–12675.

Verma, S, and RK Dutta. 2015. A facile method of synthesizing ammonia modified graphene oxide for efficient removal of uranyl ions from aqueous medium. *RSC Advances* 5 (94):77192–77203.

Vipin, AK, S Ling, and B Fugetsu. 2016. Removal of Cs^+ and Sr^{2+} from water using MWCNT reinforced zeolite-A beads. *Microporous and Mesoporous Materials* 224:84–88.

Volli, V, and RK Singh. 2012. Production of bio-oil from de-oiled cakes by thermal pyrolysis. *Fuel* 96:579–585.

Vu, TA, GH Le, CD Dao, et al. 2015. Arsenic removal from aqueous solutions by adsorption using novel MIL-53 (Fe) as a highly efficient adsorbent. *Rsc Advances* 5 (7):5261–5268.

Wagner, TV, JR Parsons, HH Rijnaarts, P de Voogt, and AA Langenhoff. 2020. Benzotriazole removal mechanisms in pilot-scale constructed wetlands treating cooling tower water. *Journal of Hazardous Materials* 384:121314.

Wan, D, H Liu, R Liu, J Qu, S Li, and J Zhang. 2012. Adsorption of nitrate and nitrite from aqueous solution onto calcined (Mg–Al) hydrotalcite of different Mg/Al ratio. *Chemical Engineering Journal* 195:241–247.

Wang, C, J Liu, Z Zhang, B Wang, and H Sun. 2012. Adsorption of Cd (II), Ni (II), and Zn (II) by tourmaline at acidic conditions: Kinetics, thermodynamics, and mechanisms. *Industrial & Engineering Chemistry Research* 51 (11):4397–4406.

Wang, H, B Gao, J Fang, et al. 2018. Engineered biochar derived from eggshell-treated biomass for removal of aqueous lead. *Ecological Engineering* 121:124–129.

Wang, H, B Gao, S Wang, J Fang, Y Xue, and K Yang. 2015. Removal of Pb(II), Cu(II), and Cd(II) from aqueous solutions by biochar derived from $KMnO_4$ treated hickory wood. *Bioresource Technology* 197:356–362.

Wang, H, X Yuan, Y Wu, et al. 2013. Adsorption characteristics and behaviors of graphene oxide for Zn(II) removal from aqueous solution. *Applied Surface Science* 279:432–440.

Wang, J, C Cheng, X Yang, C Chen, and A Li. 2013. A new porous chelating fiber: Preparation, characterization, and adsorption behavior of Pb(II). *Industrial & Engineering Chemistry Research* 52 (11):4072–4082.

Wang, J, X Peng, Z Luan, and C Zhao. 2010. Regeneration of carbon nanotubes exhausted with dye reactive red 3BS using microwave irradiation. *Journal of Hazardous Materials* 178 (1):1125–1127.

Wang, L, C Shi, L Pan, X Zhang, and J-J Zou. 2020. Rational design, synthesis, adsorption principles and applications of metal oxide adsorbents: A review. *Nanoscale* 12 (8):4790–4815.

Wang, L, C Shi, L Wang, L Pan, X Zhang, and J-J Zou. 2020. Rational design, synthesis, adsorption principles and applications of metal oxide adsorbents: A review. *Nanoscale* 12 (8):4790–4815.

Wang, L, W Liu, T Wang, and J Ni. 2013. Highly efficient adsorption of Cr (VI) from aqueous solutions by amino-functionalized titanate nanotubes. *Chemical Engineering Journal* 225:153–163.

Wang, L, Y Li, and Y Liu. 2017. Reduced graphene oxide@TiO_2 nanorod@reduced graphene oxide hybrid nanostructures for photoelectrochemical hydrogen production. *Micro & Nano Letters* (7), https://digital-library.theiet.org/content/journals/10.1049/mnl.2016.0747.

Wang, LP, and YJ Chen. 2019. Sequential precipitation of iron, copper, and zinc from wastewater for metal recovery. *Journal of Environmental Engineering* 145 (1):04018130.

Wang, M, X Yu, C Yang, et al. 2017. Removal of fluoride from aqueous solution by Mg-Al-Zr triple-metal composite. *Chemical Engineering Journal* 322:246–253.

Wang, Q, J Li, C Chen, X Ren, J Hu, and X Wang. 2011. Removal of cobalt from aqueous solution by magnetic multiwalled carbon nanotube/iron oxide composites. *Chemical Engineering Journal* 174 (1):126–133.

Wang, S, and Y Peng. 2010. Natural zeolites as effective adsorbents in water and wastewater treatment. *Chemical Engineering Journal* 156 (1):11–24.

Wang, S-y, Y-k Tang, C Chen, et al. 2015. Regeneration of magnetic biochar derived from eucalyptus leaf residue for lead (II) removal. *Bioresource Technology* 186:360–364.

Wang, T, L Zhang, C Li, et al. 2015. Synthesis of core–shell magnetic Fe_3O_4@ poly (m-phenylenediamine) particles for chromium reduction and adsorption. *Environmental Science & Technology* 49 (9):5654–5662.

Wang, T, Z Cheng, B Wang, and W Ma. 2012. The influence of vanadate in calcined Mg/Al hydrotalcite synthesis on adsorption of vanadium (V) from aqueous solution. *Chemical Engineering Journal* 181:182–188.

Wang, W-S, Y-B Li, B-J Gao, et al. 2013. Effective removal of Fe (II) impurity from rare earth solution using surface imprinted polymer. *Chemical Engineering Research and Design* 91 (12):2759–2764.

Wang, X, P Lv, H Zou, Y Li, X Li, and Y Liao. 2016. Synthesis of poly (2-aminothiazole) for selective removal of Hg (II) in aqueous solutions. *Industrial & Engineering Chemistry Research* 55 (17):4911–4918.

Wang, Y-S, C-C Cheng, J-K Chen, F-H Ko, and F-C Chang. 2013. Bioinspired supramolecular fibers for mercury ion adsorption. *Journal of Materials Chemistry A* 1 (26):7745–7750.

Wang, Z, D Shen, F Shen, and T Li. 2016. Phosphate adsorption on lanthanum loaded biochar. *Chemosphere* 150:1–7.

Wang, Z, H Guo, F Shen, et al. 2015. Biochar produced from oak sawdust by Lanthanum (La)-involved pyrolysis for adsorption of ammonium (NH^{4+}), nitrate (NO^{3-}), and phosphate (PO_4^{3-}). *Chemosphere* 119:646–653.

Watts, P, and P Howe. 2010. *Strontium and Strontium Compounds*. Geneva (Switzerland): World Health Organization.

Wei, Y, X Li, R Zhang, et al. 2016. Periodic mesoporous organosilica nanocubes with ultrahigh surface areas for efficient CO_2 adsorption. *Scientific Reports* 6 (1):20769.

Wen, T, X Wu, M Liu, Z Xing, X Wang, and A-W Xu. 2014. Efficient capture of strontium from aqueous solutions using graphene oxide-hydroxyapatite nanocomposites. *Dalton Transactions* 43 (20):7464–7472.

Weng, X, J Wu, L Ma, G Owens, and Z Chen. 2019. Impact of synthesis conditions on Pb(II) removal efficiency from aqueous solution by green tea extract reduced graphene oxide. *Chemical Engineering Journal* 359:976–981.

West, JR. 1945. Some industrial aspects of adsorption. *Journal of Chemical Education* 22 (8):398.

WHO. 2017. WHO Guidelines Approved by the Guidelines Review Committee. In *Guidelines for Drinking-Water Quality: Fourth Edition Incorporating the First Addendum*. Geneva: World Health Organization Copyright (c) World Health Organization 2017: page no. 1–564.

Wilbur, S, H Abadin, M Fay, et al. 2012. Agency for Toxic Substances and Disease Registry (ATSDR) Toxicological Profiles. In *Toxicological Profile for Chromium*. Atlanta (GA): Agency for Toxic Substances and Disease Registry (US): page no. 1–502.

Woolf, D, JE Amonette, F Alayne Street-Perrott, J Lehmann, and S Joseph. 2010. Sustainable biochar to mitigate global climate change. *Nature Communications* 1 (1):1–9.

Worch, E. 2012a. Adsorbents and Adsorbent Characterization. In *Adsorption Technology in Water Treatment: Fundamentals, Processes, and Modeling*, edited by E Worch. Germany: Walter de Gruyter. Chapt. *2*: page no. 11–40.

Worch, E. 2012b. Adsorption Dynamics in Fixed-bed Adsorbers. In *Adsorption Technology in Water Treatment: Fundamentals, Processes, and Modeling*, edited by E Worch. Germany: Walter de Gruyter. Chapt. *6*: page no. 171(169–195).

Worch, E. 2012c. Adsorption Equilibrium I: General Aspects and Single-solute Adsorption. In *Adsorption Technology in Water Treatment: Fundamentals, Processes, and Modeling*, edited by E Worch. Germany: Walter de Gruyter. Chapt. *3*: page no. 68 (41–76).

Worch, E. 2012d. Adsorption Kinetics. In *Adsorption Technology in Water Treatment: Fundamentals, Processes, and modeling*, edited by E Worch. Germany: Walter de Gruyter. Chapt. *5*: page no. 127(123–168).

Worch, E. 2012e. Introduction. In *Adsorption Technology in Water Treatment: Fundamentals, Processes, and Modeling*, edited by E Worch. Berlin: Walter de Gruyter. Chapt. *1*: page no. 1–8.

Wosnitza, TMA, and JG Barrantes. 2006. Utilization of seaweed Ulva sp. in Paracas Bay (Peru): Experimenting with compost. *Journal of Applied Phycology* 18 (1):27.

Wu, P, J Wu, L Xia, Y Liu, L Xu, and S Song. 2017. Adsorption of fluoride at the interface of water with calcined magnesium–ferri–lanthanum hydrotalcite-like compound. *RSC Advances* 7 (42):26104–26112.

Wu, S, K Zhang, J He, et al. 2016. High efficient removal of fluoride from aqueous solution by a novel hydroxyl aluminum oxalate adsorbent. *Journal of Colloid and Interface Science* 464:238–245.

Wu, W, Y Yang, H Zhou, et al. 2012. Highly efficient removal of Cu(II) from aqueous solution by using graphene oxide. *Water, Air, & Soil Pollution* 224 (1):1372.

Wu, Y, H Luo, H Wang, C Wang, J Zhang, and Z Zhang. 2013. Adsorption of hexavalent chromium from aqueous solutions by graphene modified with cetyltrimethylammonium bromide. *Journal of Colloid and Interface Science* 394:183–191.

Wu, Y, S Zhang, X Guo, and H Huang. 2008. Adsorption of chromium(III) on lignin. *Bioresource Technology* 99 (16):7709–7715.

Xia, X, L Ling, and W-x Zhang. 2017. Genesis of pure Se(0) nano- and micro-structures in wastewater with nanoscale zero-valent iron (nZVI). *Environmental Science: Nano* 4 (1):52–59.

Xiao, W, B Yan, H Zeng, and Q Liu. 2016. Dendrimer functionalized graphene oxide for selenium removal. *Carbon* 105:655–664.

Xie, Z, W Xu, X Cui, and Y Wang. 2017. Recent progress in metal–organic frameworks and their derived nanostructures for energy and environmental applications. *ChemSusChem* 10 (8):1645–1663.

Xu, B, Y Zhai, Y Zhu, et al. 2016. The adsorption mechanisms of ClO^{4-} onto highly graphited and hydrophobic porous carbonaceous materials from biomass. *RSC Advances* 6 (96):93975–93984.

Xu, J-h, N-y Gao, D-y Zhao, et al. 2015. Comparative study of nano-iron hydroxide impregnated granular activated carbon (Fe–GAC) for bromate or perchlorate removal. *Separation and Purification Technology* 147:9–16.

Xu, J-h, N-y Gao, Y Deng, and S-q Xia. 2013. Nanoscale iron hydroxide-doped granular activated carbon (Fe-GAC) as a sorbent for perchlorate in water. *Chemical Engineering Journal* 222:520–526.

Xu, M, and G McKay. 2017. Removal of Heavy Metals, Lead, Cadmium, and Zinc, Using Adsorption Processes by Cost-Effective Adsorbents. In *Adsorption Processes for Water Treatment and Purification*, edited by A Bonilla-Petriciolet, DI Mendoza-Castillo, and HE Reynel-Ávila. Cham, Switzerland: Springer International Publishing. Chapt. 5: page no. 109–138.

Xu, M, J Wu, L L, et al. 2018. The factors affecting biochar application in restoring heavy metal-polluted soil and its potential applications. *Chemistry and Ecology* 34 (2):177–197.

Yadav, VB, R Gadi, and S Kalra. 2019. Clay based nanocomposites for removal of heavy metals from water: A review. *Journal of Environmental Management* 232:803–817.

Yakout, AA, RH El-Sokkary, MA Shreadah, and OG Abdel Hamid. 2016. Removal of Cd (II) and Pb (II) from wastewater by using triethylenetetramine functionalized grafted cellulose acetate-manganese dioxide composite. *Carbohydrate Polymers* 148:406–414.

Yamani, JS, AW Lounsbury, and JB Zimmerman. 2014. Adsorption of selenite and selenate by nanocrystalline aluminum oxide, neat and impregnated in chitosan beads. *Water Research* 50:373–381.

Yamazaki, Y, Y Tachibana, T Kaneshiki, M Nomura, and T Suzuki. 2015. Adsorption behavior of uranium ion using novel phenol-type resins in contaminated water containing seawater. *Progress in Nuclear Energy* 82:74–79.

Yan, L, L Kong, Z Qu, L Li, and G Shen. 2015. Magnetic biochar decorated with ZnS nanocrystals for Pb (II) removal. *ACS Sustainable Chemistry & Engineering* 3 (1):125–132.

Yan, L, Q Zhao, T Jiang, et al. 2015. Adsorption characteristics and behavior of a graphene oxide–Al 13 composite for cadmium ion removal from aqueous solutions. *RSC Advances* 5 (83):67372–67379.

Yang, D, S Sarina, H Zhu, et al. 2011. Capture of radioactive cesium and iodide ions from water by using titanate nanofibers and nanotubes. *Angewandte Chemie International Edition* 50 (45):10594–10598.

Yang, G, L Tang, X Lei, et al. 2014. Cd (II) removal from aqueous solution by adsorption on α-ketoglutaric acid-modified magnetic chitosan. *Applied Surface Science* 292:710–716.

Yang, H, H Yu, J Sun, et al. 2017. Facile synthesis of mesoporous magnetic AMP polyhedric composites for rapid and highly efficient separation of Cs^+ from water. *Chemical Engineering Journal* 317:533–543.

Yang, H, L Sun, J Zhai, H Li, Y Zhao, and H Yu. 2014. In situ controllable synthesis of magnetic Prussian blue/graphene oxide nanocomposites for removal of radioactive cesium in water. *Journal of Materials Chemistry A* 2 (2):326–332.

Yang, J, B Hou, J Wang, et al. 2019. Nanomaterials for the removal of heavy metals from wastewater. *Nanomaterials* 9 (3):424.

Yang, P, Q Liu, J Liu, et al. 2017. Bovine serum albumin-coated graphene oxide for effective adsorption of uranium (VI) from aqueous solutions. *Industrial & Engineering Chemistry Research* 56 (13):3588–3598.

Yang, X, DK Debeli, G Shan, and P Pan. 2020. Selective adsorption and high recovery of La^{3+} using graphene oxide/poly (N-isopropyl acrylamide-maleic acid) cryogel. *Chemical Engineering Journal* 379:122335.

Yang, X, Y Wan, Y Zheng, et al. 2019. Surface functional groups of carbon-based adsorbents and their roles in the removal of heavy metals from aqueous solutions: A critical review. *Chemical Engineering Journal* 366:608–621.

Yang, Y, N Gao, W Chu, Y Zhang, and Y Ma. 2012. Adsorption of perchlorate from aqueous solution by the calcination product of Mg/(Al–Fe) hydrotalcite-like compounds. *Journal of Hazardous Materials* 209:318–325.

Yang, Y, N Gao, Y Deng, and S Zhou. 2012. Adsorption of perchlorate from water using calcined iron-based layered double hydroxides. *Applied Clay Science* 65:80–86.

Yang, Y, Y Xie, L Pang, et al. 2013. Preparation of reduced graphene oxide/poly (acrylamide) nanocomposite and its adsorption of Pb (II) and methylene blue. *Langmuir* 29 (34):10727–10736.

Yang, Z, O Xiao, B Chen, et al. 2013. Perchlorate adsorption from aqueous solution on inorganic-pillared bentonites. *Chemical Engineering Journal* 223:31–39.

Yao, Y, B Gao, J Chen, et al. 2013. Engineered carbon (biochar) prepared by direct pyrolysis of Mg-accumulated tomato tissues: Characterization and phosphate removal potential. *Bioresource Technology* 138:8–13.

Yazdani, M, T Tuutijärvi, A Bhatnagar, and R Vahala. 2016. Adsorptive removal of arsenic(V) from aqueous phase by feldspars: Kinetics, mechanism, and thermodynamic aspects of adsorption. *Journal of Molecular Liquids* 214:149–156.

Ye, N, N Cimetiere, V Heim, N Fauchon, C Feliers, and D Wolbert. 2019. Upscaling fixed bed adsorption behaviors towards emerging micropollutants in treated natural waters with aging activated carbon: Model development and validation. *Water research* 148:30–40.

Yeddou, N, and A Bensmaili. 2007. Equilibrium and kinetic modelling of iron adsorption by eggshells in a batch system: Effect of temperature. *Desalination* 206 (1–3):127–134.

Yi, R, G Ye, F Wu, M Wen, X Feng, and J Chen. 2014. Highly efficient removal of 137 Cs in seawater by potassium titanium ferrocyanide functionalized magnetic microspheres with multilayer core–shell structure. *RSC Advances* 4 (71):37600–37608.

Yin, Q, B Zhang, R Wang, and Z Zhao. 2017. Biochar as an adsorbent for inorganic nitrogen and phosphorus removal from water: A review. *Environmental Science and Pollution Research* 24 (34):26297–26309.

Yoshihara, K, M Aikawa, and M Machida. 2009. Influence of solution pH and background electrolytes on heavy metals ion adsorption by activated carbons from aqueous solutions. *Carbon* 9 (47):2301.

Yu, H-S, C-H Lee, and G-S Chen. 2002. Peripheral vascular diseases resulting from chronic arsenical poisoning. *The Journal of Dermatology* 29 (3):123–130.

Yu, J-G, X-H Zhao, L-Y Yu, F-P Jiao, J-H Jiang, and X-Q Chen. 2014. Removal, recovery and enrichment of metals from aqueous solutions using carbon nanotubes. *Journal of Radioanalytical and Nuclear Chemistry* 299 (3):1155–1163.

Yu, L, Y Ma, CN Ong, J Xie, and Y Liu. 2015. Rapid adsorption removal of arsenate by hydrous cerium oxide–graphene composite. *Rsc Advances* 5 (80):64983–64990.

Yu, S, H Mei, X Chen, et al. 2015. Impact of environmental conditions on the sorption behavior of radionuclide 90 Sr (II) on Na-montmorillonite. *Journal of Molecular Liquids* 203:39–46.

Yu, X, S Tong, M Ge, and J Zuo. 2013. Removal of fluoride from drinking water by cellulose@ hydroxyapatite nanocomposites. *Carbohydrate Polymers* 92 (1):269–275.

Yu, Z, H Han, P Feng, et al. 2020. Recent advances in the recovery of metals from waste through biological processes. *Bioresource Technology* 297:122416.

Yuan, S, J Zhang, Z Yang, S Tang, B Liang, and SO Pehkonen. 2017. Click functionalization of poly(glycidyl methacrylate) microspheres with triazole-4-carboxylic acid for the effective adsorption of Pb(ii) ions. *New Journal of Chemistry* 41 (14):6475–6488.

Yuan, S, L Zou, J-S Qin, et al. 2017. Construction of hierarchically porous metal–organic frameworks through linker labilization. *Nature Communications* 8 (1):15356.

Yuliarto, B, NLW Septiani, YV Kaneti, et al. 2019. Green synthesis of metal oxide nanostructures using naturally occurring compounds for energy, environmental, and bio-related applications. *New Journal of Chemistry* 43 (40):15846–15856.

Yusuf, M, FM Elfghi, SA Zaidi, EC Abdullah, and MA Khan. 2015. Applications of graphene and its derivatives as an adsorbent for heavy metal and dye removal: A systematic and comprehensive overview. *RSC Advances* 5 (62):50392–50420.

Zagulski, I, L Pawlowski, and A Cichocki. 1980. Method for Recovery of Water and Vanadium Compounds from Wastewater. In *Physicochemical Methods for Water and Wastewater Treatment*,edited by Lucjan Pawlowski. Oxford (UK): Pergamon: page no. 229–236.

Zare, EN, A Motahari, and M Sillanpää. 2018. Nanoadsorbents based on conducting polymer nanocomposites with main focus on polyaniline and its derivatives for removal of heavy metal ions/dyes: A review. *Environmental Research* 162:173–195.

Zare, EN, and MM Lakouraj. 2014. Biodegradable polyaniline/dextrin conductive nanocomposites: Synthesis, characterization, and study of antioxidant activity and sorption of heavy metal ions. *Iranian Polymer Journal* 23 (4):257–266.

Zare, EN, MM Lakouraj, and A Ramezani. 2016. Efficient sorption of Pb(ii) from an aqueous solution using a poly(aniline-co-3-aminobenzoic acid)-based magnetic core–shell nanocomposite. *New Journal of Chemistry* 40 (3):2521–2529.

Zare, EN, MM Lakouraj, and N Kasirian. 2018. Development of effective nano-biosorbent based on poly m-phenylenediamine grafted dextrin for removal of Pb (II) and methylene blue from water. *Carbohydrate Polymers* 201:539–548.

Zelmanov, G, and R Semiat. 2011. Iron (Fe^{+3}) oxide/hydroxide nanoparticles-based agglomerates suspension as adsorbent for chromium (Cr^{+6}) removal from water and recovery. *Separation and Purification Technology* 80 (2):330–337.

Zeng, G, Y Liu, L Tang, et al. 2015. Enhancement of Cd (II) adsorption by polyacrylic acid modified magnetic mesoporous carbon. *Chemical Engineering Journal* 259:153–160.

Zeng, L, and CY Cheng. 2009. A literature review of the recovery of molybdenum and vanadium from spent hydrodesulphurisation catalysts: Part I: Metallurgical processes. *Hydrometallurgy* 98 (1–2):1–9.

Zha, R, R Nadimicherla, and X Guo. 2014. Cadmium removal in waste water by nanostructured TiO_2 particles. *Journal of Materials Chemistry A* 2 (34):13932–13941.

Zhang, A, X Li, J Xing, and G Xu. 2020. Adsorption of potentially toxic elements in water by modified biochar: A review. *Journal of Environmental Chemical Engineering*:104196.

Zhang, A, Y Wang, and J Li. 2017. Cesium and strontium uptake utilizing a new ternary solid-state supramolecular adsorbent under a framework of group partitioning. *Journal of Chemical & Engineering Data* 62 (4):1440–1447.

Zhang, C-Z, B Chen, Y Bai, and J Xie. 2018. A new functionalized reduced graphene oxide adsorbent for removing heavy metal ions in water via coordination and ion exchange. *Separation Science and Technology* 53 (18):2896–2905.

Zhang, G, Z Ren, X Zhang, and J Chen. 2013. Nanostructured iron(III)-copper(II) binary oxide: A novel adsorbent for enhanced arsenic removal from aqueous solutions. *Water Research* 47 (12):4022–4031.

Zhang, H, Q Liu, J Wang, et al. 2015. Preparation of magnetic calcium silicate hydrate for the efficient removal of uranium from aqueous systems. *RSC Advances* 5 (8):5904–5912.

Zhang, H, X Zhao, J Wei, and F Li. 2015. Removal of cesium from low-level radioactive wastewaters using magnetic potassium titanium hexacyanoferrate. *Chemical Engineering Journal* 275:262–270.

Zhang, J-Y, N Zhang, L Zhang, et al. 2015. Adsorption of uranyl ions on amine-functionalization of MIL-101(Cr) nanoparticles by a facile coordination-based post-synthetic strategy and X-ray absorption spectroscopy studies. *Scientific Reports* 5:13514.

Zhang, L, J Wei, X Zhao, et al. 2016. Removal of strontium (II) and cobalt (II) from acidic solution by manganese antimonate. *Chemical Engineering Journal* 302:733–743.

Zhang, L, W Xia, X Liu, and W Zhang. 2015. Synthesis of titanium cross-linked chitosan composite for efficient adsorption and detoxification of hexavalent chromium from water. *Journal of Materials Chemistry A* 3 (1):331–340.

Zhang, L, X Guo, H Li, Z Yuan, X Liu, and T Xu. 2011. Separation of trace amounts of Ga and Ge in aqueous solution using nano-particles micro-column. *Talanta* 85 (5):2463–2469.

Zhang, L, X Jing, R Li, et al. 2015. Magnesium carbonate basic coating on cotton cloth as a novel adsorbent for the removal of uranium. *RSC Advances* 5 (30):23144–23151.

Zhang, L, X Liu, W Xia, and W Zhang. 2014. Preparation and characterization of chitosan-zirconium (IV) composite for adsorption of vanadium (V). *International Journal of Biological Macromolecules* 64:155–161.

Zhang, L, Y Zeng, and Z Cheng. 2016. Removal of heavy metal ions using chitosan and modified chitosan: A review. *Journal of Molecular Liquids* 214:175–191.

Zhang, L, Y Zhu, H Li, N Liu, X Liu, and X Guo. 2010. Kinetic and thermodynamic studies of adsorption of gallium(III) on nano-TiO_2. *Rare Metals* 29 (1):16–20.

Zhang, M, and B Gao. 2013. Removal of arsenic, methylene blue, and phosphate by biochar/AlOOH nanocomposite. *Chemical Engineering Journal* 226:286–292.

Zhang, M, B Gao, S Varnoosfaderani, A Hebard, Y Yao, and M Inyang. 2013. Preparation and characterization of a novel magnetic biochar for arsenic removal. *Bioresource Technology* 130:457–462.

Zhang, M, B Gao, Y Yao, Y Xue, and M Inyang. 2012. Synthesis of porous MgO-biochar nanocomposites for removal of phosphate and nitrate from aqueous solutions. *Chemical Engineering Journal* 210:26–32.

Zhang, N, S Liu, L Jiang, M Luo, C Chi, and J Ma. 2015. Adsorption of strontium from aqueous solution by silica mesoporous SBA-15. *Journal of Radioanalytical and Nuclear Chemistry* 303 (3):1671–1677.

Zhang, Q, J Wu, and X Luo. 2016. Facile preparation of a novel Hg (II)-ion-imprinted polymer based on magnetic hybrids for rapid and highly selective removal of Hg (II) from aqueous solutions. *RSC Advances* 6 (18):14916–14926.

Zhang, Q, Q Yang, P Phanlavong, et al. 2017. Highly efficient lead(II) sequestration using size-controllable polydopamine microspheres with superior application capability and rapid capture. *ACS Sustainable Chemistry & Engineering* 5 (5):4161–4170.

Zhang, Q, Y Li, P Phanlavong, et al. 2017. Highly efficient and rapid fluoride scavenger using an acid/base tolerant zirconium phosphate nanoflake: Behavior and mechanism. *Journal of Cleaner Production* 161:317–326.

Zhang, T, Q Li, H Xiao, H Lu, and Y Zhou. 2012. Synthesis of Li–Al layered double hydroxides (LDHs) for efficient fluoride removal. *Industrial & Engineering Chemistry Research* 51 (35):11490–11498.

Zhang, X, J Wang, R Li, et al. 2013. Preparation of Fe_3O_4@C@layered double hydroxide composite for magnetic separation of uranium. *Industrial & Engineering Chemistry Research* 52 (30):10152–10159.

Zhang, X, T Wu, Y Zhang, DH Ng, H Zhao, and G Wang. 2015. Adsorption of Hg^{2+} by thiol functionalized hollow mesoporous silica microspheres with magnetic cores. *RSC Advances* 5 (63):51446–51453.

Zhang, Y, and K Huang. 2019. Grape pomace as a biosorbent for fluoride removal from groundwater. *RSC Advances* 9 (14):7767–7776.

Zhang, Y, M Fu, D Wu, and Y Zhang. 2017. Immobilization of selenite from aqueous solution by structural ferrous hydroxide complexes. *RSC Advances* 7 (22):13398–13405.

Zhang, Y, X Liu, Y Wang, Z Lou, W Shan, and Y Xiong. 2019. Polyacrylic acid-functionalized graphene oxide for high-performance adsorption of gallium from aqueous solution. *Journal of Colloid and Interface Science* 556:102–110.

Zhang, Y, Y Li, L-q Yang, X-j Ma, L-y Wang, and Z-F Ye. 2010. Characterization and adsorption mechanism of Zn^{2+} removal by PVA/EDTA resin in polluted water. *Journal of Hazardous Materials* 178 (1–3):1046–1054.

Zhang, Z, H Liu, L Wu, H Lan, and J Qu. 2015. Preparation of amino-Fe(III) functionalized mesoporous silica for synergistic adsorption of tetracycline and copper. *Chemosphere* 138:625–632.

Zhao, T, and T Feng. 2016. Application of modified chitosan microspheres for nitrate and phosphate adsorption from aqueous solution. *RSC Advances* 6 (93):90878–90886.

Zhao, Y, Z Shao, C Chen, J Hu, and H Chen. 2014. Effect of environmental conditions on the adsorption behavior of Sr (II) by Na-rectorite. *Applied Clay Science* 87:1–6.

Zhao, Z, Y Baba, W Yoshida, F Kubota, and M Goto. 2016. Development of novel adsorbent bearing aminocarbonylmethylglycine and its application to scandium separation. *Journal of Chemical Technology and Biotechnology* 91 (11):2779–2784.

Zhao, Z, Y Yang, Y Xiao, and Y Fan. 2012. Recovery of gallium from Bayer liquor: A review. *Hydrometallurgy* 125–126:115–124.

Zheng, Q, Z Li, X Miao, et al. 2017. Preparation and characterization of novel organic chelating resin and its application in recovery of Zn (II) from aqueous solutions. *Applied Organometallic Chemistry* 31 (1):e3546.

Zheng, S, Z-S Wu, S Wang, et al. 2017. Graphene-based materials for high-voltage and high-energy asymmetric supercapacitors. *Energy Storage Materials* 6:70–97.

Zheng, Y, B Cheng, W You, J Yu, and W Ho. 2019. 3D hierarchical graphene oxide-NiFe LDH composite with enhanced adsorption affinity to Congo red, methyl orange and Cr(VI) ions. *Journal of Hazardous Materials* 369:214–225.

Zhong, L-B, J Yin, S-G Liu, Q Liu, Y-S Yang, and Y-M Zheng. 2016. Facile one-pot synthesis of urchin-like Fe–Mn binary oxide nanoparticles for effective adsorption of Cd (II) from water. *RSC Advances* 6 (105):103438–103445.

Zhou, J, Y Cheng, J Yu, and G Liu. 2011. Hierarchically porous calcined lithium/aluminum layered double hydroxides: Facile synthesis and enhanced adsorption towards fluoride in water. *Journal of Materials Chemistry* 21 (48):19353–19361.

Zhou, J, Y Wang, J Wang, W Qiao, D Long, and L Ling. 2016. Effective removal of hexavalent chromium from aqueous solutions by adsorption on mesoporous carbon microspheres. *Journal of Colloid and Interface Science* 462:200–207.

Zhou, Y, X Hu, M Zhang, X Zhuo, and J Niu. 2013. Preparation and characterization of modified cellulose for adsorption of Cd (II), Hg (II), and acid fuchsin from aqueous solutions. *Industrial & Engineering Chemistry Research* 52 (2):876–884.

Zhou, Z, Y-g Liu, S-b Liu, et al. 2017. Sorption performance and mechanisms of arsenic(V) removal by magnetic gelatin-modified biochar. *Chemical Engineering Journal* 314:223–231.

Zhu, H, Y Fu, R Jiang, J Yao, L Xiao, and G Zeng. 2014. Optimization of copper(II) adsorption onto novel magnetic calcium alginate/maghemite hydrogel beads using response surface methodology. *Industrial & Engineering Chemistry Research* 53 (10):4059–4066.

Zhu, J, X Lin, P Wu, and X Luo. 2016. Pectin/Al_2O_3–ZrO_2 core/shell bead sorbent for fluoride removal from aqueous solution. *RSC Advances* 6 (33):27738–27749.

Zhu, M, P Northrup, C Shi, SJL Billinge, DL Sparks, and GA Waychunas. 2013. Structure of sulfate adsorption complexes on ferrihydrite. *Environmental Science & Technology Letters* 1 (1):97–101.

Zhu, Q, and Z Li. 2015. Hydrogel-supported nanosized hydrous manganese dioxide: Synthesis, characterization, and adsorption behavior study for Pb^{2+}, Cu^{2+}, Cd^{2+} and Ni^{2+} removal from water. *Chemical Engineering Journal* 281:69–80.

Zhu, T, T Zhu, J Gao, L Zhang, and W Zhang. 2017. Enhanced adsorption of fluoride by cerium immobilized cross-linked chitosan composite. *Journal of Fluorine Chemistry* 194:80–88.

Zhu, Y, and EJ Elzinga. 2015. Macroscopic and spectroscopic assessment of the cosorption of Fe(II) with As(III) and As(V) on Al-oxide. *Environmental Science & Technology* 49 (22):13369–13377.

Zhu, Z, X Yang, L-N He, and W Li. 2012. Adsorption of Hg^{2+} from aqueous solution on functionalized MCM-41. *Rsc Advances* 2 (3):1088–1095.

Zong, Y, Y Zhang, X Lin, D Ye, X Luo, and J Wang. 2017. Preparation of a novel microsphere adsorbent of prussian blue capsulated in carboxymethyl cellulose sodium for Cs (I) removal from contaminated water. *Journal of Radioanalytical and Nuclear Chemistry* 311 (3):1577–1591.

Zuo, R, S Jin, Jie Y, et al. 2019. Removal of strontium from aqueous solutions by sodium dodecyl sulfate-modified palygorskite. *Journal of Radioanalytical and Nuclear Chemistry* 321 (1):151–159.

Zuo, W-Q, C Chen, H-J Cui, and M-L Fu. 2017. Enhanced removal of Cd (ii) from aqueous solution using $CaCO_3$ nanoparticle modified sewage sludge biochar. *RSC Advances* 7 (26):16238–16243.

Index

Printed in the United States
By Bookmasters